Advanced Technologies for Solid, Liquid, and Gas Waste Treatment

Advanced Technologies for Solid, Liquid, and Gas Waste Treatment presents the potential of using advanced and emerging technologies to effectively treat waste. This book uniquely addresses treatment techniques for waste in all three phases, solid, liquid, and gas, with the goals of mitigating negative impacts of waste and producing value-added products, such as biogas and fertilizer, as well as the use of artificial intelligence in the field.

- Covers a wide range of advanced and emerging treatment technologies such as photocatalysis processing, adsorptive membranes, pyrolysis, advanced oxidation process, electrocoagulation, composting technologies, etc.
- Addresses issues associated with wastes in different phases.
- Discusses the pros and cons of treatment technologies for handling different wastes produced by different industrial processes, such as agricultural biomass, industrial/domestic solid wastes, wastewater, and hazardous gas.
- Includes application of artificial intelligence in treatment of electronic waste.

This book will appeal to chemical, civil, and environmental engineers working on waste treatment, waste valorization, and pollution control.

Advanced Technologies for Solid, Liquid, and Gas Waste Treatment

Edited by Yeek-Chia Ho, Woei Jye Lau,
Sudip Chakraborty, N. Rajamohan,
and Saleh Al Arni

CRC Press
Taylor & Francis Group
Boca Raton London

CRC Press is an imprint of the
Taylor & Francis Group, an **informa** business

First edition published 2023
by CRC Press
6000 Broken Sound Parkway NW, Suite 300, Boca Raton, FL 33487–2742

and by CRC Press
4 Park Square, Milton Park, Abingdon, Oxon, OX14 4RN

CRC Press is an imprint of Taylor & Francis Group, LLC

ISBN: 978-1-032-19759-3 (hbk)
ISBN: 978-1-032-19763-0 (pbk)
ISBN: 978-1-003-26073-8 (ebk)

DOI: 10.1201/9781003260738

Typeset in Times
by Apex CoVantage, LLC

Contents

Preface.. vii
Editor biographies... ix
Contributors xi

Chapter 1   Composting of Food Wastes for Soil Amendment 1

*A. Sánchez, A. Artola, R. Barrena, T. Gea, X. Font,
and A.J. Moral-Vico*

Chapter 2   Integrated Management of Electronic and Electric Waste
(EEW) with the Application of Artificial Intelligence (AI):
Future and Challenges 23

Giorgio L. Russo

Chapter 3   Thermal Conversion of Solid Waste via Pyrolysis
to Produce Bio-Oil, Biochar and Syngas.. 41

Saleh Al Arni

Chapter 4   Waste Tyre Recycling: Processes and Technologies 57

Saleh Al Arni and Mahmoud M. Elwaheidi

Chapter 5   Electrochemical Removal of Organic Compounds
from Municipal Wastewater 73

R. Elkacmi

Chapter 6   Photocatalytic Membrane for Emerging Pollutants Treatment 95

*Nur Hashimah Aliasa, Nur Hidayati Othman, Woei Jye Lau,
Fauziah Marpani, Muhammad Shafiq Shayuti, Zul Adlan Mohd
Hir, Juhana Jaafar, and Mohd Haiqal Abd Aziz*

Chapter 7   Membrane and Advanced Oxidation Processes for
Pharmaceuticals and Personal Care Products Removal 115

Ryosuke Homma and Haruka Takeuchi

Chapter 8   Membrane Bioreactor for Wastewater Treatment............................ 133

Mengying Yang and Xinwei Mao

Chapter 9   Integration of Advanced Oxidation Processes as Pre-Treatment
for Anaerobically Digested Palm Oil Mill Effluent........................... 161

*E. L. Yong, Z. Y. Yong, M. H. D. Othman,
and H. H. See*

Chapter 10   Electrocoagulation and Its Application in Food Wastewater
Treatment.. 179

Mohammed J.K. Bashir and Koo Li Sin

Chapter 11   Advanced Oxidation Processes (AOPs) on the Removal of
Different Per- and Polyfluoroalkyl Substances (PFAS) Types
in Wastewater 203

Z. Y. Yong, H. Y. Tey, E. L. Yong, M. H. D. Othman, and H. H. See

Chapter 12   Photocatalysis for Oil Water Treatment 221

Baskaran Sivaprakash and N. Rajamohan

Chapter 13   Integrated Treatment Process for Industrial Gas Effluent 239

Daniel Dobslaw

Chapter 14   De-NOx SCR: Catalysts and Process Designs in the
Automotive Industry.. 267

Gerardo Coppola, Valerio Pugliese, and Sudip Chakraborty

Chapter 15   Advanced Technology for Cleanup of Syngas Produced
from Pyrolysis/Gasification Processes 289

Saleh Al Arni

Index.. 305

Preface

Advanced Technologies for Solid, Liquid, and Gas Waste Treatment has been prepared to harness significant environmental issues through a comprehensive up-to-date overview on the recent development of different treatment technologies. This book is developed not only for practicing environmental and chemical engineers, but also for civil engineers who are involved in the processes related to solid, liquid, and gas waste treatment. The book can also be used as a reference source for undergraduate and postgraduate courses in waste treatments.

This is the first book that is aimed to address the wastes in all three phases, i.e., solid, liquid, and gas. Statistics released by the BCC Research indicated the healthy market growth of the treatment technologies in remediating different wastes. For instance, the global market for scrap tires management and rubber remediation applications (i.e., solid wastes) will grow from $7.6 billion in 2017 to nearly $9.5 billion by 2022, with a compound annual growth rate (CAGR) of 4.5% for the period of 2017–2022. The global market for wastewater technologies and air pollution control equipment meanwhile was reported at $64.4 billion (2018) and $13.3 billion (2015), respectively with a CAGR of at least 5% over the next 5-year period.

This book contains the chapters that were contributed by well-known scientists/researchers from different parts of the world and covers many different types of treatment technologies such as photocatalysis, adsorptive membranes, pyrolysis, advanced oxidation process, electrocoagulation, and emerging composting methods for the treatment of agricultural biomass, industrial/domestic solid wastes, wastewater, and hazardous gas. Other technologies included in this book are related to artificial intelligence applications, value-added product formation as new energy and nutrient sources as well as hybrid processes to curb incalcitrant pollutants present in the environment.

The advances in solid, liquid, and gaseous emission are discussed in breadth and depth and summarized in 15 chapters. To illustrate waste management through different technologies, individual graphical figures, tables, and diagrams are depicted. Additional drawings and photographs are included to aid the understanding of the advanced technologies presented.

To make this book more valuable, discharge from various industries is considered. These include automotive, agriculture, electronics, food, oil and gas industry, and pharmaceuticals, just to name a few. These advanced technologies are aimed not only to reduce negative impacts caused by the wastes on the environment and humans but also to produce value-added products.

We hope this book will be of great help for practitioners and researchers in the real-world application – industry.

Editors

Yeek-Chia Ho
Universiti Teknologi PETRONAS, Malaysia

Woei Jye Lau
Universiti Teknologi Malaysia, Malaysia

Sudip Chakraborty
University of Calabria, Italy

N. Rajamohan
Sohar University, Oman

Saleh Al Arni
Hail University, Saudi Arabia

Editor biographies

Yeek-Chia Ho (Ph.D.) is Programme Manager for Master in Industrial Environmental Engineering in Universiti Teknologi PETRONAS. Additionally, she is in the Leadership Team under the Institute of Self-Sustainable Building (ISB). Also noteworthy is that she is Secretary in the International Water Association (IWA) for Design, Operation and Maintenance of Drinking Water Treatment Plants. Her current research interest is closely related to environmental engineering, specifically, Water-Food-Energy nexus. YC is involved in research projects which includes renewable energy, microalgae harvesting, water and wastewater treatments, and life cycle assessment. She has published several works in invited review papers, invited book chapters, international conference papers and international indexed journal papers. Also, YC has won a few national awards in publications and international awards for her research works; two patents and two trademarks have been filed so far. Lastly, she has been selected to be one of the invited professionals in the Leaders Innovation Fellowships Programme by the Royal Academy of Engineering, UK.

Woei Jye Lau is currently an associate professor at Faculty of Chemical and Energy Engineering, Universiti Teknologi Malaysia (UTM). He obtained his Bachelor of Engineering in Chemical-Gas Engineering and Doctor of Philosophy (Ph.D.) in Chemical Engineering from UTM. Dr Lau has a very strong research interest in the field of membrane science and technology for water applications. He has published more than 250 scientific papers and 30 reviews with a total number of Scopus citations exceeding 10,000 and h-index of 52. Currently, Dr Lau serves as a subject editor for Chemical Engineering Research and Design (Elsevier) and associate editor for Water Reuse (International Water Association). Dr Lau has been named among the top 2% scientists in the world according to the Stanford Report for 2019 published in the journal PLOS Biology.

Sudip Chakraborty has a Ph.D. in Chemical Engineering from University of Calabria, Italy. He is Abilitazione Scientifica Nazionale (ASN) – full professor in sector – ING-IND 24 at the Laboratory of Transport Phenomena and Biotechnology, University of Calabria. He was also a visiting researcher at Massachusetts Institute of Technology (MIT). His major fields of interest are membrane separation, plasmonic nanoparticles, composite materials, energy, and process intensification. Dr. Chakraborty has an h-index of 36 and has published more than 120 research publications in international journals, books, and conference proceedings.

N. Rajamohan is a senior academic researcher, industrial consultant, program manager and research administrator, currently affiliated with Sohar University as Dean (acting), Faculty of Engineering Sohar, Oman. He has completed Ph.D., Master's and Bachelor's degrees in Chemical Engineering. He has published more than 125 research articles in various indexed journals of international repute. He has delivered several major research grants including the UK-Gulf Institutional Links grant. His fields of specialization are environmental chemical engineering, heavy metal pollution control, biological treatment of toxic gases and Sustainable technologies. He has amassed rich research experience over a period of two decades. He was awarded "Outstanding Achievement in Research and Knowledge Transfer" at his institution in February 2020.

Saleh Al Arni is Associate Professor of Chemical Engineering, Hail University (UOH), Kingdom of Saudi Arabia. He received his Master's degree (Laurea) in Chemical Engineering from University of Genoa, Italy, in 1996. In 2000, he received his first Ph.D. degree in "Technologies and Economics of the Processes and Products to Safeguard the Environment," from Catania University, Italy, and in 2008, he received the second Ph.D. degree in "Chemical Sciences, Technologies and Processes," from the University of Genoa. His research and teaching activities deal with biotechnological processes, and he has published several scientific articles in international journals.

Contributors

Nur Hashimah Aliasa
Universiti Teknologi MARA
Shah Alam, Selangor, Malaysia

Saleh Al Arni
Department of Chemical Engineering –
  Engineering College
University of Ha'il
Ha'il, Saudi Arabia

A. Artola
Universitat Autònoma de Barcelona
Bellaterra, Barcelona, Spain

Mohd Haiqal Abd Aziz
Universiti Tun Hussein Onn Malaysia
Panchor, Johor, Malaysia

R. Barrena
Universitat Autònoma de Barcelona
Bellaterra, Barcelona, Spain

Mohammed J.K. Bashir
Universiti Tunku Abdul Rahman
Kampar, Perak, Malaysia

Sudip Chakraborty
Dipartimento di DIMES
Rende (CS), Cosenza, Italy

Gerardo Coppola
Dipartimento di DIMES
Rende (CS), Cosenza, Italy

Daniel Dobslaw
University of Stuttgart
Stuttgart, Germany

R. Elkacmi
University Sultan Moulay Slimane
Beni-Mellal, Morocco

Mahmoud M. Elwaheidi
Department of Geology and
  Geophysics – College of Science
King Saud University
Riyadh, Saudi Arabia

X. Font
Universitat Autònoma de Barcelona
Bellaterra, Barcelona, Spain

T. Gea
Universitat Autònoma de Barcelona
Bellaterra, Barcelona, Spain

Zul Adlan Mohd Hir
Universiti Teknologi MARA
  Pahang
Bandar Tun Abdul Razak, Jengka,
  Malaysia

R. Homma
Kyoto University
Kyoto, Japan

Juhana Jaafar
Universiti Teknologi Malaysia
Skudai, Johor, Malaysia

Woei Jye Lau
Universiti Teknologi Malaysia
Skudai, Johor, Malaysia

Xinwei Mao
Stony Brook University and New
  York State Center for Clean Water
  Technology
New York, USA

Fauziah Marpani
Universiti Teknologi MARA
Shah Alam, Selangor, Malaysia

A.J. Moral-Vico
Universitat Autònoma de Barcelona
Bellaterra, Barcelona, Spain

M. H. D. Othman
Universiti Teknologi Malaysia
Skudai, Johor, Malaysia

Nur Hidayati Othman
Universiti Teknologi MARA
Shah Alam, Selangor, Malaysia

Valerio Pugliese
Dipartimento di DIMES
Rende (CS), Cosenza, Italy

N. Rajamohan
Sohar University
Sohar, Oman

Giorgio L. Russo
University of Genoa
Genoa, Italy

A. Sánchez
Universitat Autònoma de
   Barcelona
Bellaterra, Barcelona, Spain

H.H. See
Universiti Teknologi Malaysia
Skudai, Johor, Malaysia

Muhammad Shafiq Shayuti
Universiti Teknologi MARA
Shah Alam, Selangor, Malaysia

Koo Li Sin
Universiti Tunku Abdul Rahman
Kampar, Perak, Malaysia

Baskaran Sivaprakash
Annamalai University
Annamalai Nagar, India

H. Takeuchi
Kyoto University
Kyoto, Japan

H. Y. Tey
Universiti Teknologi Malaysia
Skudai, Johor, Malaysia

Mengying Yang
Stony Brook University and Pall
   Corporation
New York, USA

E.L. Yong
Universiti Teknologi Malaysia
Skudai, Johor, Malaysia

Z.Y. Yong
Universiti Teknologi Malaysia
Skudai, Johor, Malaysia

1 Composting of Food Wastes for Soil Amendment

A. Sánchez, A. Artola, R. Barrena, T. Gea,
X. Font, and A.J. Moral-Vico

CONTENTS

1.1 Introduction...1
1.2 Food Waste, Domestic Waste, Organic Fraction of Municipal Solid
     Waste or Biowaste?..3
1.3 Particularities of Food Waste 6
     1.3.1 Overall Characteristics..6
     1.3.2 Heterogeneity and Seasonality 6
     1.3.3 Impurities and Pollutants 6
     1.3.4 Microplastics and Microbioplastics...7
1.4 Food Waste Composting 8
     1.4.1 Process Conditions..8
     1.4.2 Microbiology 9
     1.4.3 Gaseous Emissions 10
1.5 Food Waste Compost 11
     1.5.1 Stability and Maturity 11
     1.5.2 Pollutants 12
1.6 Uses of Food Waste Compost 13
     1.6.1 Use as Organic Amendment 13
     1.6.2 Soil Bioremediation 14
     1.6.3 Landfill Cover..15
     1.6.4 Suppressor Effect...16
1.7 Conclusions..16
References..17

1.1 INTRODUCTION

The global generation of municipal solid wastes (MSW) is 2.01 billion tonnes per year, and it is expected to increase to approximately 3.40 billion tonnes per year by 2050 [1]. However, per capita generation averages are wide, with calculated rates varying considerably by region, country, city and even within neighborhoods. MSW

DOI: 10.1201/9781003260738-1

1

generation rates are influenced by economic development, degree of industrialisation, public habits, and local climate [2]. In general, the higher the economic development and urbanisation rate, the higher the amount of solid waste produced. Each person in developed countries produces an average of 500 kg of solid waste per year and this is halved in developing countries. Income level and urbanisation are highly correlated with waste production. For example, urban residents produce about twice as much waste as their rural counterparts [1].

Given that the world population is increasing and being concentrated in urbanized areas, MSW management (collection, recycling, and valorizing) is becoming increasingly important. Indeed, it is estimated by the United Nations that 57% of the population live in urban areas and that by 2050 this percentage will increase to 68%. Being generated either in rural or in urban areas, big amounts of MSW are produced and must be properly managed to avoid negative impacts on the environment, reuse organic nutrients to agriculture and increase resilience of cities. Besides, collection of organic waste is increasing, for example in low-income countries it has increased from 22% to 39% [1], thus more efforts are needed to recycle MSW.

Municipal waste is defined by the Organization for Economic Co-operation and Development (OECD) as waste collected and treated by or for municipalities. According to OECD, MSW includes waste from: households, similar waste from commerce and trade, office buildings, institutions, and small businesses, as well as yard and garden waste, street sweepings, the contents of litter containers, and market cleansing waste if managed as household waste. OECD excludes waste from municipal sewage networks and treatment, and waste from construction and demolition activities from the flow of MSW. This definition can vary depending on the source, for example, the World Bank includes industrial wastes and construction and demolition wastes within MSW flows [3]. Other countries include or exclude some different materials from MSW.

Whatever the definition of MSW, this residue will contain biodegradable organic matter. Indeed, biodegradable organic waste accounts for 28 to 56% of MSW [1, 4]. This biodegradable content is also known as organic fraction of MSW (OFMSW) and it consists of: (1) food manufacturing waste; (2) household waste from food preparation, leftovers and expired food; and (3) waste from restaurants and food outlets [5]. These sources comprise 39, 42 and 19% of the total European Union (EU) food waste stream, respectively [6]. In the overall flow of MSW, the biodegradable organic fraction is responsible for some of the negative environmental impacts associated with MSW collection and treatment. Specifically, odour emission, leachate production, emission of greenhouse gases among other impacts can be appointed.

Administrations are promoting policies to reduce or avoid these environmental impacts. For example, the Landfill Directive by the European Union [7] requires its member states to reduce the amount of untreated biodegradable waste managed in landfills by adopting measures to increase and improve waste reduction, recovery, and recycling. To facilitate MSW fractions recycling and valorisation, separation at source is the best practice to obtain the minimum percentage of impurities in each fraction, including the case of OFMSW [8].

In the framework of Circular Economy, the future of OFMSW valorisation should go beyond traditional low-value recycling processes [9]. For example, in the case of the European Union, its Green Deal strategy is boosting the transformation of European society to a modern, resource-efficient, and competitive economy, claiming to recover the economic value of waste when its generation cannot be avoided. In this sense, anaerobic digestion and composting are the main recycling processes for OFMSW currently implemented.

The history of composting goes back to early civilizations in South America, India, China, and Japan, where extensive agriculture was practiced. The fields were fertilized with animal, human and agricultural waste. It is known that such waste was kept in pits or heaps to obtain the final soil amendment [10]. Since then, the technology has evolved, but there are still some aspects to improve and deepen the understanding of the composting process.

In addition to the evolution of the composting technology (improving aeration, mechanization, and automatization), the use of compost has also evolved. In addition to its use as a soil amendment, the compost obtained is being used for soil bioremediation processes and for landfill cover. Its final use depends on its quality. The quality of the compost obviously depends on the process operation, but also on the composted waste that can contain impurities. In the case of food waste, the main impurities that can be found are plastics, microplastics and glass that can finally be present in the compost.

In this chapter we will revise the entire chain of the composting process. Starting with the definition and particularities of food waste, followed by the composting processes itself and the properties of the compost obtained from food waste. Finally, the different uses of the obtained product (i.e., soil amendment, bioremediation, and landfill cover) will be revised.

1.2   FOOD WASTE, DOMESTIC WASTE, ORGANIC FRACTION OF MUNICIPAL SOLID WASTE OR BIOWASTE?

There are some confusions with the terminology used to designate the organic fraction of municipal solid waste in a general sense, which can come from households, restaurants, markets, etc. Molina-Peñate et al. [11] presents a complete study of different terms that have been used throughout scientific literature. Firstly, the term "biomass", that is, mass of living organisms [12], has been used to define all natural carbonaceous resources that can be used to generate fuels [13]. For this reason, it is an unspecific overused term. On the other hand, and according to Molina-Peñate et al., terms such as "organic waste", "biowaste", or "food waste" do not precisely describe the origin of the waste (industrial, agricultural or municipal). Contrarily, the terms "municipal waste" or "household waste" fail to describe accurately the type of material. It is the opinion of these authors that the most accurate term is "the organic fraction of municipal solid waste", or its acronym OFMSW, although it is less frequently used. Sometimes, OFMSW is also used to simulate food waste from restaurants using fresh food waste. However, this analogy cannot be representative of the complexity of the OFMSW collected in municipalities and managed in municipal waste treatment facilities [14]. In this chapter, the term "OFMSW" will be used. Several pictures related to the OFMSW are presented in Figure 1.1. Herein, food

(a)

(b)

FIGURE 1.1   Pictures of the Organic Fraction of Municipal Solid Waste (OFMSW) a)
OFMSW as received in a composting plant from a source-separated collection
system, b) Details of the entire OFMSW, c) Details of the impurities found in
the OFMSW, d) Compost from the OFMSW (not refined).

(*Source*: the authors).

(c)

(d)

FIGURE 1.1 (Continued)

waste will refer to mainly OFMSW and will include food waste from markets and food industry for a more general perspective.

1.3   PARTICULARITIES OF FOOD WASTE

1.3.1   Overall Characteristics

Food waste has unique characteristics and it is probably the most challenging biomass to manage because of the heterogeneous biochemical composition and physical properties dependent on a broad range of technical and socioeconomic factors and seasonality [15, 16]. It is a valuable source of nitrogen and phosphorous and unsuitable management leads to significant losses of resources [17]. For this reason, food waste composting and anaerobic digestion are largely encouraged under the circular economy paradigm to close organic and nutrient cycles and return them to soil, avoiding the use of chemical fertilizers. However, concern is arising on the potential risk of critical pollutants' distribution in the food-supply chain, where food waste plays a key role [18].

1.3.2   Heterogeneity and Seasonality

OFMSW is a very heterogeneous material since it contains multiple ingredients such as rests of fruits, vegetables, meat and fish leftovers, cereals, processed food, etc. These ingredients differ in biochemical composition, water content, particle size, density, and resistance to compaction. They also differ among regions and countries due to different culture, food habits or climate [15]. The level of development of the region/country directly affects OFMSW composition and amount [17].

Moreover, OFMSW composition varies along the year according to certain foods availability in different seasons. In addition, the amount of garden waste collected with OFMSW is almost negligible in winter and abundant in spring or autumn. This directly affects overall OFMSW composition (i.e., moisture or lignin content) [16]. For this reason, some regions implement separate collections for garden waste.

Food waste streams from markets will show the same heterogeneity and seasonality as household or kitchen waste, while food waste from the food industry may be more homogeneous and constant in time.

This has some serious implications when designing the composting process because the time needed to produce a good quality compost depends on the mentioned parameters: size, composition, etc. Variations in biodegradability will require different aeration rates. Variations in water content and porosity will lead to modifications in the bulking agent ratio. Thus, some operational adjustments may be necessary along the year to maintain compost quality standards.

1.3.3   Impurities and Pollutants

One of the main constraints in OFMSW management is the serious amount of non-compostable materials that are collected with food waste. They include packaging material, glass or cardboard, but also batteries, electronic devices or clothes.

The level of impurities can exceed 30% w/w [16, 19] and this fact complicates the OFMSW management and makes it more expensive. In addition, these non-compostable materials transfer pollutants into the organic fraction, lowering its quality as feedstock for composting and eventually contaminating the final compost.

The number of impurities present in OFMSW has been repeatedly linked to OFMSW selection and collection systems [19–21]. Source-selected OFMSW has much higher quality than mechanically-sorted OFMSW [8] and is less likely to exceed the regulatory limits for heavy metal concentration in the compost produced from it. The type of source-selection systems also affects OFMSW quality. Door-to-door collection systems provide much higher quality OFMSW than anonymous systems based in street containers, which present a higher non-compostable content (4-fold increase, according to [16]). The presence of these impurities has serious implications since they affect the composting process and the quality of the final compost, not only visually (plastic and glass fragments), but also agronomically (impurities alter porosity and water holding capacity). In addition, and as mentioned before, they can potentially transfer chemical pollutants to the organics.

However, chemical pollutants make their way to food and food waste much before the collection system. In the farm where food is produced, chemical pesticides are applied and remain in food even at trace level. Later, several opportunities arise for pollutants to reach food along manufacturing, packaging, and distribution. Isenhour et al. [22] reflect on the impossible separation of technical and biological cycles and the consequences for the conceptualization of the circular economy. Besides classical physical contaminants, heavy metals, and pathogens, some other pollutants are of arising concern and have been detected in food waste samples [18, 23]. They include chemicals such as pesticides, halogenated compounds or aromatic hydrocarbons, antibiotic resistance genes and microplastics. Pesticides and halogenated organics (i.e., per-and polyfluoroalkyl substances, PFAS) are highly persistent and tend to accumulate in the environment. When composting food waste, these contaminants are not degraded and end up in the compost. Antibiotic resistance genes dispersed in the environment increase risk of gene transfer to human pathogens [23]. Further research is needed to understand and assess risk related to these pollutants in the food supply chain to ensure food security [18].

1.3.4 MICROPLASTICS AND MICROBIOPLASTICS

Microplastics (MPs) contamination has arisen as one of the major environmental problems in recent years, receiving much attention by the scientific community. MPs usually refer to plastic fragments smaller than 2- or 5-mm. MPs have been found ubiquitous in water bodies and soils and are present in food waste too.

There is a lack of research data on the presence of MPs in food waste. At the moment, there does not exist a standard method to quantify MPs or nanoplastics in complex matrices such as food waste samples. Thus, many of the current studies focus on developing an efficient methodology [24, 25]. Obviously, the presence of plastics and microplastics in food waste directly affects compost quality.

It is important to notice that MPs can be produced from conventional plastics and also from biodegradable plastics. Although some studies report the absence of microbioplastics (MBPs) in compost samples [21], thus confirming their compostability, other authors reported the release of MBPs into compost because the wrong process conditions can reduce MBPs degradation [15]. Presence of MBPs in food waste has been estimated as 5.4–8.4 MBPs/10 g OFMSW after shredding pre-treatment before anaerobic digestion [15]. More often, MPs have been assessed in compost from OFMSW with results ranging from 20 to >2500 MPs/kg dry weight according to [15] and 5–20 items per g dry weight according to [21]. The presence of MPs in food waste affects product quality and modifies the microbial communities present.

In conclusion, food waste quality is essential, and efforts must be addressed to avoid food and food waste contamination in the different stages of the food-supply chain, from farm practices to collection and management systems. Otherwise, compost from food waste applied to soils as an amendment may be polluted with the contaminants mentioned in this section thus endangering the food chain [20]. When closing the nutrients cycle in the loop food-food waste-compost-soil-food there is a risk of concentrating persistent pollutants and MPs and spreading antibiotic resistance genes that should be understood and properly managed.

1.4   FOOD WASTE COMPOSTING

1.4.1   Process Conditions

Composting is a fully aerobic process, which requires oxygen, moisture and porosity to stabilize organic wastes, and their common control variables are temperature, oxygen and moisture [26]. A complex metabolic process is responsible for the decomposition and fractional humification of the organic matter, which ultimately transforms it into a soil amendment (compost), in which stability and maturity plays the main role as quality parameters [27].

Among these variables, ensuring the necessary oxygen supply is probably the most important parameter to consider in food waste composting. Accordingly, aeration is critical, especially when using the OFMSW, which requirements of oxygen supply are among the higher ones found in organic waste, as observed using respiration measurements [28–30]. In fact, the efficiency of the composting process is strongly affected by the oxygen level because the composting process is directly associated with microbial population dynamics [31]. Rasapoor et al. [32] compared different aeration systems in the OFMSW composting. Both forced aeration and pile turning are shown to be efficient in terms of final compost quality, although the latter showed better results for agricultural applications. Other studies evaluated the performance of different systems for composting the source-selected OFMSW [33]. In this case, turned pile, static forced-aerated pile and turned forced-aerated pile were analyzed at full scale. The results demonstrated that static systems showed worse values in terms of stability as the compaction of material resulted in an evident loss of porosity. In conclusion, besides oxygen demand, the porosity of the solid matrix to be composted is a key parameter for the successful composting of OFMSW [34].

Porosity is influenced by several parameters such as particle size and moisture content and it has a direct influence in aeration distribution and the oxygen content across the organic matrix. In the case of the OFMSW, water content is typically high [35], whereas porosity is not sufficient, which can result in anaerobic zones, undesirable gaseous emissions, or unpleasant odors. Achieving proper porosity levels ensures correct air circulation through the solid matrix and provides full aerobic conditions. In general, this issue is approached by composters and researchers by including a known volumetric ratio of bulking agent to OFMSW to adjust the moisture content, C/N ratio and porosity, although this is not always possible [36, 37].

In OFMSW composting, several bulking agents have been evaluated. Among them, lignocellulosic waste from agricultural activities, pruning waste, residual pellets and wood chips have resulted in good porosity conditions [35, 38]. Most of the studies reported an adequate porosity range of 30–50% for OFMSW composting, expressed as FAS (free air space). However, these values are obtained for the initial mixture. In general, there is a lack of information about the evolution of porosity during the entire composting process. When using a suitable bulking agent with enough resistance to compaction, FAS levels are maintained near initial values. It is also frequent to observe a reduction in FAS in the first days of the process due to compaction and substrate size reduction due to biodegradation. Later, FAS increases again as biodegradation proceeds [33, 34]. Yu et al. [39] assessed the effect of porosity during the curing stage by using passive aeration and confirmed and modelled the direct positive effect of FAS on microbial kinetics. Anyway, lack of FAS could limit the understanding of the entire process and have a negative impact on the compost quality.

Another important parameter in any composting process is the C/N ratio. This variable is critical for several aspects of composting such as ammonia emissions, but it is particularly crucial for the development of microorganisms during composting because both elements are required for cellular growth. Limiting the content of nitrogen is undesirable because it generates a reduction in the C consumption rate, whereas an excess can generate the release of ammonia in gaseous form [40]. The recommended initial C/N ratio at the start of the composting process ranges from 25–30. However, many other authors have used slightly different C/N ratios [41, 42], with good results. Regarding C/N ratio, it is important to note that OFMSW can present slow or non-biodegradable carbon sources depending on the presence of impurities such as plastics, textiles, wood, etc. In this sense, the use of a ratio based on the biodegradable organic carbon should be more adequate, as pointed out by Puyuelo et al. [43].

Other properties have traditionally been used for the monitoring of composting processes. However, biological and biochemical parameters have been reported as excellent indicators to know the biological activity of the process [44]. Biological methods are typically based on respiration activities. Other biologically-related parameters such as enzymatic activities or maturity tests are also emerging parameters to monitor the composting process [45, 46].

1.4.2 MICROBIOLOGY

During the last decades, a large variety of mesophilic, thermotolerant and thermophilic aerobic microorganisms including bacteria, actinomycetes, yeasts and other fungi have been extensively reported in compost and composting [47–49]. In fact, composting is a process performed by a series of microorganisms associated with different biodegradation systems [50].

In the case of the OFMSW, several authors have monitored the microbial communities and succession in the composting process [47]. Most of the studies report the importance of maintaining microbial diversity by controlling variables such as moisture, oxygen and turning.

One of the main problems found in OFMSW composting is to have a homogenous and reproducible process. Indeed, the operational parameters are responsible for the microbial fluctuations of the process. In this sense, the composting microbial communities act on a succession of different microorganisms that are strongly dependent on each other and are conditioned by biotic and abiotic factors [49]. Traditionally, it has been reported the succession of microbial communities from mesophilic to thermophilic along the composting process, in parallel to temperature profile. Recently, an exhaustive study on composting microbiome has challenged this perspective when demonstrating huge levels of thermotolerance by 90% of microbial strains involved in composting [50].

Composting is a process that occurs spontaneously due to autochthonous microorganisms present in the organic waste, thus no inoculation would be necessary. Despite this, several studies have reported that the addition of inoculating agents can result in enhancement of the composting process. These inoculants can be a sole strain [51–53] or a commercial mixture of several strains and enzymes [54–56]. The objective of these studies is to reveal a considerable reduction in the operation time of the composting process, odour reduction or the improvement of compost quality. However, other studies show different results, where inoculation does not present any significant improvement [57]. In general, the studies conducted on the suitability of different inoculants are inconclusive and scarce for the OFMSW. Moreover, home and community composting perform perfectly without the use of inoculum, which reinforces the idea that the autochthonous microbial communities in the OFMSW are active enough to start the process [58].

1.4.3 GASEOUS EMISSIONS

Although composting is generally considered an environmentally friendly technology, it also has negative environmental impacts. This is the case of atmospheric emissions, especially in the case of greenhouse gases (GHG) and volatile organic compounds (VOC).

Among the released GHGs, they can be attributed to energy requirements of the composting plant operation and to biochemical reactions, which produces carbon dioxide (CO_2), methane (CH_4), and nitrous oxide (N_2O) in the biodegradation of organic matter [59–60]. Regarding VOCs, the rates and specific forms of these emissions highly depend on the feedstock materials and composting phases, considering that aeration of the composting mixture plays a considerable role in releasing these compounds [61–62].

Undesirable gaseous emissions are formed due to inadequate aerobic conditions of composting [63]. This results in CH_4 emissions, whereas nitrogen transformation and loss (NH_3 and N_2O) are related to ammonification, nitrification, and denitrification [64]. The rate of gaseous emissions strongly depends on the initial content of carbon and nitrogen [65]. It is also important to note that the emissions flow is also influenced by the composting technology [66]. To reduce the impact of gaseous emissions from composting, biofilters effectively reduced ammonia and VOCs emission, being one of the most extended technologies in composting plants [67].

Finally, when comparing composting with other technologies for the treatment of organic solid waste, different studies have demonstrated that it has less impact on global warming, as it produces lower amounts of GHGs. This fact was concluded and documented by different studies, which emphasized that composting produces lower amounts of emissions than incineration or landfill in terms of g CO_2-eq/ton of waste [68–69]. The sole exception is vermicomposting, which can be defined as the process by which worms are used to convert organic materials (usually wastes) into a humus-like material known as vermicompost. In this case, when composting and vermicomposting were compared, it was found that the vermicomposting process caused lower GHG emissions compared to traditional composting [70].

1.5 FOOD WASTE COMPOST

The production of good quality compost is a crucial aspect to guarantee its safe use in agriculture. Essentially, it should present low levels of trace elements and organic contaminants, (near) absence of pathogenic organisms, and be mature and stable enough.

1.5.1 STABILITY AND MATURITY

Numerous techniques have been reported to assess compost quality [71]. Among them, respiration indices are one of the most used and recognized methods in the evaluation of compost stability [44] as an indicator of the extent to which biodegradable organic matter is being broken down within a specified time period. Respirometric activity can be determined directly from the O_2 consumption or CO_2 production and indirectly through the heat released during the process [72]. Among the numerous existing methodologies, two main aspects could be used for its classification. The first one is related to the way that the index is calculated and expressed: as an uptake rate or a cumulative (both O_2 consumption and CO_2 production). The second aspect is whether oxygen uptake measurement is made in the absence (static respiration index, SRI) or the presence (dynamic respiration index, DRI) of continuous aeration of the biomass [73].

Respiration indices of food waste compost have been widely reported using different methodologies from plants treating OFMSW presenting differences in collection systems, treatment technology and density of served population [73–76]. Nowadays, different composting facilities treating OFMSW are well-implemented around the world. Table 1.1 presents input and output values of DRI from samples of treatment plants using different technologies. As observed, a stable compost could be produced with the current technology if the composting process is accurate, and the length of maturation is respected. Most of them achieved DRI values lower than 1 g O_2 kg^{-1} OM h^{-1} [73] and close to the valued 0.8 g O_2 kg^{-1} OM h^{-1} suggested in the EU Regulation for fertilizing products (Regulation (EU) 2019/1009) [77]. In this regulation, the compost shall meet at least one of the following stability criteria: (a) Oxygen uptake rate: (maximum 25 mmol O_2/kg organic matter/h); or (b) Self-heating: (minimum Rottegrad III).

Maturity is usually related to the absence of phytotoxic substances and is generally determined by plant bioassays [72]. The germination test is the most reported method, although the seed selection is important for a sensitive and reliable indicator [78]. Food waste compost has been reported to have high salt concentrations [79] that could inhibit seed germination and plant growth. Hence, electrical conductivity should be carefully considered in the application of compost.

1.5.2 POLLUTANTS

For a long time, the main concern about the use of compost derived from MSW has been the presence of heavy metals (Cd, Cu, Ni, Pb, Zn, Hg and Cr). However, the amount of them in compost has been clearly related to the system of collection [80]. The most effective method of reducing heavy metals in compost is the source separation of the organic fraction. Nowadays, European regulations only allow food waste resulting from separate collections at the source as suitable input material to obtain compost (Regulation (EU) 2019/1009). Mixed municipal waste is excluded. According to its use as a component of organic fertiliser, soil improver or growing media, the limits are detailed in the regulation. Also, the content of pathogens (usually *Salmonellas, Salmonella* spp. and *Escherichia coli* or *Enterococcaceae*) is usually analyzed and delimited in compost regulations.

Other organic toxins detected in MSW compost have been related to the quality of the organic fraction, being higher in mixed MSW [79]. Although composting can be an effective way to reduce levels of these compounds enhancing its degradation (per example PAHs and short-chain phthalates [81]) their presence is limited when organic fraction is separately collected. However, there is a rising concern about the presence of pollutants in compost because some of them can persist in the environment and could be accumulated in the food supply chain [18, 23]. More data is required about the content of microplastics, heavy metals, pesticides, polychlorinated biphenyls (PCBs), polycyclic aromatic hydrocarbons (PAHs), per- and polyfluoroalkyl substances (PFAS), and pathogens in compost and digestates derived from food waste [18]. It is convenient to know the presence and concentration of toxic and dangerous compounds to assess the risk that compost handling, treatment, processing, and use can contribute to human health and the environment, helping to prevent many unwanted risks.

Compost quality also depends on the presence of non-compostable materials. Especially attention has been paid in the last years to plastic and emerging contaminants of plastics impurities [24,25]. In 2022, Edo et al. [21] found that the concentration of small fragments and fibers (diameter <5 mm) in compost was in the 5–20 items/g of dry weight. Five polymers represented 94% of the plastic items found: polyethylene, polystyrene, polyester, polypropylene, polyvinyl chloride.

TABLE 1.1

Stability of OFMSW Compost From Facilities Adopting Different Technologies

OFMSW source	OFMSW treatment technology		DRI (g O_2 kg⁻¹ TS h⁻¹)	
	Decomposition phase	Maturation stage	Input	Output
Street bin collection	Thermophilic anaerobic digestion (21 days) + Tunnel (1 week)	Static turned windrow (1–2 weeks)	3.4 ± 0.7	0.4 ± 0.1
	Composting tunnel (2 weeks)	Aerated windrow (8 weeks)	3.6 ± 0.1	0.9 ± 0.1
	Turned windrow (2 weeks)	Turned windrow (2 months)	–	1.0 ± 0.1
	Mesophilic anaerobic digestion (22 days) + Tunnel (3 weeks)	–	3.0 ± 0.9	0.3 ± 0.1
Door to door collection	Home composter (3 months)	–	5.5 ± 1.0	0.1 ± 0.1
	Aerated module, covered by a Gore-Tex® layer (5 weeks)	Static windrow (4 weeks)	4.8 ± 0.8	1.1 ± 0.2
	Vermicomposting (3 months)	–	3.1 ± 0.2	0.2 ± 0.1
Mechanically separated (street bin mixed waste)	Channels (4 weeks)	–	2.7 ± 0.5	1.6 ± 0.2

Created by the authors from [28, 44, 76]. OFMSW: organic fraction of municipal solid waste, DRI: dynamic respiration index, TS: total solids.

Interestingly, the presence of plastic was fewer in compost from simple composting plants but treating door-to-door collected organic fraction. In that sense, the use of compostable bioplastics seems to be a useful way to eliminate the presence of plastic in compost [21].

1.6 USES OF FOOD WASTE COMPOST

1.6.1 USE AS ORGANIC AMENDMENT

Closing the loop of nutrients between food consumption and agriculture is in the aim of the whole composting process, the use of compost as organic amendment being the summit. Compost application for agronomic purposes is regulated by each country's legislation. In the case of the European Union, a directive was published in 2019 [82] to ensure safe use of compost relying on its biological stability, maturity and agronomical value, the later evaluated mainly by means of compost content on macro and micronutrients and lack of phytotoxicity [83]. Benefits of compost application to soil have extensively been reported [84]. Among them, increase in soil fertility, water holding capacity and structure must be highlighted, resulting in reduced water needs. Other properties improved are cation exchange and buffering capacity mainly due to compost content in humic acid [85]; erosion prevention being also reported.

As organic amendment, compost application has been pointed to perform better than mineral fertilizers in terms of soil organic carbon content, also improving plant growth and yield strongly depending on compost application rate [86]. In this sense, repeated application of compost for agricultural purposes increases organic matter content of soil as well as C/N ratio, helping in retaining nutrients that in case of mineral fertilizers are easily drained by rain and irrigation [87].

On the other hand, the literature also reports some drawbacks on compost use as organic amendment such as the presence of heavy metals that can be absorbed by crops after accumulating in soil due to repeated applications, despite the chelating effect of humic substances [79]. In addition to heavy metals' entrance to the trophic chain, the presence of Zn, Cu and Pb can also diminish the activity of some enzymes at soil level [88]. Although pathogen inactivating capacity of the composting process has been recognized when provided proper conditions, active *Listeria spp.* and *Salmonella spp.* presence in compost has been detected [87]. This is also the case of some organic pollutants whose origin has been attributed to hazardous household and industrial wastes present as impurities in the OFMSW.

Thus, establishing compost quality to be used as organic amendment has been one of the priorities of national and international organizations. Compost quality depends on the origin of the waste and the composting process itself. As mentioned before, door to door collection systems have been pointed as the most adequate strategy to ensure a very low to null presence of impurities [89]. Barrena et al. [76], after analyzing compost samples from 25 different treatment plants, concluded that in terms of stability, current composting technologies demonstrate the potential for producing a high-quality product depending on the proper control of the composting process. As previously described, stability and maturity are complementary when assessing compost quality, measuring transformation of unstable organic matter in the raw waste to stable organic matter in the final product and the effects of compost on plants respectively [90]. pH, C/N ratio, moisture and organic matter content, cation exchange capacity (CEC), humic to fulvic acids ratio, heavy metals content, respiration index, self-heating capacity, germination index, are measured assessing compost quality. However, not to forget, attention should be given to the presence of emerging contaminants and microplastics.

1.6.2 SOIL BIOREMEDIATION

As widely known, compost is commonly used as soil fertilizer. However, its role in soil bioremediation has been discovered in the last few decades as very effective [91]. Contaminated soils normally do not have enough degrading capacity by themselves since the physicochemical conditions do not allow a satisfactory elimination of the pollutants. The presence of compost permits the boosting of the biodegrading process by providing the system with a satisfactory metabolic activity on one side, and providing the microbes already present in the soil with nutrients. Hence, it can help in two soil improvement tasks: bioremediation and soil amendment [92]. Compost can degrade several organic pollutants normally present in soil coming from very different sources. Among them, pesticides are the most abundant in soils dedicated to agriculture, and it becomes essential to degrade these contaminants for the recovery of these soils. Besides pesticides, other usual pollutants can be found in soils, such as heavy metals and petroleum derivatives, which cause an important damage in soil in terms of physicochemical and microbiological characteristics [93], significantly increasing the interest in the use of compost thanks to its mentioned capacity to trigger the pollutants biodegradation.

The techniques normally used for the remediation of soils are often expensive, making compost addition a very promising technique in economic terms [94]. Besides, the bioremediation of soils is an attractive technology that permits the transformation of contaminants to biomass and innocuous final products [95], which results in a much more environmentally friendly technique to treat polluted soils. However, some drawbacks should be considered, for instance, the composition of compost can vary significantly depending on its origin, which has a very important influence on the microbial and physical conditions of the compost. Hence, it becomes important to know if the selected compost can be applied for the degradation of the contaminants present in the soil to be treated, as some compounds degradation can result easier than others. In this sense, it is essential that the microbes have an easy access to contaminants, and the concentration of these should be previously known as high concentrations could result in toxicity. All these factors make the relationship between compost amendments and contaminants quite complex, and research is still needed to clarify this topic [96].

1.6.3 LANDFILL COVER

It is necessary to apply a daily cover on landfills to reduce their odors and gases emissions, which generate much inconvenience or are highly pollutant. Landfills have evolved significantly from being open dumps with no kind of treatment, or being covered with any type of soil, to modern facilities where a proper and specific treatment is applied to residues [97]. One of the main problems that landfills present is that they are considered as one of the great sources of methane emissions nowadays, a problem that needs to be addressed urgently due to global warming. Even though covering landfills with ordinary soils can effectively reduce gases emissions [98], it results much more useful to use compost as landfill cover in terms of specific mitigation of methane emissions, since it contains an increased microbiological activity and diversity. In fact, compost can act both as a biofilter, thanks to its high capacity to permeate gas flows [99] and as an active source for methane oxidation [100]. In particular, it has been reported that composts are excellent matrices for the growth and activity of methanotrophs, which are the bacteria that have the ability to oxidize methane as their only source of carbon, generating water and carbon dioxide. Besides, compost contains a high nutrient quantity for the satisfactory development of these bacteria [101], and it is an economical and environmentally friendly material to obtain high methane oxidation rates [102]. It has also been described how compost of specific materials can also be used not only in methane mitigation, but also in reducing landfills emissions of odors, widening its environmental advantages. For instance, biochar has been reported as a good material for this purpose [103]. Thus, further research in this field is needed for the development of different types of compost materials which may prove very efficient for the reduction of the several species of gases involved in landfills emissions.

1.6.4 SUPPRESSOR EFFECT

Pest and disease suppression have also been associated with compost application presumably reducing the need for pesticides [83]. Suppression capacity has been associated with the microbiota present in compost [104] which interact with the organic matter already present in soil and the target crops/plants regulates the rhizosphere microbial community. The mechanisms responsible for the compost suppressive effects can be found in more detail in the review presented by De Corato [105] where pros and cons of compost repeated application on soil are analyzed in terms of enhancement of natural soil suppressiveness. Specifically, compost from OFMSW can contribute to the control of *P. ultimum* in cucumber, *Fusarium oxysporum f.* sp. *basilici* in basil or *Sclerotinia sclerotiorum* in lettuce [105]. In fact, the influence of raw materials in the suppressive effect of compost has been reported as well as composting process operation [104]. The type of bulking agent used and its particle size seem also to play a role in defining final compost suitability to act on specific soil-borne plant pathogens [106]. Bonanomi et al. [107] studied data from more than 250 papers to ascertain the key parameters in predicting the beneficial effect of different organic amendments in disease suppression as results presented lead to inconsistent conclusions depending on the organic matter used and target plant disease. Microbiological and enzymatic characteristics were found to provide more information than the physico-chemical ones. FDA activity, substrate respiration, microbial biomass, total culturable bacteria, fluorescent *Pseudomonas* and *Trichoderma* populations were among the most useful parameters with not a single variable considered sufficient to determine disease suppression properties.

1.7 CONCLUSIONS

Composting of the OFMSW, in its different technologies and scales, is a highly valuable process for waste management that permits, by means of a biological robust process, obtaining an end product with a high potential for soil amendment, although other alternative uses as pathogen suppressor or soil bioremediation enhancer are being considered in modern literature.

The composting operational conditions must be carefully defined for a successful process: moisture, biodegradability, C/N ratio and porosity are the main parameters to be adjusted. Particularly, in the composting of the OFMSW the presence of impurities must be also considered, being plastics (macro and micro) one of the main problems. These impurities can contaminate the final compost, in which stability and maturity are also crucial.

Finally, the determination and further mitigation of gaseous emissions is very important for ensuring the sustainability of the composting process in the framework of a circular economy. Further research on food waste composting should be focused on its use in innovative applications. It also should include a holistic approach to avoid the introduction of pollutants in the food supply chain. Special attention must be given to chemical pollutants, antibiotic resistance genes and microplastics.

REFERENCES

[1] Kaza S, Yao L, Bhada-Tata P, Van Woerden F. *What a waste 2.0: A global snapshot of solid waste management to 2050*. World Bank, Washington, DC (License: Creative Commons Attribution CC BY 3.0 IGO). 2018.

[2] Hoornweg D, Bhada-Tata P. *What a waste: A global review of solid waste management*. World Bank, Washington, DC. 2012.

[3] Kawai K, Tasaki T. Revisiting estimates of municipal solid waste generation per capita and their reliability. *Waste Manage* 2016; 18:1–13.

[4] Hoornweg D, Bhada-Tata P, Kennedy C. Waste production must peak this century. *Nature* 2013; 502:615–17.

[5] Braguglia CM, Gallipoli A, Gianico A, Pagliaccia P. Anaerobic bioconversion of food waste into energy: A critical review. *Bioresource Technol* 2018; 248:37–56.

[6] European Commission. Assessment of the management of bio-waste. Working document: Accompanying the communication from the commission on future steps in bio-waste management in the European Union. 2010. Available from: https://eur- lex.europa.eu/legal-content/ES/ALL/?uri=CELEX:52010SC0577

[7] Council Directive 1999/31/EC of 26 April 1999 on the landfill of waste https://eur-lex.europa.eu/legal-content/EN/TXT/?uri=celex%3A31999L0031

[8] Angouria-Tsorochidou E, Thomsen M. Modelling the quality of organic fertilizers from anaerobic digestion – Comparison of two collection systems. *J Clean Prod* 2021; 304:127081.

[9] Korhonen J, Honkasalo A, Seppälä J. Circular economy: The concept and its limitations. *Ecol Econ* 2018; 143:37–46.

[10] Diaz LF, de Bertoldi M. Chapter 2: History of composting, in *Waste management series*. Editors: LF. Diaz M. de Bertoldi W. Bidlingmaier, E. Stentiford, Elsevier, London, 2007, pp. 7–24.

[11] Molina-Peñate E, Artola A, Sánchez A. Organic municipal waste as feedstock of biorefineries. Bioconversion technologies integration and challenges. *Rev Environ Sci Biotechnol* 2022; 21:247–67.

[12] Houghton RA. Biomass, in *Encyclopedia of ecology*. Elsevier Inc., London, 2008, pp. 448–53.

[13] Pang S. Fuel flexible gas production: Biomass, coal and bio-solid wastes, in *Fuel flexible energy generation: Solid, liquid and gaseous fuels*. Elsevier Inc., London, 2016, pp. 241–69.

[14] Alibardi L, Cossu R. Composition variability of the organic fraction of municipal solid waste and effects on hydrogen and methane production potentials. *Waste Manage* 2015; 36:147–55.

[15] Bandini F, Taskin E, Bellotti G, Vaccari F, Misci C, Guerrieri MC, Cocconcelli PS, Puglisi E. The treatment of the organic fraction of municipal solid waste (OFMSW) as a possible source of micro- and nano-plastics and bioplastics in agroecosystems: A review. *Chem Biol Technol Agric* 2022; 9:4.

[16] Pecorini I, Rossi E, Iannelli R. Bromatological, proximate and ultimate analysis of OFMSW for different seasons and collection systems. *Sustainability* 2020; 12:2639.

[17] Niu Z, Jin Ng S, Li B, Han J, Wu X, Huang Y. Food waste and its embedded resources loss: A provincial level analysis of China. *Sci Total Environ* 2022; 823:153665.

[18] O'Connor J, Mickan BS, Siddique KHM, Rinklebe J, Kirkham MB, Bolan NS. Physical, chemical, and microbial contaminants in food waste management for soil application: A review. *Environ Pol* 2022; 300:118860.

[19] Alvarez MD, Sans R, Garrido N, Torres A. Factors that affect the quality of the bio-waste fraction of selectively collected solid waste in Catalonia. *Waste Manage* 2008; 28:359–66.

[20] Friege H, Eger Y. Best practice for bio-waste collection as a prerequisite for high-quality compost. *Waste Manage Res* 2022; 40:104–10.

[21] Edo C, Fernández-Piñas F, Rosal R. Microplastics identification and quantification in the composted organic fraction of municipal solid waste. *Sci Total Environ* 2022; 813:151902.

[22] Isenhour C, Haedicke M, Berry B, MacRae J, Blackmer T, Horton S. Toxicants, entanglement, and mitigation in New England's emerging circular economy for food waste. *J Environ Stud Sci* 2022. https://doi-org.are.uab.cat/10.1007/s13412-021-00742-w

[23] Thakali A, MacRae JD, Isenhour C, Blackmer T. Composition and contamination of source separated food waste from different sources and regulatory environments. *J Environ Manage* 2022; 314:115043.

[24] Ruggero F, Gori R, Lubello C. Methodologies for microplastics recovery and identification in heterogeneous solid matrices: A review. *J Polym Environ* 2020; 28:739–48.

[25] Ruggero F, Porter AE, Voulvoulis N, Carretti E, Lotti T, Lubello C, Gori R. A highly efficient multi-step methodology for the quantification of micro-(bio) plastics in sludge. *Waste Manag Res* 2021; 39:956–65.

[26] Haug RT. *The practical handbook of compost engineering.* Lewis Publishers, Boca Raton, FL, 2003.

[27] Bertoldi M, de Vallini G, Pera A. The biology of composting: A review. *Waste Manage Res* 1983; 1:157–76.

[28] Barrena R, Gea T, Ponsá S, Ruggieri L, Artola A, Font X, Sánchez A. Categorizing raw organic material biodegradability via respiration activity measurement: A review. *Compost Sci Util* 2011; 19:105–13.

[29] Almeira N, Komilis K, Barrena R, Gea T, Sánchez A. The importance of aeration mode and flowrate in the determination of the biological activity and stability of organic wastes by respiration indices. *Bioresource Technol* 2015; 196:256–62.

[30] Mejías L, Komilis D, Gea T, Sánchez A. The effect of airflow rates and aeration mode on the respiration activity of four organic wastes: Implications on the composting process. *Waste Manage* 2017; 65:22–8.

[31] Nakasaki K, Hirai H. Temperature control strategy to enhance the activity of yeast inoculated into compost raw material for accelerated composting. *Waste Manage* 2017; 65:29–36.

[32] Rasapoor M, Adl M, Pourazizi B. Comparative evaluation of aeration methods for municipal solid waste composting from the perspective of resource management: A practical case study in Tehran, Iran. *J Environ Manage* 2016; 184:528–34.

[33] Ruggieri L, Gea T, Mompeó M, Sayara T, Sánchez A. Performance of different systems for the composting of the source-selected organic fraction of municipal solid waste. *Biosyst Eng* 2008; 101:78–86.

[34] Ruggieri L, Gea T, Artola A, Sanchez A. Air filled porosity measurements by air pycnometry in the composting process: A review and a correlation analysis. *Bioresource Technol* 2009; 100:2655–66.

[35] Adhikari B, Barrington S, Martinez J, King S. Characterization of food waste and bulking agents for composting. *Waste Manage* 2008; 28:795–804.

[36] Külcü R. Determination of the relationship between FAS values and energy consumption in the composting process. *Ecol Eng* 2015; 81:444–50.

[37] Mu D, Horowitz N, Casey M, Jones K. Environmental and economic analysis of an in-vessel food waste composting system at Kean University in the U.S. *Waste Manage* 2017; 59:476–86.

[38] Schwalb M, Rosevear C, Chin R, Barrington S. Food waste treatment in a community center. *Waste Manage* 2011; 31:1570–5.

[39] Yu S, Grant Clark O, Leonard J. Influence of free air space on microbial kinetics in passively aerated compost. *Bioresource Technol* 2009; 100:782–90.

[40] Zhang H, Li G, Gu J, Wang G, Li Y, Zhang D. Influence of aeration on volatile sulfur compounds (VSCs) and NH_3 emissions during aerobic composting of kitchen waste. *Waste Manage* 2016; 58:369–75.

[41] Maulini-Duran C, Artola A, Font X, Sanchez A. Gaseous emissions in municipal wastes composting: Effect of the bulking agent. *Bioresource Technol* 2014; 172:260–68.

[42] Zhang L, Sun X. Improving green waste composting by addition of sugarcane bagasse and exhausted grape marc. *Bioresource Technol* 2016; 218:335–43.

[43] Puyuelo B, Ponsá S, Gea T, Sánchez A. Determining C/N ratios for typical organic wastes using biodegradable fractions. *Chemosphere* 2011; 85:653–9.

[44] Barrena R, d'Imporzano G, Ponsá S, Gea T, Artola A, Vázquez F, Sánchez A, Adani F. In search of a reliable technique for the determination of the biological stability of the organic matter in the mechanical-biological treated waste. *J Hazard Mater* 2009; 162:1065–72.

[45] Zhang S, Wang J, Chen X, Gui J, Sun Y, Wu D. Industrial-scale food waste composting: Effects of aeration frequencies on oxygen consumption, enzymatic activities and bacterial community succession. *Bioresource Technol* 2021; 320A:124357.

[46] Siles-Castellano AB, López MJ, López-González JA, Suárez-Estrella F, Jurado MM, Estrella-González MJ, Moreno J. Comparative analysis of phytotoxicity and compost quality in industrial composting facilities processing different organic wastes. *J Cleaner Prod* 2020; 252:119820.

[47] Antunes L, Martins L, Pereira R, Thomas A, Barbosa D, Lemos L, Silva G, Moura L, Epamino G, Digiampietri L, Lombardi K, Ramos P, Quaggio R, de Oliveira J, Pascon R, Cruz J, da Silva A, Setubal J. Microbial community structure and dynamics in thermophilic composting viewed through metagenomics and metatranscriptomics. *Sci Rep* 2016; 6:38915.

[48] Kinet R, Destain J, Hiligsmann S, Thonart P, Delhalle L, Taminiau B, Daube G, Delvigne F. Thermophilic and cellulolytic consortium isolated from composting plants improves anaerobic digestion of cellulosic biomass: Toward a microbial resource management approach. *Bioresource Technol* 2015; 189:138–44.

[49] López-González J, Vargas-García M, López M, Suárez-Estrella F, Jurado M, Moreno J. Biodiversity and succession of mycobiota associated to agricultural lignocellulosic waste-based composting. *Bioresource Technol* 2015; 187:305–13.

[50] Moreno J, López-González JA, Arcos-Nievas MA, Suárez-Estrella F, Jurado MM, Estrella-González MJ, López MJ. Revisiting the succession of microbial populations throughout composting: A matter of thermotolerance. *Sci Total Environ* 2021; 773:145587.

[51] Hou N, Wen L, Cao H, Liu K, An X, Li D, Wang H, Du X, Li C. Role of psychrotrophic bacteria in organic domestic waste composting in cold regions of China. *Bioresource Technol* 2017; 236:20–8.

[52] Tsai C, Chen M, Ye A, Chou M, Shen S, Mao I. The relationship of odor concentration and the critical components emitted from food waste composting plants. *Atmos Environ* 2008; 42:8246–51.

[53] Zhao Y, Lu Q, Wei Y, Cui H, Zhang X, Wang X, Shan S, Wei Z. Effect of actinobacteria agent inoculation methods on cellulose degradation during composting based on redundancy analysis. *Bioresource Technol* 2016; 219:196–203.

[54] Ke G, Lai C, Liu Y, Yang S. Inoculation of food waste with the thermo-tolerant lipolytic actinomycete *Thermoactinomyces vulgaris* A31 and maturity evaluation of the compost. *Bioresource Technol* 2010; 101:7424–31.

[55] Manu MK, Kumar R, Garg A. Performance assessment of improved composting system for food waste with varying aeration and use of microbial inoculum. *Bioresource Technol* 2017; 234:167–77.

[56] Nair J, Okamitsu K. Microbial inoculants for small scale composting of putrescible kitchen wastes. *Waste Manage* 2010; 30:977–82.

[57] Karnchanawong S, Nissaikla S. Effects of microbial inoculation on composting of household organic waste using passive aeration bin. *Int J Recycl Org Waste Agric* 2014; 34:113–9.

[58] Sánchez A. Decentralized composting of food waste: A perspective on scientific knowledge. *Front Chem Eng* 2022; 4.

[59] Friedrich E, Trois C. GHG emission factors developed for the recycling and composting of municipal waste in South African municipalities. *Waste Manage* 2013; 33:2520–31.

[60] Luo W, Yuan J, Luo YM, Li GX, Nghiem L, Price WE. Effects of mixing and covering with mature compost on gaseous emissions during composting. *Chemosphere* 2014; 117:14–9.

[61] Delgado-Rodríguez M, Ruiz-Montoya M, Giráldez I, Cabeza IO, López R, Diaz MJ. Effect of control parameters on emitted volatile compounds in municipal solid waste and pine trimmings composting. *J Environ Sci Health Part A* 2010; 45:855–62.

[62] Font X, Artola A, Sánchez A. Detection, composition and treatment of volatile organic compounds from waste treatment plants. *Sensors* 2011; 11:4043–59.

[63] Dhamodharan K, Varma VS, Veluchamy C, Pugazhendhi A, Rajendran K. Emission of volatile organic compounds from composting: A review on assessment, treatment and perspectives. *Sci Total Environ* 2019; 695:133725.

[64] Yang F, Li G, Shi H, Wang Y. Effects of phosphogypsum and superphosphate on compost maturity and gaseous emissions during kitchen waste composting. *Waste Manage* 2015; 36:70–6.

[65] Ba S, Qu Q, Zhang K, Groot JCJ. Meta-analysis of greenhouse gas and ammonia emissions from dairy manure composting. *Biosyst Eng* 2020; 193:126–37.

[66] Colón J, Cadena E, Pognani M, Barrena R, Sánchez A, Font X, Artola A. Determination of the energy and environmental burdens associated to the biological treatment of source-separated municipal solid wastes. *Energy Environ* Sci 2012; 5:5731–41.

[67] Colón J, Martínez-Blanco J, Gabarrell X, Rieradevall J, Font X, Artola A, Sánchez A. Performance of an industrial biofilter from a composting plant in the removal of ammonia and VOCs after material replacement. *J Chem Technol Biotechnol* 2009; 84:1111–7.

[68] Güereca LP, Gassó S, Baldasano JM, Jiménez-Guerrero P. Life cycle assessment of two biowaste management systems for Barcelona, Spain. *Resour Conserv Recycl* 2006; 49:32–48.

[69] Yay ASE. Application of life cycle assessment (LCA) for municipal solid waste management: A case study of Sakarya. *J Cleaner Prod* 2015; 94:284–93.

[70] Lleó T, Albacete E, Barrena R, Font X, Artola A, Sánchez A. Home and vermicomposting as sustainable options for biowaste management. *J Cleaner Prod* 2013; 47:70–6.

[71] Komilis, DP. Compost quality: Is research still needed to assess it or do we have enough knowledge? *Waste Manage* 2015; 38:1–2.

[72] Wichuk KM, McCartney D. Compost stability and maturity evaluation—A literature review. *Can J Civ Eng* 2010; 37:1505–23.

[73] Scaglia B, Acutis M, Adani F. Precision determination for the dynamic respirometric index (DRI) method used for biological stability evaluation on municipal solid waste and derived products. *Waste Manage* 2011; 31:2–9.

[74] Binner E, Böhm K, Lechner P. Large scale study on measurement of respiration activity (AT4) by Sapromat and OxiTop. *Waste Manage* 2012; 32:1752–59.

[75] Aspray TJ, Dimambro ME, Wallace P, Howell G, Frederickson J. Static, dynamic and inoculum augmented respiration based test assessment for determining in-vessel compost stability. *Waste Manage* 2015; 42:3–9.

[76] Barrena R, Font X, Gabarrell X, Sánchez A. Home composting versus industrial composting: Influence of composting system on compost quality with focus on compost stability. *Waste Manage* 2014; 34:1109–16.

[77] Regulation (EU) 2019/1009, 2019. Regulation (EU) 2019/1009 of the European Parliament and of the Council of 5 June 2019, laying down rules on the making available on the market of EU fertilising products and amending Regulations (EC) No 1069/2009 and (EC) No 1107/2009 and repealing Regul Off J Eur Union 2019; 1–114.

[78] Yang Y, Wang G, Li G, Ma R, Kong Y, Yuan J. Selection of sensitive seeds for evaluation of compost maturity with the seed germination index. *Waste Manage* 2021; 136:238–43.

[79] Hargreaves JC, Adl MS, Warman PR. A review of the use of composted municipal solid waste in agriculture. *Agric Ecosyst Environ* 2008; 123:1–14.

[80] Huerta-Pujol O, Gallart M, Soliva M, Martínez-Farré FX, López M. Effect of collection system on mineral content of biowaste. *Resour Conserv Recycl* 2011; 55:1095–9.

[81] Graça J, Murphy B, Pentlavalli P, Allen CCR, Bird E, Gaffney M, Duggan T, Kelleher B. Bacterium consortium drives compost stability and degradation of organic contaminants in in-vessel composting process of the mechanically separated organic fraction of municipal solid waste (MS-OFMSW). *Bioresour Technol Rep* 2021; 13:100621.

[82] European Union, 2019. European Union Regulation (EU) 2019/1009. Available at: https://eur-lex.europa.eu/legal-content/EN/TXT/?uri=celex%3A32019R1009

[83] Le Pera A, Sellaro M, Bencivenni E. Composting food waste or digestate? Characteristics, statistical and life cycle assessment study based on an Italian composting plant. *J Clean Prod* 2022; 350:131552.

[84] Martínez-Blanco J, Lazcano C, Christensen TH, Muñoz P, Rieradevall J, Møller J, Antón A, Boldrin A. Compost benefits for agriculture evaluated by life cycle assessment. A review. *Agron Sustain Dev* 2013; 33:721–32.

[85] Garcia-Gil JC, Ceppi SB, Velasco MI, Polo A, Senesi N. Long-term effects of amendment with municipal solid waste compost on the elemental and acidic functional group composition and pH-buffer capacity of soil humic acids. *Geoderma* 2004; 121:135–42.

[86] Baldi E, Cavani L, Margon A, Quartieri M, Sorrenti G, Marzadori C, Toselli M. Effect of compost application on the dynamics of carbon in a nectarine orchard ecosystem. *Sci Total Environ* 2018; 637:918–25.

[87] Srivastava V, Ferreira de Araujo AS, Vaish B, Bartelt-Hunt S, Singh P, Singh RP. Biological response of using municipal solid waste compost in agriculture as fertilizer supplement. *Rev Environ Sci Biotechnol* 2016; 15:677–96.

[88] García-Gil JC, Plaza C, Soler-Rovira P, Polo A. Long term effects of municipal solid waste compost application on soil enzyme activities and microbial biomass. *Soil Biol Biochem* 2000; 32:1907–13.

[89] Colón J, Mestre-Montserrat M, Puig-Ventosa I, Sánchez A. Performance of compostable baby used diapers in the composting process with the organic fraction of municipal solid waste. *Waste Manage* 2013; 33:1097–103.

[90] Azim K, Soudi B, Boukhari S, Perissol C, Roussos S, Thami Alami I. Composting parameters and compost quality: A literature review. *Org Agric* 2018; 8:141–58.

[91] Kästner M, Miltner A. Application of compost for effective bioremediation of organic contaminants and pollutants in soil. *App Microbiol Biotechnol* 2016; 100:3433–49.

[92] Taiwo AM, Gbadebo AM, Oyedepo JA, Ojekunle ZO, Alo OM, Oyenirana AA, Onalaja OJ, Ogunjimia D, Taiwo OT. Bioremediation of industrially contaminated soil using compost and plant technology. *J Hazard Mater* 2016; 304:166–72.

[93] Gonçalves Sales da Silva I, Gomes de Almeida FC, Padilha da Rocha e Silva NM, Casazza AA, Converti A, Sarubbo LA. Soil bioremediation: Overview of technologies and trends. *Energies* 2020; 13:4664.

[94] Michels J, Track T, Gehrke U, Sell D. *Leitfaden—biologische verfahren zur bodensanierung*. Veröffentlichungen des BMBF (Grün-Weiße-Reihe), Umweltbundesamt, Berlin. 2000.

[95] Sayara T, Sarra M, Sanchez A. Effects of compost stability and contaminant concentration on the bioremediation of PAHs-contaminated soil through composting. *J Hazard Mater 2010*; 1–3:999–1006.

[96] Bastida F, Jehmlich N, Lima K, Morris BEL, Richnow HH, Hernández T, von Bergen M, García C. The ecological and physiological responses of the microbial community from a semiarid soil to hydrocarbon contamination and its bioremediation using compost amendment. *J Proteomics* 2016; 165:162–9.

[97] Chetri JK, Reddy KR. Advancements in municipal solid waste landfill cover system: A review. *J Indian Inst Sci* 2021; 101:557–88.

[98] Sadasivam BY, Reddy KR. Landfill methane oxidation in soil and bio-based cover systems: A review. *Rev Environ Sci Biotechnol* 2014; 13:79–107.

[99] Scheutz C, Pedersen GB, Costa G, Kjeldsen P. Biodegradation of methane and halocarbons in simulated landfill biocover systems containing compost materials. *J Environ Qual* 2009; 38:1363–71.

[100] Tienen YMdSV, de Lima GM, Mazur DL, Martins KG, Stroparo EC, Schirmer WN. Methane oxidation biosystem in landfill fugitive emissions using conventional cover soil and compost as alternative substrate – A field study. *Research Square*, 2021.

[101] Niemczyk M, Berenjkar P, Wilkinson N, Lozecznik S, Sparling R, Yuan Q. Enhancement of CH_4 oxidation potential in bio-based landfill cover materials. *Process Saf Environ* 2021; 146:943–51.

[102] Chiemchaisri C, Chiemchaisri W, Kumar S, Wicramarachchi PN. Reduction of methane emission from landfill through microbial activities in cover soil: A brief review. *Crit Rev Environ Sci Technol* 2012; 42:412–34.

[103] Ding Y, Xiong J, Zhou B, Wei J, Qian A, Zhang H, Zhu W, Zhu J. Odor removal by and microbial community in the enhanced landfill cover materials containing biochar-added sludge compost under different operating parameters. *Waste Manage* 2019; 87:679–90.

[104] Hernández-Lara A, Ros M, Cuartero J, Bustamante MA, Moral R, Andreu-Rodríguez FJ, Fernández JA, Egea-Gilabert C, Pascual JA. Bacterial and fungal community dynamics during different stages of agro-industrial waste composting and its relationship with compost suppressiveness. *Sci Total Environ* 2022; 805:150330.

[105] De Corato U. Disease-suppressive compost enhances natural soil suppressiveness against soil-borne plant pathogens: A critical review. *Rhizosphere* 2010; 13:100192.

[106] Morales AB, Bustamante MA, Marhuenda-Egea FC, Moral R, Ros M, Pascual JA. Agrifood sludge management using different co-composting strategies: Study of the added value of the composts obtained. *J Clean Prod* 2016; 121:186–97.

[107] Bonanomi G, Antignani V, Capodilupo M, Scala F. Identifying the characteristics of organic soil amendments that suppress soilborne plant diseases. *Soil Biol Biochem* 2010; 42:136–44.

2 Integrated Management of Electronic and Electric Waste (EEW) with the Application of Artificial Intelligence (AI)
Future and Challenges

Giorgio L. Russo

CONTENTS

2.1 Introduction..23
2.2 Global Production of EEW..25
2.3 Regulation Frameworks of Chemical Risk in EEW 27
2.4 Classification of EEW..28
2.5 Recycling Batteries: Example of E-Waste Stream Management..................31
2.6 Ecological Design for Minimizing Carbon Footprint in EEW.....................32
    2.6.1 Chemical Management...32
    2.6.2 Energy Efficiency..33
    2.6.3 Materials Management 33
    2.6.4 Design for Recycling and End-of-Life 33
    2.6.5 Product Expandability and Longevity 33
2.7 Artificial Intelligence (AI) in EEW Management 33
2.8 Outline of EEW Management from an Environmental Sustainability
    Point of View 35
2.9 Conclusions..38
References...38

2.1 INTRODUCTION

The improper disposal and informal treatment of electronic and electric waste (EEW) have raised serious concerns for the environment and human health and over time (Figure 2.1). Several national legislative frameworks have been implemented to regulate the management of EEW through the recycling and reuse of parts and components. This is both to prevent environmental pollution and adopt the reuse

DOI: 10.1201/9781003260738-2

FIGURE 2.1   Different types of electronic and electric waste.

of resources as the most advanced logic of saving and conserving environmental resources of raw materials, energy, and the environment in the context of the green economy.

EEW legislations in several countries mainly include restrictions on the import/export of electronic waste with updating of blacklists on components and electronic parts that can be updated with materials and compounds containing substances considered to be of concern to human health. The supply of new components for new products in the electronics supply chain has also been regulated to exclude components, paints, or other parts, which may contain metals or other substances that once they become waste or directly during the use of the equipment, may harm the population or fall as waste in the natural biological cycle. An example is the Substances of Concern In articles (SCIP) database that provides information on substances of interest in articles as such or in complex objects (products) [1].

Substance information is updated by ECHA—European Chemical Chemicals Agency, for information submitted on substances of concern in articles, on their own or in complex objects (products) established under the Waste Framework Directive (WFD) [1]. This requires that any supplier of an article must provide information on that article containing substances of very high concern (SVHC) in the Candidate List in a concentration of more than 0.1% w/w (Figure 2.2).

Therefore, the regulations for the recycling of specific categories of electronic waste and responsibilities over time have been extended by the producer, user, or final collector of the waste. Although each country's EEW legislation is structured to address the specifics of the individual country, the legislation is not always harmonized. The non-harmonization leads to several management problems still evolving under the pressure of new environmental paradigms when it comes to waste and up to the useful life of a product [2]. This implies recycling of raw materials at the end of their life cycle for reuse of individual parts and components, minimizing the phases of reduction of waste volume.

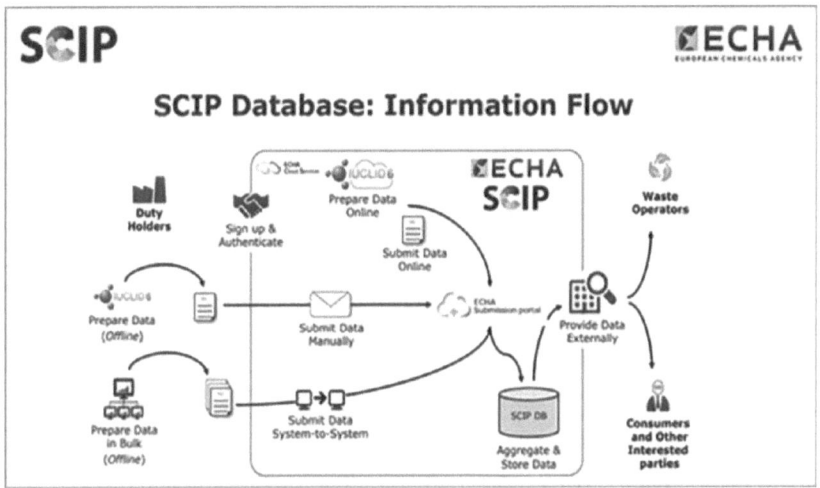

FIGURE 2.2   Information flow in the SCIP database [1].

(*Source:* European Chemicals Agency (ECHA), http://echa.europa.eu/.)

2.2   GLOBAL PRODUCTION OF EEW

The consumption of EEW is strongly related to widespread global economic development. In fact, EEW has become indispensable in modern societies and is improving living standards, but its production and use can be very demanding in terms of resources, which also illustrates a contrast to this same improvement in living standards. Higher levels of income, increasing urbanization and mobility, and further industrialization in some parts of the world are leading to increasing amounts of EEW. On average, the total weight (excluding photovoltaic panels) of global EEW consumption increases annually by 2.5 million tons [1, 2]. After its use, EEW is disposed of, generating a waste stream that contains hazardous and valuable materials. This waste stream is referred to as electronic waste, or waste electrical and electronic equipment (WEEE), a term mainly used in Europe.

In 2019, the world generated an astonishing 53.6 Mt of electronic waste, an average of 7.3 kg per capita. Global EEW production has grown by 9.2 Mt since 2014 and is expected to grow to 74.7 Mt by 2030, almost doubling in just 16 years. The increasing amount of EEW is mainly fueled by higher EEW consumption rates, short life cycles and few repair options. Asia generated the highest amount of EEW in 2019 at 24.9 Mt, followed by the Americas (13.1 Mt) and Europe (12 Mt), while Africa and Oceania generated 2.9 Mt and 0.7 Mt respectively. Europe ranked first in the world in terms of EEW. Oceania followed by the Americas (13.3 kg per capita), while Asia and Africa generated only 5.6 and 2.5 kg per capita respectively (Figure 2.3) [2].

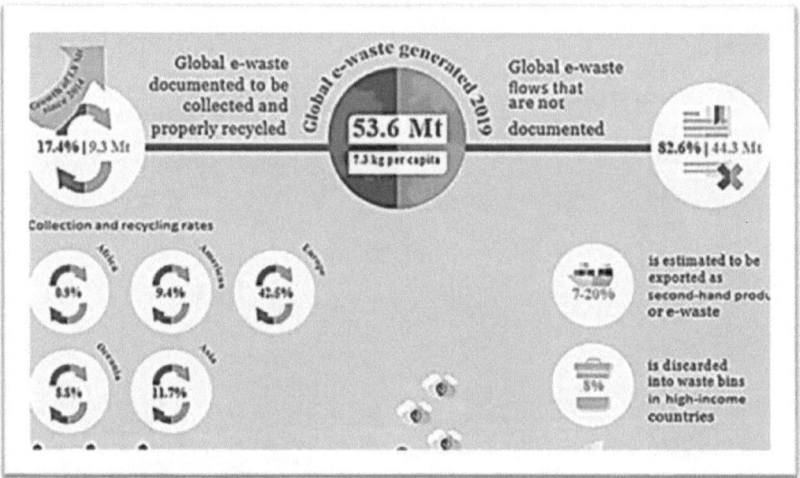

FIGURE 2.3   Global e-waste generated in 2019 [2].

In 2019, formally documented collection and recycling amounted to 9.3 Mt, i.e., 17.4% compared to the electronic waste generated. It has grown 1.8 Mt since 2014, an annual growth of almost 0.4 Mt. However, the total production of electronic waste increased by 9.2 Mt, with an annual growth of almost 2 Mt. Therefore, recycling activities are not keeping pace with the global growth of EEW. Statistics show that in 2019, the continent with the highest collection and recycling rate was Europe with 42.5%, Asia in second place with 11.7%, the Americas and Oceania were similar to 9.4% and 8.8%, respectively, and Africa had the lowest rate at 0.9% [1, 2].

The fate of 82.6% (44.3 Mt) of EEW generated in 2019 is uncertain, and its location is whereabouts, and the environmental impact varies across the different regions. In high-income countries, a waste recycling infrastructure is usually developed and About 8% of EEW is discarded in garbage cans and subsequently landfilled or incinerated. This is mostly composed of small equipment and small IT [1, 2].

Discarded products can sometimes still be reconditioned and reused, and therefore are usually shipped as second-hand products from high-income countries to low- or middle-income countries. However, a significant amount of EEW is still being exported illegally or under the pretext of being reused or pretending to be scrap metal. It can be assumed that the volume of transboundary movements of used EEW or electronic waste varies from 7 to 20% of the EEW generated [1].

Most undocumented household and commercial EEW are likely mixed with other waste streams, such as plastic waste and metal waste. This means that easily recyclable fractions could be recycled but often in lower conditions without depollution and without the recovery of all valuable materials. Therefore, such recycling is not preferred.

In middle- and low-income countries, the EEW management infrastructure is not yet fully developed or, in some cases, is absent altogether. Therefore, EEW is mainly managed by the informal sector. In this case, EEW is often handled under inferior conditions, causing serious effects on the health of workers and children who often live, work, and play near EEW management activities.

2.3 REGULATION FRAMEWORKS OF CHEMICAL RISK IN EEW

Chemicals in products and hazardous substances in the life cycle of electrical and electronic products have long been emerging policy issues under the (SAICM) Strategic Approach International Chemical Management, since 2019. Among others, the activities within the project to track and control chemicals along the value chains of electrical and electronic products, stakeholders must first identify the relevant chemicals to be addressed. Given the complexity of the chemical world as well as the complexity of value chains in the electronics sector. National and international regulatory bodies have established reference regulatory frameworks for the management of hazardous chemicals. Some of these frameworks relate specifically to chemicals related to electrical and electronic products, while others relate to chemicals. The framework presented maps the existing regulatory frameworks including specific substances provisions for electronic and electrical products [3].

TABLE 2.1

Regulatory Frameworks Comprising Substance Specific Provisions and Lists With Regards to Electronic and Electrical Products and Batteries

Country/ region	Regulation name	Brief description	Reference
China	Administrative Measures on the Restriction of the Use of Certain Hazardous Substances in Electrical and Electronic Products	Restricts the use of certain substances and substance groups for their use in certain electrical and electronic products and establishes labelling requirements. Substances regulated include lead, mercury, cadmium and PBDEs.	[4]
China	Limitation of Mercury, Cadmium and Lead Contents for Alkaline and Non-alkaline Zinc Manganese Dioxide Batteries (GB24427–2009)	–	[5]
EU	Directive 2011/65/EU of the European Parliament and of the Council of 8 June 2011 on the restriction of the use of certain hazardous substances in electrical and electronic equipment (RoHS)	Restricts the use of ten substances and substance groups for their use in the manufacture of various types of electronic and electrical equipment. Substances regulated include lead, mercury, PBDEs and certain phthalates. The regulation also provides a list of specific applications exempted from the restrictions.	[6]

(Continued)

TABLE 2.1
(Continued)

Country/region	Regulation name	Brief description	Reference
EU	Directive 2006/66/EC of the European Parliament and of the council of 6 September 2006 on batteries and accumulators and waste batteries and accumulators and repealing Directive 91/157/EEC (Batteries directive)	Regulates the manufacturing and disposal of batteries and accumulators in the EU and contains provisions for the use and labelling requirements of certain chemicals in batteries.	[7]
Norway	Regulations on restrictions on the use of hazardous chemicals and other products (product regulations, FOR-2004-06-01-922)	Restricts or prohibits the use of certain substances in products, including certain electronic and electrical products and batteries. List of substances provided within the regulatory document.	[8]
USA/CA	Health and safety code – Division 20. Miscellaneous health and safety provisions [24000–26250] – chapter 6.5. Hazardous Waste Control [25100–25259] (California RoHS)	Regulations prohibiting a covered electronic device from being sold or offered for sale in California if that device is prohibited from being sold or offered for sale in the EU due to the presence of lead, mercury, cadmium, or hexavalent chromium above certain maximum concentration values (MCVs).	[9]
USA/NY	Environmental conservation act – Article 27: Collection, Treatment and Disposal of Refuse and Other Solid Waste – Title 7: Solid Waste Management and Resource Recovery Facilities – Section 27–0719: Battery management and disposal	Regulations on the placing of the market of batteries containing certain levels of hazardous substances. Substances addressed include mercury, lead, and cadmium.	[10]

2.4   CLASSIFICATION OF EEW

EEW refers to all items of electrical and electronic equipment and its parts that have been discarded by its owner as waste without the intent of re-use. EEW is also known as waste electrical and electronic equipment (WEEE), or in short called electronic waste (e-waste). EEW includes both household appliances such as refrigerators, washing machines and microwaves, as well as brown products consisting of

TVs, radios and computers that have achieved their purposes for their current owner. Obsolete electrical and electronic products represent a high environmental risk [2].

A classification system for EEW must consider the composition of comparable materials, in terms of hazardous substances and industrially useful materials and their end-of-life attributes. Products within the same category must have, to simplify classifications, a homogeneous average weight, and a distribution over lifetime, for quantitative evaluation for similar products. Finally, large, or environmentally relevant EEW, where a lot of data is potentially available, should be treated separately. Currently, there is a classification system that meets these criteria: the classification developed by the UNU [11]. This classification is called UNU-KEYS. UNU-KEYS are constructed in such a way that product groups share comparable average weights, material compositions, end-of-life characteristics, and lifetime distributions. The full list of UNU-KEYS is presented in Table 2.2.

TABLE 2.2
UNU Keys Used in EEW Classification [11]

UNU KEY	WEEE	UNU Key Description
0101	4	Professional Heating & Ventilation (excl. cooling equipment)
0102	4	Dishwashers
0103	4	Kitchen (f.i. large furnaces, ovens, cooking equipment)
0104	4	Washing Machines (incl. combined dryers)
0105	4	Dryers (wash dryers, centrifuges)
0106	4	Household Heating & Ventilation (f.i. hoods, ventilators, space heaters)
0108	1	Fridges (incl. combi-fridges)
0109	1	Freezers
0111	1	Air Conditioners (household installed and portable)
0112	1	Other Cooling (f.i. dehumidifiers, heat pump dryers)
0113	1	Professional Cooling (f.i. large air conditioners, cooling displays)
0114	5	Microwaves (incl. combined, excl. grills)
0201	5	Other Small Household (f.i. small ventilators, irons, clocks, adapters)
0202	5	Food (f.i. toaster, grills, food processing, frying pans)
0203	5	Hot Water (f.i. coffee, tea, water cookers)
0204	5	Vacuum Cleaners (excl. professional)
0205	5	Personal Care (f.i. tooth brushes, hair dryers, razors)
0301	6	Small IT (f.i. routers, mice, keyboards, external drives & accessories)
0302	6	Desktop PCs (excl. monitors, accessories)
0303	2	Laptops (incl. tablets)
0304	6	Printers (f.i. scanners, multifunctionals, faxes)
0305	6	Telecom (f.i. (cordless) phones, answering machines)
0306	6	Mobile Phones (incl. smartphones, pagers)

(Continued)

TABLE 2.2
(Continued)

UNU KEY	WEEE	UNU Key Description
0307	4	Professional IT (f.i. servers, routers, data storage, copiers)
0308	2	Cathode Ray Tube Monitors
0309	2	Flat Display Panel Monitors (LCD, LED)
0401	5	Small Consumer Electronics (f.i. headphones, remote controls)
0402	5	Portable Audio & Video (f.i. MP3, e-readers, car navigation)
0403	5	Music Instruments, Radio, HiFi (incl. audio sets)
0404	5	Video (f.i. Video recorders, DVD, Blue Ray, set-top boxes)
0405	5	Speakers
0406	5	Cameras (f.i. camcorders, photo & digital still cameras)
0407	2	Cathode Ray Tube TVs
0408	2	Flat Panel Display TVs (LCD, LED, PDP)
0501	5	Lamps (f.i. pocket, Christmas, excl. LED & incandescent)
0502	3	Compact fluorescent lamps
0503	3	Straight tube fluorescent lamps
0504	3	Special (Hg, high & low pres. Na, other prof. lamps
0505	3	LED
0506	5	Luminaires (incl. HH incandescent fittings)
0507	5	Luminaires
0601	5	Tools (all HH saws, drills, cleaning, garden, etc.)
0602	4	Tools (Professional tools, excl. dual use)
0701	5	Toys (small toys, vehicles, small music)
0702	6	Game Consoles (video games and consoles)
0703	4	Toys and sound beds (exercising, large music instr.)
0801	5	Medical (small HH thermometers, blood pressure meters)
0802	4	Medical (hospital, dentist, diagnostics, etc.)
0901	5	Monitoring (alarm, heat, smoke, security, excluding screens)
0902	4	Monitoring (prof. M&C, garage, diagnostic, etc.)
1001	4	Dispenser (non-cooled vending, coffee, tickets, etc.)
1002	1	Dispenser (cooled vending, bottles, candy, etc.)

In addition to the established WEEE categories, the United Nations University (UNU) has developed an additional classification that includes 54 categories. These categories are related to the 10 original EU WEEE categories (the first two digits of the UNU keys) but are further divided into superficial categories. UNU keys are chosen in such a way that each key describes a single product or a small range of products that have a uniform average weight, environmental relevance, and market behavior.

The UNU-KEYS are used in the implementing act to describe the common methodology for calculating the WEEE collection targets for Article 7 of the European Commission 2017/2018 [3].

TABLE 2.3

Schematic Diagram of the Generic EEW Management Model for Implementing Effective EEW Legislation [3]

E—WASTE LEGISLATION				
↓				
Amendments to Regulators	Law Enforcement Bodies		System Feedback	
Identified Productors	Production Abiding Legislation Guidelines		Declaration of Components and Hazardous Substances	Consumers
↓				
E – WASTE GENERATED				
Separate Collection of E-waste	E – Waste disposal with other wastes			
E—Waste Recycling	Landfilling and Incineration			Informal processing
Materials Recovered	Risk Assessment			
↑↓				
E—WASTE: ILLEGAL IMPORT EXPORT				

2.5 RECYCLING BATTERIES: EXAMPLE OF E-WASTE STREAM MANAGEMENT

Batteries Li and Co: important and useful technologies for mobility and new ecological problems and challenges. Like other components of modern electronics, lithium, and cobalt batteries, they are everywhere: from micro appliances to electric vehicles and smartphones. Yet they are not counted in global e-waste streams. Batteries normally contain mainly Lithium, which is the fastest growing market. Battery costs are falling, and demand is increasing, driven by the demand for smartphones and electric vehicles: a revolution in non-polluting transport due to the absence of combustion. Electric vehicle batteries often contain the same amount of lithium as 1,000 smartphones [10]. The EU and the People's Republic of China have introduced laws making car manufacturers responsible for recycling batteries. There is also the potential for a large market for second-life batteries. The renewable energy grids of the future will need vast amounts of storage, which could be filled by batteries that are too old for cars, but useful for static uses.

Over 11 million tons of used lithium-ion batteries are expected to be discarded by 2030, representing a significant challenge, but also given the opportunity for dramatic increase in demand for materials such as lithium and 11-fold cobalt [12, 13]. In electronics, the collection of devices remains critical and, as with all components, will be important for the increased collection of batteries for recycling. When a battery reaches the end of its life, it will be essential to ensure that the constituent components are intended for selective recovery. Already today it will have to be created eco-sustainable right from the design phase, which also includes instructions for disassembly at the end of its life and a simpler identification of the materials to be used for recycling or other forms of recovery.

At the management level, producers and recyclers will have to be connected to ensure that the batteries are disposed of properly. One of the most important materials for recent electric batteries is also cobalt. From a strategic and geographic point of view, 66% of the world's cobalt is extracted in the Democratic Republic of Congo (DRC). About 90% of the cobalt produced in the DRC comes from large-scale and mechanized mining operations. However, 10% are it is estimated to come from small-scale mines, often in dangerous working conditions, where child labor is widespread also due to the extraction of this metal which is dangerous for humans as well as for the environment [14]. Efforts are underway to address these challenges, which embrace the life cycle and value chain of battery technology.

Lithium batteries are not all the same; in fact, the six construction types of lithium accumulators can be classified as follows: Lithium cobalt oxide (LCO) with cobalt oxide ($LiCoO_2$); LMO with manganese oxide ($LiMn_2O_4$); NMC with nickel manganese cobalt oxide; LFP with iron phosphate ($LiFePO_4$); and lithium titanium LTO (Li_2TiO_3) [12, 13]. The $LiCoO_2$ battery, thanks to its high specific energy, is the most common on smartphones, laptops and digital cameras. It is made up of a cobalt oxide cathode and a graphite anode. The cathode has a microscopic structure with superimposed planes, between which the lithium ions flow. During operation, the lithium ions go from the anode to the cathode, while in the charging phase the flow is reversed. The defects of this type of cell are a relatively short life, low thermal stability, and low specific power. These batteries are subject to the phenomenon of SEI (solid electrolyte interface), that is, the deposit on the anode of a layer of solid lithium which increases its resistance, especially due to rapid recharging at low temperatures. In the newer batteries nickel, manganese and aluminum have been added to improve life and lower costs. The lithium-cobalt battery is giving way to lithium-manganese due to the high cost of cobalt and improved performance of the batteries that use cathodes with different materials [12, 13].

2.6 ECOLOGICAL DESIGN FOR MINIMIZING CARBON FOOTPRINT IN EEW

Environmental sustainability is an important factor that companies, suppliers, and consumers are considering more each day when designing, manufacturing, and buying products, including electronics. Designers, companies, and suppliers have the responsibility to create sustainable products that are up to the standards that both regulators and consumers demand, and which will reduce social and environmental negative impacts [15]. The principles phases to Design Environmentally Sustainable Electronics are provided below:

2.6.1 CHEMICAL MANAGEMENT

A lot of products use chemicals and metals that could be harmful for the environment and human health. Now, manufacturers are aiming to drastically reduce or completely remove these materials in the production of electronic products with the reduction of elimination of sensitive substances from all products [16].

2.6.2 ENERGY EFFICIENCY

Reducing energy consumption is important to reduce the emission of greenhouse gasses and of course, save significant costs in electricity. This includes power management features, such as low-power standby mode, auto power savings, and user-activated power options via the Energy Star features [16].

2.6.3 MATERIALS MANAGEMENT

Although the miniaturization of components and devices has helped to reduce the use of resources needed to produce electronics, there are still opportunities to improve the use and management of resources. Design, when possible, smaller products that will need less materials and resources in manufacturing and which will also generate less waste at the end of their lifecycle. Replace difficult-to-recycle plastics with lighter and more durable materials, such as aluminum and magnesium, that make the products more valuable to recycle. Consider the use of vegetable-based and recycled plastics, recycled glasses, and propylene in your products, packaging, and shipping materials. Design the product for reusable and long-life batteries instead of single use batteries [16].

2.6.4 DESIGN FOR RECYCLING AND END-OF-LIFE

Electronics contain an important variety of materials that can be recovered through responsible recycling. These reclaimed materials can be used in the manufacturing of new products, with a significantly smaller footprint than traditional mining and extraction. Consumers are more committed to buying electronics that can be recycled at the end of their life cycle and this is driving processes and design principles that make electronics easier to dismantle, sort, and recycle [17–19].

2.6.5 PRODUCT EXPANDABILITY AND LONGEVITY

Consumers want to purchase products that are multi-functional units and that will have a longer lifecycle. Consider adopting a part-update approach. Instead of constantly releasing new versions of the product, offer the option to only change the part that was updated. Design products with features that will keep the device useful and relevant longer, extending the life of the device and avoiding it entering the waste stream early. Consider repair procedures as part of your mechanical design process so that one small failure does not lead to the disposal of the whole product. This kind of check may also lead to the discovery of additional assembly process efficiencies [18, 19].

2.7 ARTIFICIAL INTELLIGENCE (AI) IN EEW MANAGEMENT

Designers, manufacturers, investors, traders, miners, commodity producers, consumers, policymakers, and others have a crucial role to play in reducing waste, maintaining value within the system, prolonging the economic and physical life of an item,

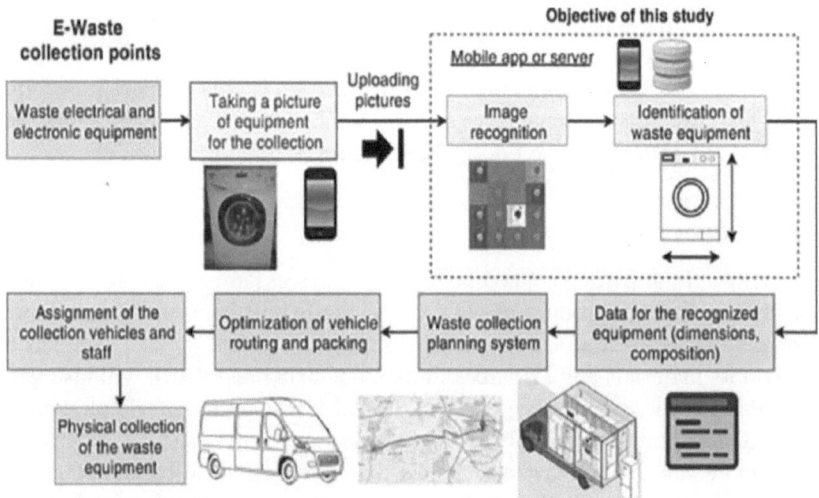

FIGURE 2.4   A typical procedure of e-waste collection based on AI technology in image recognition [21].

(Source: Under a Creative Commons license https://s100.copyright.com/AppDispatchServlet? publisherName=ELS&contentID=S0956053X20302105&orderBeanReset=true.)

as well as its ability to be repaired, recycled, and reused (Figure 2.4). Technological changes such as cloud computing and the Internet of Things (IoT) could have the potential to "dematerialize" the electronics industry [20]. Increased service business models and better monitoring and recall of products could lead to global circular value chains. Material efficiency, recycling infrastructure, and increasing the volume and quality of recycled materials to meet the needs of electronics supply chains will all be essential. The term "AI waste management" or "AI-based waste management" refers to the use of artificial intelligence in relation to waste management. Artificial intelligence is already being used in various ways to manage the infrastructure around us, and this includes e-waste management. For example, AI can be used to inform waste management companies when garbage bins are full, meaning collection routes can be optimized, labor reduced, and fuel saved. Alternatively, AI can be used to match operational resources with nearby waste services and sites, ensuring that more waste is sent to recycling centers than to landfills. There are various ways in which artificial intelligence can be used within the e-waste management industry, it is the separate collection of waste based on artificial intelligence that can really be of help to the sector [20, 21]. When the e-waste is thrown away, it is sorted according to the rules of the law and the criteria useful for recycling. This is done separately by the user of the waste or in a material recovery plant. The increasing economic value of the number of recyclable materials increasingly justifies the strengthening of this chain of recovery of quasi-raw materials as they can be considered waste to be recycled. Recycled materials are prevented from being sent to landfill, with the consequent associated harmful environmental impact. E-waste sorting is usually undertaken by a combination of human workers and sorting mechanics: the use of artificial intelligence, combined with robotics, can simplify the task, and increase the speed and accuracy with which waste can be sorted.

AI-driven robots can work long hours at a constant speed that allows separate collection to continue for an extended period, with an AI-powered waste detection capability that is as accurate as the human eye.

If e-waste is placed in the wrong bin or waste is mixed, for example, the additional burden of sorting and separating the waste can be considered excessive in terms of cost and time. When AI is used for e-waste segregation, however, material recovery facilities can separate waste with the added benefits of AI robots that are able to work consistently for long periods [20, 21]. A circular economy promotes economic development through the reuse, repair, recycling, and renewal of existing materials, significantly reducing the negative environmental impact and energy consumption for collective hygiene from waste production. How we treat waste and the amount of waste we recycle will be a key factor in a circular economy and making sustainable waste management cost-effective through separate collection of AI waste brings us to this reality. Advanced technology, such as artificial intelligence, computer vision and machine learning, is already being used in our daily life (Figure 2.5).

Complex nonlinear processes are difficult to model, predict and optimize using conventional methods. Artificial intelligence (AI) techniques have gained momentum in offering alternative computational approaches to solving e-waste management problems. AI has been efficient at addressing problems by learning from experience and handling uncertainty and incomplete data. AI technology deals with the design of computer systems and programs that can mimic human traits such as problem-solving, learning, perception, understanding, reasoning, and awareness of the surrounding environment [21].

Models such as artificial neural networks (ANN), expert systems, genetic algorithms and fuzzy logic (FL) have the ability to solve ill-defined problems, configure complex mappings and predict results. Each model or branch of AI performs a specific function; for example, ANN models can train data for classification and prediction. In addition, NNs can be used to manage big data in urban geographies and perform geographic analysis. Expert systems, such as FL, can acquire human cognitive and reasoning skills and possess a knowledge base. These systems have a simple linguistic syntax that is adept at handling complex operations and qualitative attributes. On the other hand, evolutionary algorithms such as GA, adopt the concept of natural selection to obtain optimal results by selecting the most suitable data to manage unforeseen conditions [22].

2.8 OUTLINE OF EEW MANAGEMENT FROM AN ENVIRONMENTAL SUSTAINABILITY POINT OF VIEW

The reduction of the total quantity of waste that must be managed locally and worldwide, passes from a high efficiency of the management of the life cycle of the electronic equipment. The balance will be given between the sources of the materials used, the energy used to produce it, distribute them, on the other hand, the increase in reuse, repairability, extension of the life of products and the recycling of electronics. A life cycle approach could be the key element in the sustainable management of materials in the sustainable future globally [18]. The life cycle of electronic products that includes several phases and paradigms of sustainable growth can be introduced, indicating the direction to be taken in industrial and collective policies from the producer-designer to the consumer, passing from the responsibility of the seller and local communities for the collection of EEW at the end of life (Table 2.4).

TABLE 2.4

Relationship Among Life Cycle Steps of EE Equipment, Sustainability of EES and EEW, and Stakeholders

Life cycle steps of EE equipment	Economic and Ecological Sustainability (EES)	Sustainability objectives for EEW	Stakeholders
Raw Materials	Raw or virgin materials such as oil, iron, gold, palladium, platinum, copper, are in high-tech electronics. These materials are due mined, transported and processed. These activities use large amounts of energy and produce greenhouse gas emissions, pollution, and a drain on our natural resources.	Source reducing raw materials can save natural resources, conserve energy, and reduce pollution.	Loop by collections of refuse on EEW, productors of EE, productors of raw materials. Others: authority for the landscape protections, authority for the works protections. Final stakeholder: global community-consumers.
Supply/ Manufacturing	In analyzing the life cycle of a product, the reduction of sources is important in production as it is preferable from an environmental point of view.	Electronics will use fewer materials overall, use more recycled materials, and be more durable and recyclable. In turn, the overall impact of the product on human health and the environment will be reduced.	Productors (for their own policy of sustainability always request on market and buyers-consumers attentive to ecological sustainability. Sustainability planners, Authority, University.
Sale	The first step in using electronics sustainably involves purchasing equipment that has been designed with environmentally preferable attributes.	News ecological standards for products for new targets of consumers-users, sustainability oriented.	Consumers, users.
Use	Products with best longevity, durability, reusability, and recyclability	News vision-direction for users and consumers, how quality and environmental savings of the product they are purchasing.	Consumers, productors oriented at sustainability markets (e.g., carbon footprint, carbon free) governments and institutions, agency educative oriented to sustainability policy.

Life cycle steps of EE equipment	Economic and Ecological Sustainability (EES)	Sustainability objectives for EEW	Stakeholders
Collection	End-of-life electronics are normally collected at delivery points (retailers) or in areas dedicated to public access. The strengthening of this important system for the recovery of materials must be increased by improving, already in the design phase, both the dismantling for recycling and the identification of the components for recycling (design of sustainability, util and simply information on labeling)	Reductions of components to landfill. Energy and raw materials saving.	Designer, manufacturer, retailer, user.
Landfill	Today they are designed (when foreseen) to minimize the possibility of the release of hazardous waste into the environment.	For an optimization of the recovery of resources from discarded materials, the design of the landfills may include a further selection and separation of the waste to improve the total yield of the separate collection of waste from EE.	Local authority, citizens (of natural resources, human health protection).
Reuse/Refurbish	Refurbished electronics are electronics that have been updated and repaired for resale.	Reuse of used electronics extends product lifespans and contributes to the source reduction of raw materials. Will be studying communications techniques to enhance these new products.	Engineering designer, manufacturer, retailer, users, advertising.
Recycling	The recycling includes sorting, dismantling, mechanical separation, and recovery of valuable materials. Over time, better and more efficient recycling technologies can be developed.	Recycling of used electronics can yield materials that can be to the supply chain to be used again, reducing raw materials used and the overall need for disposal.	Users, retailers, local authority, diffusions of these educational centers for diffusion of these practices.

2.9 CONCLUSIONS

EEW is the fastest growing household waste stream in the world, fueled primarily by higher consumption rates of electrical and electronic equipment, short life cycles, and few options for repair. Only 17.4% of e-waste in 2019 was collected and recycled. This means that gold, silver, copper, platinum, and other high-value recoverable materials were mostly dumped or burned rather than being collected for treatment and reuse.

Waste electrical and electronic equipment include a wide range of devices such as computers, refrigerators, and mobile phones at the end of their lives. This type of waste contains a complex mixture of materials, some of which are hazardous. These can cause serious environmental and health problems if discarded devices are not managed properly. In addition, modern electronics contain rare and expensive resources, which can be recycled and reused if waste is managed effectively. Improving the collection, treatment, and recycling of EEE at the end of its life cycle can improve sustainable production and consumption, increase resource efficiency, and contribute to the circular economy. Recycling e-waste reduces chemical emissions during the disposal of devices and within the process of creating new devices. When companies use recycled materials to make new products, they use less energy than they would if they used brand new materials. When e-waste is buried in landfills, these toxic chemicals can penetrate the soil and contaminate local water supplies. This endangers the health of any nearby community and can lead to environmental catastrophes.

The increase in the production of electronic products and the growing intrinsic need to possess the latest available technologies has led to a significant increase in the amount of electronic waste produced each year. Therefore, proper management and recycling of e-waste is critical to the sustainability of any modern society. While the industrial and commercial collection of e-waste has been in the spotlight, solutions for collecting e-waste from individual households are still small and limited. The implementation of AI allows automated machines – collection bins, vehicles, conveyor belts, storage – to identify common electronic waste based on learning about transfer from images and other specifications that will accelerate the collection and recovery of this important economic resource which is electronic waste. Complex identification systems based on convolutional neural networks are now used to classify electronic waste with high accuracy. The system will relieve unskilled labor from the dangerous sorting process, while providing cost reduction.

REFERENCES

[1] ECHA: https://echa.europa.eu/en/scip. Accessed on June 25, 2022.
[2] Forti V., Baldé C.P., Kuehr R., Bel G. *The Global E-waste Monitor 2020: Quantities, flows and the circular economy potential. United Nations University (UNU)/United Nations Institute for Training and Research (UNITAR) – co-hosted SCYCLE Programme,* International Telecommunication Union (ITU) & International Solid Waste Association (ISWA), Bonn/Geneva/Rotterdam.
[3] Patil R.A., Ramakrishna S. A comprehensive analysis of e-waste legislation worldwide. *Environmental Science and Pollution Research* 27(13):14412–14431 (2020).

[4] China administrative measures on the restriction of the use of certain hazardous substances in electrical and electronic products. http://www.miit.gov.cn/n1146285/n1146352/n3054355/n3057254/n3057260/c4608532/content.html. Accessed on June 25, 2022.

[5] China GB 24427–2009. http://c.gb688.cn/bzgk/gb/showGb?type=online&hcno=F-4196DA15745273FD55C1D8952B12747. Accessed on June 25, 2022.

[6] EU RoHS. https://ec.europa.eu/environment/waste/rohs_eee/legis_en.html. Accessed on June 25, 2022.

[7] EU Batteries Directive. https://ec.europa.eu/environment/waste/batteries/. Accessed on June 25, 2022.

[8] Norwegian Product Regulation. https://lovdata.no/dokument/SF/forskrift/2004-06-01-922. Accessed on June 25, 2022.

[9] Ro HS. https://dtsc.ca.gov/restrictions-on-the-use-of-certain-hazardous-substances-rohs-in-electronic-devices/. Accessed on June 25, 2022.

[10] New York environmental conservation law, battery management and disposal §27–0719. www.nysenate.gov/legislation/laws/ENV/27-0719.

[11] Wang F., Huisman J., Baldé K., Stevels A. A systematic and compatible classification of WEEE. In *Electronics Goes Green*, Berlin: IEEE (2012).

[12] Wu B.Y., et al. Assessment of toxicity potential of metallic elements in discarded electronics: A case study of mobile phones in China. *Journal of Environmental Sciences* 20(11):1403–1408 (2008).

[13] Li Z., et al. Comparative life cycle analysis for value recovery of precious metals and rare earth elements from electronic waste. *Resources, Conservation and Recycling* 149:20–30 (2019).

[14] Mentore V., Fabiola Z., Margaret B., Terry T., Teklit A. Application of an integrated assessment scheme for sustainable waste management of electrical and electronic equipment: The case of Ghana. *Sustainability* 12:8:3191 (2020).

[15] Minimizing Your Carbon Footprint by Feroz Balsara: https://neuronicworks.com/blog/environmentally-sustainable-design/

[16] www.epa.gov/smm-electronics/basic-information-about-electronics-stewardship#01

[17] www.epa.gov/smm-electronics/assessment-tools-electronics-stewardship#01

[18] Isabel C.N., Eloiza K., Fernanda H.B., Alexandre B.F., Delcio P. Life cycle analysis of electronic products for a product-service system. *Journal of Cleaner Production* 314:127926 (2021). https://doi.org/10.1016/j.jclepro.2021.127926.

[19] Fiore S., et al. Improving waste electric and electronic equipment management at full-scale by using material flow analysis and life cycle assessment. *Science Total Environment* 659:928–939 (2019).

[20] Jueru H., Dmitry D.K. Artificial intelligence for planning of energy and waste management. *Sustainable Energy Technologies and Assessments* 47:101426 (2021).

[21] Nowakowski P., Pamuła T. Application of deep learning object classifier to improve e-waste collection planning. *Waste Management* 109:1–9 (2020).

[22] Yetilmezsoy K., Ozkaya B., Cakmakci M. Artificial intelligence-based prediction models for environmental engineering. *Neural Network World* 21:3 (2011).

3 Thermal Conversion of Solid Waste via Pyrolysis to Produce Bio-Oil, Biochar and Syngas

Saleh Al Arni

CONTENTS

3.1 Introduction..41
3.2 Solid Waste 42
3.3 Waste Conversion into Energy...44
3.4 Thermochemical Processes...45
3.5 Pyrolysis Processes...47
    3.5.1 Types, Mechanisms and Conditions 47
    3.5.2 Reactor Types..49
    3.5.3 Products of Pyrolysis Processes..49
3.6 Conclusion 51
References...52

3.1 INTRODUCTION

Energy can be recovered from waste using several technologies such as biochemical (anaerobic digestion, gas fermentation, carbon chain elongation) and thermochemical conversion processes (pyrolysis, gasification, hydrothermal carbonization) [1]. These technologies were applied on industrial wastes such as tyres [2–4], plastics [5–7], and municipal solid waste (MSW) [8–11]. Czajczyńska et al. [12] presented work on the benefits of using pyrolysis processes in the waste management sector. Al-Salem et al. [7] investigated the application of pyrolysis process on plastic solid waste and emphasized its advantages in mitigating environmental pollution. Umeki et al. [2] investigated the use of pyrolysis on waste tyres to produce alternative fuels. Wang et al. [11] investigated the possibility of using pyrolysis processes in North Carolina in the field of MSW treatment and disposal where five pyrolysis plants were built.

Pyrolysis plants are, also, operating in Germany and other European countries and Japan for MSW treatment [13]. In addition, Chen et al. [14] presented a review about the MSW pyrolysis. They studied the influence of operating parameters such as temperature, heating rate and residence time and the types of reactors, products and environmental impacts.

DOI: 10.1201/9781003260738-3

Waste has several types; in this chapter the intended type of waste is limited to the waste that could be treated by thermochemical conversion processes especially pyrolysis such as rubber, plastic waste, municipal solid waste, agriculture waste includes wood, and biowaste like manure, or sludge. Pyrolysis has been highlighted as a technology for treatment of plastics, rubbers, natural lignocellulosic waste and sludge [1]. It has several advantages in the application of combustion (turbines, boilers, engines, etc.) such as transportation, storing and flexibility in solicitation [15]. In addition, using this process, it is possible to obtain a clean energy that is different from the one produced by the traditional municipal solid waste incineration plants [14]. Furthermore, Chen et al. [14] presented a review on municipal solid waste treatment by pyrolysis technologies including types of pyrolysis reactors, commercial and semi-commercial pyrolysis technologies and systems available in the literature.

Energy recovery from waste is a waste management process that is done by the thermal decomposition of organic-based materials [16]. The main processes involved in recovering energy are pyrolysis and gasification processes. The recovered energy could be in the form of fuel 'synthesis gas' that can be directly combusted to generate electricity. To be emphasized, finally, that thermal treatment processes of waste can be combined such as gasification and pyrolysis or gasification and combustion or all three processes. This chapter will deal with the pyrolysis processes.

3.2 SOLID WASTE

Waste is produced by human activities and needs to be disposed of in a proper way that does not represent a risk to the environment and to public health [17]. The global waste estimation is expected to reach 2.59 and 3.40 billion tonnes per year by 2030 and 2050, respectively [18]. This huge quantity requires efficient solutions and proper management policies. Solid waste materials (SWM) are mainly composed of biomass matter, like agricultural and forestry residues, and other waste such as chicken litter and tires. The percentage of composition of SWM are 17% paper and paperboard, 12% plastics, 2% rubber and leather, 2% wood, 44% food and green and 9% miscellaneous wastes includes metals, glass and other inorganic [18]. The SWM is easy to be disposed of to be converted into useful energy, especially, the MSW. The MSW, including organics, plastic, papers, and textiles materials can be classified as organic and non-organic materials (Figure 3.1) [19] or biomass and fabricate materials that are combustible materials. The combustible materials consist of 82% biomass waste (paper, cardboard, wood, cotton, wool leather, yard trimmings, food wastes, etc.) and 18% petrochemical wastes (plastics, rubber, and fabrics) [20].

The USEPA [21] reported that the municipal solid waste generated in the United States in 2018 was approximately 265 million tonnes, 11.3% (30 million tonnes) out of this were combusted to recover energy. The composition percentage of the municipal solid waste were 23.1 paper and paperboard, 12.2 plastics, 3.1 rubber and leather, 5.8 textiles, 6.2 wood, 33.7 food and yard trimmings and 15.9 miscellaneous waste that includes metals, glass and other inorganic.

Many researchers have studied the pyrolysis of MSW [14, 22, 23]. Table 3.1 shows the transformation of waste materials by pyrolysis processes into useful products.

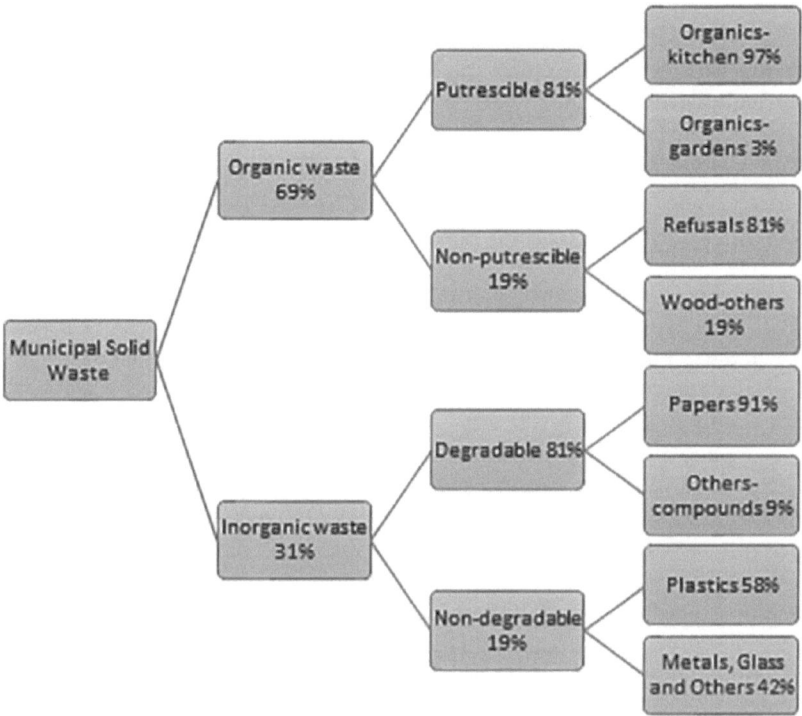

FIGURE 3.1   Composition percentage of MSW.

TABLE 3.1

Transformation of Waste Materials by Pyrolysis Processes into Useful Products

Feedstock	Process	Reactor	Ref.
Coffee wastes	Pyrolysis	Microwave fast pyrolysis	[24]
Tea waste	Pyrolysis	muffle furnace	[25, 26]
Rice husk			
Waste tyre	Pyrolysis	A reactor	[27, 28]
Waste tyre	Catalytic pyrolysis	A reactor	[27]
Plastics wastes	Catalytic pyrolysis	a small pilot scale pyrolysis reactor	[29]
Wastepaper	Vacuum pyrolysis	Tubular furnace	[30]
Wastepaper	Slow pyrolysis	Semi-batch	[31]
Paper biomass	Slow pyrolysis	Fixed bed	[32]
Paper cups	Slow pyrolysis	Semi-batch	[33]
Printing paper	Preheated lab-scale facility	Batch reactor	[34]
Cardboard	Slow pyrolysis	Packed-bed	[35]
Office paper	Fast pyrolysis	Rotating Microwave Reactor	[36]
Tetrapak ® cartons	Intermediate pyrolysis	Flow screw	[37]
Tetrapak ® cartons	Slow pyrolysis	Semi-batch	[38]

Moya et al. [19] calculated the energy potentials produced by thermochemicals to be 5970 kWh/t MSW and power generation potentials were estimated at 0.07 MW/t MSW for thermochemical processes. In the USA, there are 88 plants operated for converting waste to energy from MSW in 25 US states [20]. The MSW is composed of miscellaneous materials that need pretreatment; this includes separation of undesirable materials, size reduction and drying before feeding to pyrolysis reactor.

3.3   WASTE CONVERSION INTO ENERGY

In the strategy of waste conversion into energy, the Advanced Thermal Processes Methods (ATPM) or Advanced Thermal Treatment Technologies (ATTT) uses heat energy in the applications of waste treatment to recover materials and energies. The ATTT is commercially applied to the treatment of MSW at a larger scale in North America, Europe and Japan [39, 40].

The organic materials that are composed of solid waste materials (including municipal solid waste, agri-residues, and industrial waste) are subjected to thermal conversion. Thermal processes, that are often called heat treatment processes, use the temperature or heat to break down the solid waste material into sub-materials such as gases, oils, chars, slag and ashes. The main thermal processes are combustion or incineration, pyrolysis, and gasification. These processes are called "processes to produce energy or recover energy". The diagram in Figure 3.2 shows the main differences among the thermal processes: incineration, gasification and pyrolysis. These differences are basically in the quantity of oxygen entered to the reactor and the type of the output products. Pyrolysis takes place in the absence of oxygen and produces gas, oil, and char. In gasification, a limited supply of oxygen with a quantity less than the stoichiometric of reactions is used and it produces gas, ash, and tar. Incineration or combustion, on the other hand, is a complete oxidation process with excess supply of oxygen and produces flue gas and ash.

Other thermal conversion processes include plasma, torrefaction and hydrothermal processes. The plasma is a state of matter in which the materials are in ionization substance, sometimes called the fourth state of matter. The plasma process uses an electric arc at high voltage that passes between two electrodes, with a torch's

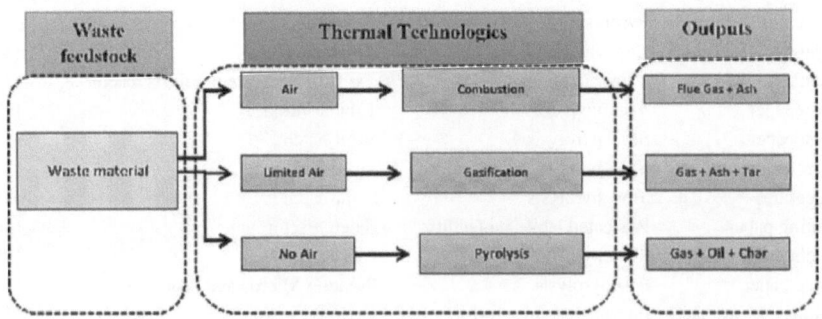

FIGURE 3.2   The main differences among the thermal processes: incineration, gasification and pyrolysis.

temperature ranging from 2,000 to 14,000 °C. This will result in a complete conversion of the waste into its original elemental components as individual atoms and/or molecules. The plasma reactor can be used for the conversion of any type of waste, such as nuclear waste, hazardous waste, and others. The plasma arc technology is used to ionize gas and catalyze raw material into syngas and slag (molten of vitrified glass, usually inorganic material such as metal, silica, gravel, soil, concrete, glass, and so on). Waste materials pass into heating, melting and vaporizing cases. The molecular dissociation is done by breaking of molecular bonds using pyrolysis (plasma pyrolysis).

Unlike pyrolysis process, torrefaction process is a thermochemical process in which the biomass is partly decomposed into bio-coal (torrefied residue), in operating conditions of oxygen absence, milder temperatures usually between 200 and 350 °C, low particle heating rates, atmospheric pressure, and a residence time in reactor of about one hour.

The hydrothermal processes are thermal processes that are used to convert wet biomass usually of more than 30 wt.%, in different operating conditions of temperatures and pressures. In the case of moderate temperatures (180°C and 250°C) and pressures (20–40 bar), the organic matter transforms into bio-coal by using water, and the process is called hydrothermal carbonization (HTC). In the case of intermediate temperatures (250 and 375°C), and pressures up to 180 bar, the product is a liquid that is called bio-crude and the process is known as hydrothermal liquefaction (HTL). In the case of higher temperatures (above 375°C) and pressures (beyond 200 bar), the product is a syngas, and the process is known as hydrothermal gasification (HTG) or supercritical water gasification (SCWG). In each of the above cases, the dedicated process generates a biofuel with a higher energy density from the used feedstock.

As an example of thermal processes, the waste from sugar factories, known as bagasse, is often used as the main source of fuel in sugar factories for sugar mills. It is used in the boilers for steam production that is used to power the process. It is also used as a primary fuel source when burned in sufficient quantity; it produces sufficient heat energy to supply all the needs of a typical sugar mill with energy to spare [41].

3.4 THERMOCHEMICAL PROCESSES

Different types of thermochemical processes have been involved to convert waste materials into values such as energy and reused materials. The following are summary of these processes:

Incineration: Incineration is a thermal treatment process that aims to achieve volume reduction of waste, reduction of the concentration of waste contamination, energy recovery and production of secondary raw materials such as ashes and/or gases. It is used to destroy a waste material in controllable conditions of oxygen (air) and change it into submaterials that are less hazardous and safer. This process passes through three main stages: drying, thermal decomposition (pyrolysis) and complete combustion (burning). These stages usually

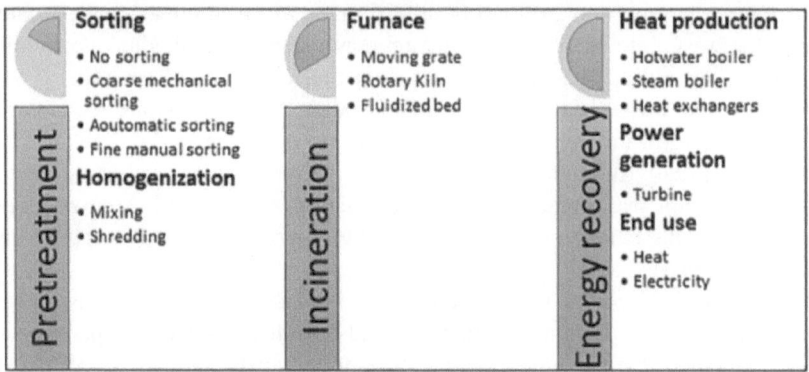

FIGURE 3.3 A diagram illustrating waste incineration facilities.

take place at a temperature around 750–1000 °C and reach up to 1600 °C [17]. The waste materials are converted into hot gases (mainly CO_2, N_2, O_2, hydrocarbons, and H_2O), ashes and heat which is used to produce steam that can be used in turbine generators to produce electricity. The typical calorific value of MSW is about 9 MJ/kg [42]. Figure 3.3 provides a technological overview of the incineration facilities and energy recovery process from waste.

Further details on the incineration process are available in the literature [17].

Gasification: the gasification is the process of thermo-chemical reduction of a waste based upon partial oxidation (limited oxygen supply) without direct combustion. The waste materials react with agents, usually air and/or steam under the application of heat. Process temperatures are in the range of 500–1500°C. The main product is typically a gas that is called "syngas" and a solid that is known as char. Syngas is a mixture of gases that contain methane (CH_4), carbon monoxide (CO), carbon dioxide (CO_2), hydrogen, nitrogen (H_2), water vapor (H_2O), nitrogen (N_2), few amounts of higher hydrocarbons such as C_nH_m and inorganic impurities and particulates as alkali metals such HCN, NH_3, H_2S, and HCl. There are several utilizations of the syngas product; it can be burned to create heat and steam to produce electricity. When syngas is directly used in internal combustion engines, it needs higher cleaning, while in other cases it needs less treatment. It also can be converted into methanol, ethanol, and other chemicals or liquid fuels. The solid residues (char) can be burned to provide heat for the gasifier reactor itself or to produce steam.

Compared to the conventional incineration of MSW, the pyrolysis processes are considered as an alternative cleaner way for treating waste since lower amounts of nitrogen oxides (NO_x) and sulphur oxides (SO_2) are produced. In addition, they are able to reduce the waste to zero in the absence of oxygen and can produce useful new products. The pyrolysis technologies have been conducted on several types of solid waste such as municipal solid waste, tyres, plastics and sewage sludge [39].

Detailed information on gasification process is widely available in the literature, e.g., on typical waste gasification reactions [17, 43], an gasifiers' designs, [44], updraft gasifier [45], downdraft gasifier and fluidized-bed gasifier [46].

3.5 PYROLYSIS PROCESSES

3.5.1 Types, Mechanisms and Conditions

Pyrolysis processes are thermal decomposition processes employed to convert waste material, which is rich with organic materials in place (reactor) in the absence of oxygen and under inert atmosphere gas flow carriers, into the three final states of products. The carrier gas employed is usually argon, helium or nitrogen. The final products of pyrolysis technology are fuels such as liquid (tar & oil), solid (char) and synthesis gas (or syngas as hydrogen, carbon monoxide and carbon dioxide, and light molecular weight hydrocarbon gases as methane).

The pyrolysis processes can be classified as slow, intermediate and fast, which depends on the operation conditions (Table 3.2). The slow pyrolysis reaction for converting biomass mainly into charcoal, while the fast pyrolysis, which maximizes liquid production rather than charcoal. In this case, the gases are the main products of ethylene-rich gases and can be condensed to produce bio-oil and alcohols. In the slow pyrolysis process, the time of heating is longer than the time of retention of biomass. The particle size is the major factor for the heating rate of the pyrolysis reaction. However, in fast pyrolysis, the initial heating time of the precursors is smaller than the final retention time at pyrolysis. Based on the medium, pyrolysis can be divided into two types: hydrous pyrolysis and hydro-pyrolysis. Slow and fast pyrolysis is usually carried out in an inert atmosphere, whereas hydrous pyrolysis is carried out in presence of water and hydro-pyrolysis is carried out in presence of hydrogen [9, 14, 47].

The residence time of vapor in the pyrolysis medium is longer for the slow pyrolysis process. This process is mainly used to produce char production. It can be further classified as carbonization and conventional. On the contrary, the vapor residence time is only for seconds or milliseconds. This type of pyrolysis, which is primarily used for the production of bio-oil and gas, is of two main types: (1) flash and (2) ultrarapid (Table 3.1). Usually, the pyrolysis process temperatures are in the range of 300–900°C. When the range of temperature is less than 500, the major portion of products are char (solid), while the major portion of products are liquid when temperature is around 500–550 °C, and when it is over than 700 °C, syngas is major product [39, 48, 49]. Short exposure to high temperatures is termed "flash pyrolysis", which maximizes the amount of liquids generated, and is normally operated at around 500–700 °C. If lower temperatures are applied for longer periods of time under vacuum, chars predominate compared to the fast pyrolysis product, and the process is called "vacuum pyrolysis".

Pyrolysis mechanism can be explained as follows: waste materials contain different types of submaterials, such as papers, plastics and others. Usually, pyrolysis processes need some mechanical sorting and separation of inert materials prior to processing the waste material [39]. These materials have complex polymers which will be degraded under the action of heat (thermal pyrolysis without oxygen) into molecules of less molecular weight. In fact, the breaking down of the natural organic molecular chains, such as cellulose, hemicellulose and lignin, and /or synthesis of large polymer (plastics) produce gases, liquid (tar) and solid, or the so-called char. The tar is a black mixture of free carbon and hydrocarbons in the form of polycyclic aromatic hydrocarbons. The initial gases produced are condensable and non-condensable. The condensable gases will have the characteristic of a liquid or oil, while non-condensable are final gases. During the employment of pyrolysis processes, the chemical reactions are very complex. Effectively, thermal decompositions of chemical bonds between the functional groups of compounds in the organic materials (large molecules to small molecules) occurs.

The pyrolysis mechanism (steps of breaking down of the natural organic molecule) are based on process types (flash, fast, slow, catalytic and vacuum pyrolysis), parameters and operating conditions, which are heating rate, resident time, temperature and pressure conditions, reactor types and configuration. The final product depends on the temperature, feedstock, heating rates, and the time of process. Summary of different types of pyrolysis processes characteristics are presented in Table 3.2 that shows the comparison between the pyrolysis processes. As an example, in the case of fast pyrolysis, the rate of heating is very high to reach the final temperature (500°C) with retention time less than two seconds, the dedicated products are bio-oil [9, 14, 47].

The general equation for thermal pyrolysis processing decompositions [14] is expressed as:

$$\text{Biomaterial } (C_x H_y O_z) + \text{Total heat } (Q_{tot}) => \text{Solid (Char)} + \text{Liquid } (H_2O + \text{Oil} + \text{Tar}) + \text{Gas (Syngas)} \quad \text{(Eq. 3.1)}$$

where: total heat is material dyer heat (the heat necessary to moisture vaporization, which depends on the material quantity and the content of moisture percentage) +

TABLE 3.2

Summary of the Characteristics of Different Types of Pyrolysis Processes

Pyrolysis Type	Rate of heating	Final temperature (°C)	Retention time (s, m, d)*	Dedicated products	Reference
Ultra-rapid	Very high	1000	<0.5 s	Chemical and gas	[9, 14, 47]
Flash	Very high	650	<1.0 s	Bio-oil, chemicals and gas	[9, 14, 47, 50]
Fast	Very high	500	<2.0 s	Bio-oil	[9, 14, 47]
Vacuum	Medium	400	2–30 s	Bio-oil	[9, 14, 47]
Hydro-pyrolysis	High	<500	<10 s	Bio-oil	[9, 14, 47]
Conventional	Low	600	5–30 min	Char, bio-oil and gas	[9, 14, 47]
Carbonization	Very low	400	days	Charcoal	[9, 14, 47]

* s = second; m = minute; d = day

heat requirement for the pyrolysis process (Σ specific heat capacities of materials x ΔT) + heat loss from the reactor to the surrounding.

3.5.2 REACTOR TYPES

There are many types of reactors that have been developed by researchers for specific purposes. In fact, reactors have been designed to satisfy specific conditions that depend on pyrolysis configurations and operating conditions, such as heating rate, temperature, residence time, required pressure, and desired product. There are several types of reactors used in pyrolysis processes for solid waste treatment that depend on the process type (e.g., drum, rotary kilns, and screw reactors are used in slow pyrolysis, while fluidized bed, rotating cones, entrained flow, vacuum and ablative reactors are used in fast pyrolysis systems). Some reactors are applied in both slow and fast pyrolysis as auger-type reactors [51].

There are several types of reactors that are being used in MSW pyrolysis such as fixed-bed reactors [11, 52, 53], rotary kiln [30], and vacuum [3]. Rotary kilns, fluidized beds, fixed-bed reactors, and other reactors are technologies used for waste pyrolysis treatment.

3.5.3 PRODUCTS OF PYROLYSIS PROCESSES

Pyrolysis is a thermal process that can be applied for treatment of organic waste (biomass) in the absence of oxygen to produce liquid (termed bio-oil or bio-crude), solid (charcoal) and gaseous fractions (syngas). The process condition depends on the end product desired. There are slow and fast pyrolysis processes [48]; this could be either the production of fuel gases or the complete destruction of the waste materials. The complete elimination of air from waste materials is very difficult in practice; therefore, the pyrolysis process is usually performed in a neutral atmosphere of argon, helium or nitrogen which acts as carrier flow [54, 55]. Figure 3.4 provides a schematic diagram of the pyrolysis process of waste.

The final products can have several uses; in the case of waste feedstock, it is usually used as a fuel for energy production. Pyrolysis oils can be used as raw materials for specialty chemicals production and/or liquid fuel (fuel oil). The pyrolysis solid

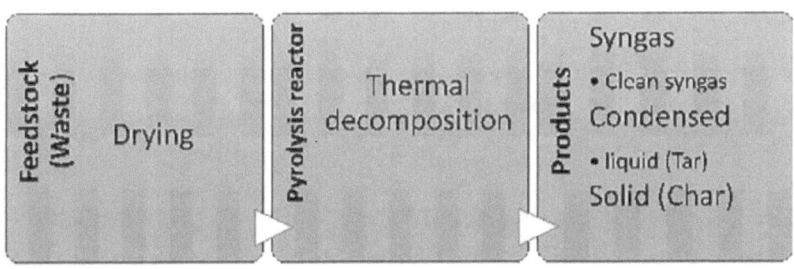

FIGURE 3.4   Schematic diagram of the pyrolysis process of waste.

char can be used as a solid fuel in power generation, in this case that is not attractive as gas and oil, and/or as an activated carbon to remove contaminants (i.e., water purification), and as a soil amendment [51].

Syngas is the main product when the wastepaper has a main component of MSW and the pyrolysis temperature is over 600°C. The gas yields are generally below 1 Nm^3kg^{-1} and have an averaged calorific value near 15 $MJNm^{-3}$ in most cases [39]. Pyrolysis gases can be used as fuel for the pyrolysis process itself and or as fuel gas that is rich with hydrogen. Pyrolysis gas contains CO, CO_2, H_2, and hydrocarbon compounds such CH_4, C_2H_8, and traces of compounds such as ammonia (NH_3), hydrogen sulphide (H_2S), and hydrogen chloride (HCl). Prior to using the syngas in recovery systems, it needs to be cleaned up. All commercial equipment for pyrolysis processes is equipped with emission abatement devices to ensure the cleanliness of the product syngas [39].

Table 3.3 provides the percentage of main products of biomass waste and various related conditions of pyrolysis processes that are collected from several literature resources.

TABLE 3.3
The Percentage of Products of Differences in Biomass Waste and Conditions of Various Pyrolysis Processes

Waste	Process	Char%	Oil%	Gas%	Ref.
Tyre	pyrolysis process	33–39	34–42	19–33	[28]
Tyre *	vacuum pyrolysis	38	41	12	[28]
Biomass	fast pyrolysis	60	20	20	[56]
Plastics waste (polyethylene, polypropylene, polystyrene, and Polyethylene terephthalate	catalytic pyrolysis	32.8	28	39.2	[29]
Food waste	pyrolysis of digestate	27.16 to 45.27	60.30	7.40	[1]
MSW (composition, V%): kitchen 38.67; fabric 1.93; plastic 12.59; paper 13.92; metal 0.25; glass 17.20; stone 15.44	pyrolysis	49.1 50.21 48.88 51.00 47.56	2.1 2.55 3.67 3.10 5.73	48.8 47.24 47.45 45.90 46.71	[23]
Biomass	slow pyrolysis	25–35	20–50	20–50	[57]
Biomass	intermediate pyrolysis	30–40	35–45	20–30	[57]
Biomass	fast Pyrolysis	10–25	50–70	10–30	[57]
Defatted coffee grounds	slow pyrolysis	28	13.7	24	[24]
Spent coffee grounds	slow Pyrolysis	27	27.2	21	[24]
Biomass	slow Pyrolysis	35	30	35	[58]
Biomass	fast Pyrolysis	20	50	30	[58]
Biomass	flash Pyrolysis	12	75	13	[29]
Biomass	fast Pyrolysis	15–25	60–75	10–20	[29]
Soft wood bark	vacuum pyrolysis	27.6	45.0	27.4	[59]

Waste	Process	Char%	Oil%	Gas%	Ref.
Hard wood bark	vacuum pyrolysis	26.2	55.9	19.9	[59]
Waste textiles	slow Pyrolysis	16.91	29.74	53.35	[60]
Waste plastics	pyrolysis	33.5	40.9	25.6	[6]
MSW	pyrolysis	30	61	9	Jiang, 2006 cited in [9 and 14]
MSW	pyrolysis	15	50	35	Jiang, 2006 cited in [9 and 14]
Waste tyre	pyrolysis	42	40	18	[27]
Dry wood	torrefaction (slow) Pyrolysis	80	0	20	[61]
Dry wood	carbonization (slow) Pyrolysis	35	30	35	[61]
Dry wood (Eucalyptus mallee)	fast Pyrolysis	12	75#	13	[61, 62]
Dry wood	intermediate pyrolysis	25	50	25	[61]
Eucalyptus mallee	slow pyrolysis	35	50##	35	[62]
Biomass	slow pyrolysis	25–35	20–50	20–50	[57]
Biomass	intermediate pyrolysis	30–40	5–45	20–30	[57]
Biomass	fast pyrolysis	10–25	50–70	10–30	[57]
Sugarcane bagasse	fast pyrolysis	25.34**	50.89**	14.12**	[48]
Sugarcane bagasse	conventional pyrolysis	37.64***	26.11***	25.1***	[48]
Waste paper	slow pyrolysis	35.23	47.03	17.74	[30]

* Steel scrap about 9%; ** Losses 9.65%; ***Losses 11.15 %; # include 25% water; ## include 50% water.

3.6 CONCLUSION

In this chapter we discussed the thermochemical conversions of biowaste into energy and useful materials. Energy can be recovered from waste materials using several technologies such as thermochemical conversion processes. Several methods are used for waste conversion into energy; these include the ATPM and ATTT. In particular, the ATTT is commercially applied in the treatment of MSW at a larger scale in North America, Europe, and Japan.

Using pyrolysis, which is a thermochemical process, is proved effective in deriving clean energy from waste materials such as rubber, plastic waste, tyres, and municipal solid waste.

The pyrolysis mechanisms are based on process type (flash, fast, slow, catalytic and vacuum pyrolysis), parameters and operating conditions, which are heating rate, resident time, temperature and pressure conditions, reactor types, and configuration. The final product depends on the temperature, feedstock, heating rates, and the time of process.

The final products of the pyrolysis process have several uses. Pyrolysis oil and gas can be used as raw materials for special chemical production and/or fuel. Furthermore, solid char can be used as a solid fuel in power generation.

REFERENCES

[1] Lü F., Hua Z., Shao L., He P. (2018) Loop bioenergy production and carbon sequestration of polymeric waste by integrating biochemical and thermochemical conversion processes: A conceptual framework and recent advances. *Renewable Energy* 124: 202–211. https://doi.org/10.1016/j.renene.2017.10.084

[2] Umeki E.R., de Oliveira C. F., Torres R.B., dos Santos R.G. (2016) Physico-chemistry properties of fuel blends composed of diesel and tire pyrolysis oil. *Fuel* 185: 236–242. http://dx.doi.org/10.1016/j.fuel.2016.07.092

[3] Williams P.T. (2013) Pyrolysis of waste tyres: A review. *Waste Management* 33: 1714–1728. http://dx.doi.org/10.1016/j.wasman.2013.05.003

[4] Alexander R. Gamboa, Ana M.A. Rocha, Leila R. dos Santos, João A. de Carvalho Jr. (2020) Tire pyrolysis oil in Brazil: Potential production and quality of fuel. *Renewable and Sustainable Energy Reviews* 120: 109614. https://doi.org/10.1016/j.rser.2019.109614

[5] Ganjar F., Is F., Imam S., Muhammad M.M., Teuku Meurah I.M., Oki M. (2021) Recent Progress in low-cost catalysts for pyrolysis of plastic waste to Fuels. *Catalysts* 11: 837. https://doi.org/10.3390/catal11070837

[6] Adrados A., de Marco I., Caballero B.M., López A., Laresgoiti M.F., Torres A. (2012) Pyrolysis of plastic packaging waste: A comparison of plastic residuals from material recovery facilities with simulated plastic waste. *Waste Management* 32(5): 826–832. https://doi.org/10.1016/j.wasman.2011.06.016

[7] Al-Salem S.M., Antelava A., Constantinou A., Manos G., Dutta A. (2017) A review on thermal and catalytic pyrolysis of plastic solid waste (PSW). *Journal of Environmental Management* 197: 177–198. http://dx.doi.org/10.1016/j.jenvman.2017.03.084

[8] José María F.-G., Carmen D.-L., Jaime M.-P., Montserrat Z. (2020) Recycling organic fraction of municipal solid waste: Systematic literature review and bibliometric analysis of research. *Trends Sustainability* 12: 4798. www.mdpi.com/journal/sustainability; http://doi:10.3390/su12114798

[9] Bosmans A., Vanderreydt I., Geysen D., Helsen L. (2013) The crucial role of Waste-to-Energy technologies in enhanced landfill mining: A technology review. *Journal of Cleaner Production* 55: 10–23. http://dx.doi.org/10.1016/j.jclepro.2012.05.032

[10] Wang H., Wang L., Shahbazi A. (2015) Life cycle assessment of fast pyrolysis of municipal solid waste in North Carolina of USA. *Journal of Cleaner Production* 87: 511–519. http://dx.doi.org/10.1016/j.jclepro.2014.09.011

[11] Wang, Y., Zhang, S.T., Zhang, Y.F., Xie, H., Deng, N., Chen, G.Y. (2005) Experimental studies on low-temperature pyrolysis of municipal household garbage–temperature influence on pyrolysis product distribution. *Renewable Energy* 30: 1133–1142.

[12] Czajczyńska D., Anguilano L., Ghazal H., Krzyżyńska R., Reynolds A.J., Spencer N., Jouhara H. (2017) Potential of pyrolysis processes in the waste management sector. *Thermal Science and Engineering Progress* 3: 171–197. http://dx.doi.org/10.1016/j.tsep.2017.06.003

[13] Bosmans A., Helsen, L. (2010) Energy from waste: Review of thermochemical technologies for refuse derived fuel (RDF) treatment. Proceedings Venice 2010, Third International Symposium on Energy from Biomass and Waste, Venice, Italy. Date: 8–11 November 2010. CISA, Environmental Sanitary Engineering Centre. Location: Venice, Italy.

[14] Chen D., Yin L., Wang H., He P. (2015) Reprint of: Pyrolysis technologies for municipal solid waste: A review. *Waste Management* 37: 116–136. http://dx.doi.org/10.1016/j.wasman.2015.01.022

[15] Chowdhury Z.Z., Kaushik P., Wageeh A.Y., Suresh S., Syed T.S., Ganiyu A.A., Emy M., Rahman F.R., Rafie B.J. (2017) Pyrolysis: A sustainable way to generate energy from waste. In *Pyrolysis*, Edited by Mohamed Samer. http://dx.doi.org/10.5772/intechopen.69036.

[16] Aliyu, A., Lee J.G.M., Harvey, A.P. (2021) Microalgae for biofuels: A review of thermochemical conversion processes and associated opportunities and challenges. *Bioresource Technology Reports* 15: 100694. https://doi.org/10.1016/j.biteb.2021.100694

[17] Al Arni, S., Elwaheidi, M. (2021) *Concise handbook of waste treatment technologies.* London: CRC Press and Taylor & Francis Group.

[18] Kaza S., Lisa Y., Perinaz B.-T., Van Woerden, F. (2018) *What a waste 2.0: A global snapshot of solid waste management to 2050. Overview booklet.* World Bank, Washington, DC (License: Creative Commons Attribution CC BY 3.0 IGO). doi: 10.1596/978-1-4648-1329-0.

[19] Moya D., Aldás C., David Jaramillo D., Játivad D., Kaparaju P. (2017) Waste-to-energy technologies: An opportunity of energy recovery from municipal solid waste, using Quito—Ecuador as case study. 9th International Conference on Sustainability in Energy and Buildings, SEB-17, 5–7 July 2017, Chania, Crete, Greece. Energy Procedia 134 (2017) 327–336. https://doi.org/10.1016/j.egypro.2017.09.537.

[20] Psomopoulos C.S., Bourka A., Themelis N.J. (2009) Waste-to-energy: A review of the status and benefits in USA. *Waste Management* 29: 1718–1724. https://doi.org/10.1016/j.wasman.2008.11.020

[21] USEPA (United States Environmental Protection Agency) (2020) *Advancing sustainable materials management: 2018 tables and figures assessing trends in materials generation and management in the United States.* Washington, DC: USEPA.

[22] Chhabra, V., Shastri, Y., Bhattacharya, S. (2016) Kinetics of pyrolysis of mixed municipal solid waste-a review. *Procedia Environmental Sciences*, 35: 513–527.

[23] Wu, D., Zhang, A., Xiao, L., Ba, Y., Ren, H., Liu, L. (2017) Pyrolysis characteristics of municipal solid waste in oxygen-free circumstance. *Energy Procedia*, 105: 1255–1262. doi: 10.1016/j.egypro.2017.03.442

[24] Limousy, L., Jeguirim, M., Labaki, M. (2005) Energy applications of coffee processing by-products, in *Handbook of coffee processing by-products* (pp. 323–367).

[25] Rajapaksha, A.U., Vithanage, M., Zhang, M., Ahmad, M., Mohan, D., Chang, S.X., Ok, Y.S. (2014) Pyrolysis condition affected sulfamethazine sorption by tea waste biochars. *Bioresource Technology* 166: 303–308.

[26] Vithanage, M., Mayakaduwa, S.S., Herath, I., Ok, Y.K., Mohan, D. (2016) Kinetics, thermodynamics and mechanistic studies of carbofuran removal using biochars from tea waste and rice husks. *Chemosphere* 150: 781–789. https://doi.org/10.1016/j.chemosphere.2015.11.002

[27] Ayanoğlu, A., Yumrutas, R. (2016) Production of gasoline and diesel like fuels from waste tyre oil by using catalytic pyrolysis. *Energy* 103: 456–468. http://dx.doi.org/10.1016/j.energy.2016.02.155

[28] Yaqoob, H., Teoh, Y.H., Jamil, M.A., Gulzar, M. (2021) Potential of tire pyrolysis oil as an alternate fuel for diesel engines: A review. *Journal of the Energy Institute* 96: 205–221. https://doi.org/10.1016/j.joei.2021.03.002

[29] Miandad, R., Rehan, M., Barakat, M.A., Aburiazaiza, A.S., Khan, H., Ismail, I.M.I., Dhavamani, J., Gardy, J., Hassanpour, A., Nizami, A.-S. (2019) Catalytic pyrolysis of plastic waste: Moving toward pyrolysis based biorefineries. *Front. Energy Res* 7: 27. doi: 10.3389/fenrg.2019.00027

[30] Li, L., Zhang, H. and Zhuang, X. (2005) Pyrolysis of waste paper: Characterization and composition of pyrolysis oil. *Energy Sources, Part A Recover Util. Environ. Eff.* 27(9): 867–873. http://dx.doi.org/10.1080/00908310490450872.

[31] Sarkar, A., Chowdhury, R. (2014) Co-pyrolysis of paper waste and mustard press cake in a semi-batch pyrolyser-optimization and bio-oil characterization. *Int. J. Green Energy* 13: 373–382. http://dx.doi.org/10.1080/15435075.2014.952423.

[32] Chattopadhyay, J., Pathak, T.S., Srivastava, R., Singh, A.C. (2016) Catalytic co-pyrolysis of paper biomass and plastic mixtures (HDPE (high density polyethylene), PP (polypropylene) and PET (polyethylene terephthalate)) and product analysis. *Energy* 103: 513–521. http://dx.doi.org/10.1016/j.energy.2016.03.015.

[33] Biswal, B., Kumar, S., Singh, R.K. (2013) Production of hydrocarbon liquid by thermal pyrolysis of paper cup waste. *J. Waste Manag* 1–7, http://dx.doi.org/10.1155/2013/731858.

[34] Zhou, C., Yang, W., Blasiak, W. (2013) Characteristics of waste printing paper and cardboard in a reactor pyrolyzed by preheated agents. *Fuel Process. Technol.* 116: 63–71, http://dx.doi.org/10.1016/j.fuproc.2013.04.023.

[35] Yang, Y. B., Phan, A.N., Ryu, C., Sharifi, V., Swithenbank, J. (2007) Mathematical modelling of slow pyrolysis of segregated solid wastes in a packed-bed pyrolyser. *Fuel* 86: 169–180. http://dx.doi.org/10.1016/j.fuel.2006.07.012.

[36] Zhang, Z., Macquarrie, D.J., De Bruyn, M., Budarin, V.L., Hunt, A.J., Gronnow, M.J., Fan, J., Shuttleworth, P.S., Clark, J. H., Matharu A.S. (2014) Low-temperature microwave-assisted pyrolysis of waste office paper and the application of bio-oil as an Al adhesive. *Green Chem.* 17: 260–270, http://dx.doi.org/10.1039/C4GC00768A.

[37] Haydary, J., Susa, D., Dudáš, J. (2013) Pyrolysis of aseptic packages (tetrapak) in a laboratory screw type reactor and secondary thermal/catalytic tar decomposition. *Waste Manag.* 33: 1136–1141, http://dx.doi.org/10.1016/j.wasman.2013.01.031.

[38] Korkmaz, A., Yanik, J., Brebu, M., Vasile, C. (2009) Pyrolysis of the tetra pak. *Waste Manag.* 29: 2836–2841, http://dx.doi.org/10.1016/j.wasman.2009.07.008.

[39] Chen D., Yin L., Wang H., He P. (2014) Pyrolysis technologies for municipal solid waste: A review. *Waste Management* 34: 2466–2486.

[40] DEFRA Department for Environment Food & Rural Affairs (2013) Report on "advanced thermal treatment of municipal solid waste, February 2013, available online on 2/4/2022, at: https://assets.publishing.service.gov.uk/government/uploads/system/uploads/attachment_data/file/221035/pb13888-thermal-treatment-waste.pdf

[41] Al Arni S., Converti, A. (2012) Conversion of sugarcane bagasse into a resource, in João F.G., Kauê, D.C. (eds.), *Sugarcane: Production, cultivation and uses.* Hauppauge, NY: Nova Science Publishers, Inc.

[42] Williams, P.T. (2005) *Waste treatment and disposal.* 2nd ed., Hoboken, NJ: John Wiley & Sons Ltd.

[43] Knoef, H. (2005) Practical aspects of biomass gasification, in H. Knoef (ed.), *Handbook of biomass gasification.* Enschede: BTG-Biomass Technology Group (BTG).

[44] Basu, P. (2006) *Combustion and gasification in fluodized beds.* London: Taylor & Francis Group, LLC.

[45] Basu, P. (2010) *Biomass gasification and pyrolysis practical design and theory.* London: Elsevier Inc.

[46] Basu, P. (2013) *Biomass gasification, pyrolysis and torrefaction—practical design and theory.* 2nd ed., London: Elsevier Inc.

[47] Zaman C.Z., Pal K., Yehye W. A., Sagadevan S., Shah S.T., Adebisi G. A., Marliana E., Rafique R.F. and Johan R.B. (2017) *Pyrolysis: A sustainable way to generate energy from waste.* http://dx.doi.org/10.5772/intechopen.69036

[48] Al Arni, S. (2018) Comparison of slow and fast pyrolysis for converting biomass into fuel. *Renewable Energy* 124: 197–201. https://doi.org/10.1016/j.renene.2017.04.060

[49] Al Arni, S., Bosio, B., Arato, E. (2010) Syngas from sugarcane pyrolysis: An experimental study for fuel cell applications. *Renewable Energy* 35: 29–3. Available online at: http://dx.doi.org/10.1016/j.renene.2009.07.005

[50] Aguado, R, Olazar, M, Gaisan, B, Prieto, R, Bilbao, J. (2002) Kinetic study of polyolefin pyrolysis in a conical spouted bed reactor. *Industrial & Engineering Chemistry Research* 41: 4559–4566

[51] Roy, P. and Dias, G. (2017) Prospects for pyrolysis technologies in the bioenergy sector: A review. *Renewable and Sustainable Energy Reviews* 77: 59–69. http://dx.doi.org/10.1016/j.rser.2017.03.136

[52] Ates, F., Miskolczi, N., Borsodi, N. (2013) Comparison of real waste (MSW and MPW) pyrolysis in batch reactor over different catalysts. Part I: Product yields, gas and pyrolysis oil properties. *Bioresour. Technol.* 133: 443–454.

[53] Miskolczi, N., Ates, F., Borsodi, N. (2013) Comparison of real waste (MSW and MPW) pyrolysis in batch reactor over different catalysts. Part II: contaminants, char and pyrolysis oil properties. *Bioresour. Technol.* 144: 370–379.

[54] Al Arni S. (2004a). An experimental investigation for gaseous products from sugarcane by fast pyrolysis. *Energy Education Science and Technology* 13(2): 89–96.

[55] Al Arni S. (2004b). Hydrogen-rich gas production from biomass via thermochemical pathways. *Energy Education Science and Technology* 13(1): 47–54.

[56] Sadaka S. 2017. Pyrolysis and BioOil. University of Arkansas, United States Department of Agriculture, and County Governments Cooperating. FSA1052-PD-4–2017RV. www.uaex.uada.edu

[57] Brownsort P. (2009) *Biomass pyrolysis processes: Review of scope, control and variability*. Edinburgh: Biochar Research Center. http://refhub.elsevier.com/S1364-0321(17)30471-9/sbref26

[58] Jahirul, M.I., Rasul, M.G., Chowdhury, A.A., Ashwath, N. (2012) Biofuels production through biomass pyrolysis—a technological review. *Energies* 5(12): 4952–5001. doi:10.3390/en5124952

[59] Torri, I.D.V., Paasikallio, V., Faccini, C.S., Huff, R., Caramão, E.B., Sacon, V., Oasmaa, A. and Zini, C.A. (2016) Bio-oil production of softwood and hardwood forest industry residues through fast and intermediate pyrolysis and its chromatographic characterization. *Bioresource Technology*, 200: 680–690. https://doi.org/10.1016/j.biortech.2015.10.086

[60] Barışçı S. and Öncel M. S. (2014) The disposal of combed cotton wastes by pyrolysis. *International Journal of Green Energy* 11(3): 255–266, http://dx.doi.org/10.1080/15435075.2013.772516

[61] Bridgwater A.V. (2012) Review of fast pyrolysis of biomass and product upgrading. Biomass and Bioenergy 3(8): 68–94. http://dx.doi.org/10.1016/j.biombioe.2011.01.048

[62] Bridgwater, A. V., Carson, P., Coulson, M. (2007) A comparison of fast and slow pyrolysis liquids from mallee. *International Journal of Global Energy Issues* 27(2): 204. doi:10.1504/ijgei.2007.013655

4 Waste Tyre Recycling
Processes and Technologies

Saleh Al Arni and Mahmoud M. Elwaheidi

CONTENTS

4.1 Introduction...57
4.2 Volume Estimation of Produced Waste Tyres............................58
4.3 Waste Tyre Management Regulations...58
4.4 Tyres Components 59
4.5 Recycling Processes...62
    4.5.1 Reducing, Reusing and Recycling of Waste Tyres 62
4.6 Thermochemical Conversion of Waste Tyres into Useful Energy 66
    4.6.1 Burning Process 66
    4.6.2 Pyrolysis Technology...66
    4.6.3 Gasification Technology 68
4.7 Tyres Disposal in Landfills 68
4.8 Conclusion 69
References...69

4.1 INTRODUCTION

Recently, waste tyre (sometimes called "off-the-road tyre" or "end-of-life tyre") is being considered as an important product worldwide that is produced by vehicles and considered a sustainable waste [1]. An inappropriate management of waste tyres may cause serious environmental and ecological issues due to non-degradability and fire hazards [2]. Specifically, waste tyres are considered a challenging source of waste due to the huge volume produced, durability and components [3]. However, the growing global emphasis on waste recycling has greatly influenced the management of waste tyres, which generated interest in developing alternative technologies of waste recycling [4].

Tyre recycling is defined as the process of recycling consumed vehicle tyres that are no longer usable. Few recycling solutions include reusing or transformation of waste tyres into useful objects for daily uses such as rockers for children, a basket for several uses, outdoor furniture, and many others (the reader can, e.g., google it using "Images for waste tyre creative arts" for further details). Other recycling solutions include reuse of waste tyres in civil construction works to reinforce engineering properties of soil [5]. In addition, waste tyres can be used as an unconventional source of fuel using thermal conversion processes [6].

DOI: 10.1201/9781003260738-4

There are several authors discussed the problem of waste tyres from different points of view, including waste tyre recycling, management, treatment, conversion, and others [7–9]. In this chapter, we investigate the management efficiency of waste tyres. Furthermore, methods and systems of recycling and exploring the potential of their reuse will be discussed.

4.2 VOLUME ESTIMATION OF PRODUCED WASTE TYRES

Estimating the volume of the waste tyres that is generated globally each year is not an easy task. In literature, however, several methods of volume estimation were reported. A reported method is based on the number of tyres produced annually. According to this method, the volume of waste tyres generated annually is estimated to exceed one billion tyres worldwide every year [7, 8].

Another estimation of the waste tyres production is based on the number of population residents in cities. In this case, it is estimated that about one scrap tyre is produced per person every year [2], and the standardized measuring unit is the Equivalent Passenger Units (EPU).

Additional various estimations were reported in the literature; Liu et al. [9] estimated an increase in the global annual waste tyres to 1.1 billion tonnes by the end of 2030. Moreover, the Australian Bureau of Statistics estimates an approximate yearly increase of waste tyres to 5 billion tonnes [10].

Estimating the volume of the produced waste tyres remains, however, the cornerstone in any waste management policy that aims to protect the environment and to overcome economic challenges in terms of the cost of recycling and disposal processes.

4.3 WASTE TYRE MANAGEMENT REGULATIONS

Waste tyres are considered a non-hazardous commercial solid waste. Effective management of this waste is governed by governmental policies, laws, regulations, and organizational framework. Treatment of waste tyres comes under local and national environmental regulations that are related to the collection, treatment, and disposal of waste. Furthermore, the financial capabilities of countries play a vital role in waste tyre management [11].

Waste tyre management regulations for recycling programs are based on the principle of the 3Rs; that is reduce, reuse, and recycle, and/or the 4Rs; that is, reduce, reuse, recycle, and recover. The systems or strategies of the programs are set to increase the recycling and resource recovery at local, regional, and national levels. Strategies, systems, and programs of waste tyres management are different from country to another. These regulations define the direct and indirect responsibilities, and the roles of stakeholders, such as generator or producer, exporter or importer, collector or transporter, waste treatment operator and local or national regulatory agencies.

In the EU, Australia, and Hong Kong, waste tyre management systems charge the waste producer with the responsibility for recycling waste tyres, while in Korea the system is based on manufacturers and importers, who pay a refunded deposit fee (a tax system) if they collect the used tyres. On the other hand, Japan, unlike many other developed countries, does not have specific laws regulating tyre recycling; instead, tyres are dealt with as part of solid wastes [7]. Other management systems are based on the free market system, which assumes the profitability of recovery and recycling of tyres by private companies that use waste tyres as a source of raw materials [12].

There are different methods and techniques for managing waste tyres that are intended for reuse and recycling in various applications, such as retreading, recycling, conversion into useful energy and disposal in landfill. However, before discussing the recycling techniques of waste tyres, it is necessary to discuss the composition of tyres.

4.4   TYRES COMPONENTS

The raw materials used in the production of automobile tyres are composed of about 40% natural rubber and about 60% synthetic rubbers [8]. These are discussed in the next paragraphs:

a) Natural rubber is a natural polymeric material that is obtained from latex sap of trees (rubber tree or *Hevea brasiliensis* tree), which are harvested in South America, Thailand, Indonesia and Africa. It is a hydrocarbons polymer with a total molecular weight range of 7.2×10^4 to 4.5×10^6 atomic mass units (amu) or Daltons (Da) [13]. The chemical mono structure unit of natural rubber is 2-methyl-1,3-butadiene, as follows (Figure 4.1). The chemical compositions of natural Rubber are poly (2-methyl-1,3-butadiene) or cis-1,4-polyisoperene, and gum elastic. Natural rubber in its state is a translucent, light yellow to dark brown, soft, and elastic material. When kept below 0 °C, rubber is hardened. On the other hand, with the increase of temperature it gets softer and weakens [14].

b) Synthetic rubber is a rubber made from various hydrocarbons that are derived from petroleum, such as isoprene, butadiene, chloroprene, isobutylene and styrene. These have a molecular weight that ranges from 5×10^4 to 5.8×10^6 amu and an average value of 2×10^6 amu for the molecular weight of synthetic rubber [8]. Synthetic rubbers include styrene-butadiene rubbers, ethylene-propylene rubbers, acrylic elastomer and silicone rubbers, and butyl and polybutadiene rubbers.

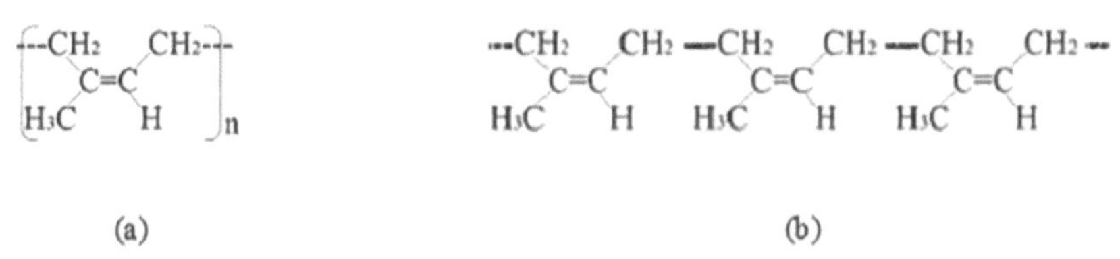

(a)   (b)

FIGURE 4.1   Chemical Formula of mono (a) and poly (b) of natural rubber.

The tyres that are made from rubbers (60–65 wt%) are mixed with other materials to achieve best reinforcement and safest tyre materials performance. These components are metals and textile materials that are used for reinforcement and to ensure rigidity, flexibility and to aid in maintaining the shape of the tyre. Furthermore, silica or silicon oxide (SiO_2) is used as an additive to the rubber to increase strength of and to resist ruptures. In addition, carbon black (25–35 wt%) is added to increase the resistance and give the distinctive black colour of the tyre [8, 15].

The percentage of the additive components is variable and depends on the type of tyre. For example, passenger car tyres are different in weight and in percentage of components from the truck tyres. Table 4.1 presents a comparison of components in percentage between passenger car tyres and truck tyres for different countries including USA, UK, Australia and EU [7, 10, 12].

To understand the feedstock contents of waste tyres, it is necessary to know the proximate and ultimate analysis of different waste tyres. The high heat value (HHV) of energy and percentage average values (on a steel free basis) of 32 records that are reported in literature by several authors [8, 15, 16] for waste tyres analysis are summarized in Table 4.2. This Table shows the average values of proximate and ultimate analysis of different waste tyres and calorific values that reported in literature.

TABLE 4.1
Components Comparison Between Passenger Car Tyres and Truck Tyres

Material	Passenger car Tyre					Truck Tyre				
	USA[a]	EU[a]	EU[b]	Australia[c]	United Kingdom[c]	USA[a]	EU[a]	EU[b]	Australia[c]	United Kingdom[c]
Natural rubber (%)	14	22	14	16	17	27	30	27	29	28
Synthetic rubber (%)	27	23	27	29	31	14	15	14	13	15
Carbon black and silica (%)	28	28	22	23	22	28	20	21.5	24	21
Metal for reinforcement (%)	14–15	13	25	16	15	14–15	25	16.5	25	27
Fillers, additives (sulphur, resin, etc.), textile for reinforcement (%)	16–17	14	8	14	13	16–17	10	15	2	6
Weight, new tyre(kg)	11	8.5	9.5	NA	NA	54	56	NA	NA	NA
Weight, scrap tyre (kg)	9	7	8	NA	NA	45	56	NA	NA	NA

NA = not available

[a] Ref. [12]

[b] Ref. [7]

[c] Ref. [10]

TABLE 4.2
The Average Values of Proximate and Ultimate Analysis of Different Waste Tyres and Calorific Values that Reported in Literature [8, 15, 16]

Proximate analysis (wt%)		Ultimate analysis (wt%)	
Fixed carbon	26.55	Carbon	81.65
Volatile matter	61.80	Hydrogen	7.19
Ash	9.46	Nitrogen	0.64
Moisture	1.08	Sulphur	1.76
HHV	35.22MJ/Kg	Oxygen	5.85

TABLE 4.3
Ash Analysis of Waste Tyres

Component	Side wall rubber[a] (Wt%)	Tyre tread rubber[a] (Wt%)	Waste tyre rubber ash[b] (Wt%)	Overall (Average value)
ZnO	0.63	1.19	NA	0.91
CaO	0.30	0.09	12.9	4.43
SiO_2	0.06	7.76	26.5	11.44
Al_2O_3	0.02	0.01	8.7	2.91
Fe_2O_3	0.01	0.07	9.3	3.13
MgO	0.01	0.02	6.4	2.14
K_2O	0.01	0.03	1.1	0.38
TiO_2	NA	0.01	1.0	0.51
SO_3	NA	NA	1.6	1.60
Na_2O	NA	NA	1.4	1.40
Cl^-	NA	NA	0.1	0.10
Zn	NA	NA	20.2	20.20
Loss on ignition	NA	NA	10.6	10.60

NA = not available
[a] Ref. [17]
[b] Ref. [18]

The ash content inorganic compounds include ZnO, CaO, SiO_2, Al_2O_3, Fe_2O_3, MgO and K_2O. These compounds can act as catalysts in pyrolysis process [17] and as a low-cost adsorbent for removal of heavy metals such as lead (II) ion from aqueous solution [18]. The analysis of ash (wt%) of waste tyres is reported in Table 4.3. Waste tyres have high contents of materials and combustible composition that are considered sources for recovery of material and energy [17, 19, 20].

4.5 RECYCLING PROCESSES

Waste tyre recycling processes start with the reuse of tyres and ends with the recovery of materials and energy. These processes are considered indispensable for environment protection and waste reduction. Furthermore, used tyres are considered as a source of raw materials or as an alternative to fossil fuels. In particular, rubber wastes can be converted into energy or into new polymer materials [21]. However, recycling waste tyres is considered a difficult task to achieve due to economic challenges.

The advantages of recycling are numerous among which we mention: reduces the amount of waste; conserves natural resources; prevents pollution by reducing the need to collect new raw materials; saves energy; reduces greenhouse gas emissions that contribute to global climate change; helps sustain the environment for future generations; helps create new well-paying jobs in the recycling and manufacturing industries.

Recycling steps start with collection, sorting and separation of waste materials; and marketing the product. Practically, tyres are usually collected from vehicle wheel tyre change stations, and then transported, treated, and recycled before moving them to their final destination.

There are several solutions for the waste tyres problem; some solutions transform waste tyres into useful objects for our life, other solutions are based on recovering the materials and energy from the waste tyres by thermal conversion. Furthermore, there are several other mixed methods that could be applied for waste tyres treatment. A brief discussion of the methods to be used in the management of waste tyres based on the principle of 4Rs is provided below.

4.5.1 REDUCING, REUSING AND RECYCLING OF WASTE TYRES

Reducing waste tyres aims to minimize the produced volume using several methods that include reusing, recycling, and converting into energy or new materials.

Reused waste tyres are classified as a controlled waste. The traditional method for reusing waste tyres is retreading that has shown limitations in its application due to safety concerns. However, the method can be used in other applications such as in creative resources for new products or as a raw material, for example, for flower or animal pots, furniture, and shoes. In this case, the process for treatment includes cutting, shaping, or moulding, weaving, or knitting and finishing.

In addition, waste tyres can be directly used as a raw material in civil constructions or shredded (crumbs) and or after separating their contents into different materials [2, 22]. In general, waste tyres can be reused as whole tyres, crumb, and separated components. Some examples are given below:

I) Reusing as a whole

There are several applications of waste tyres as whole without any treatment such as:

- used in boat bumpers to protect from scratching at the side of wharf.
- used in construction of roads for slope stability or under roads for improved stability in the seismic zones.

- used as lightweight infill materials at embankments in construction of dams.
- used as crash barriers at sides of rivers and erosion control for rainwater runoff.
- used in construction works in wetland soil in establishment and drainage around building foundations.
- used as fuel in furnaces for cement production or steam production.

Furthermore, there are other applications of whole tyres after applying superficial treatment as retreaded. *Retreaded tyres* are subjected to a process of regeneration of waste tyres not intended to be reused on a vehicle. In this process, the old wearing surface of the tyre is replaced with a new one. It is an efficient and safe process that uses heat and pressure in a similar way to the process used in the creation of a new tyre but with the advantage of using only 30–50% of new raw material [1]. The process is economically convenient for the consumer and the manufacturer, and significantly reduces the amount of tyre waste.

II) Reusing as crumbs

There are several mechanical operations using shredders, granulators and rolling mills with ribbed rollers and grinding to obtain a crumb rubber from waste tyres at ambient temperature or hot operations [8]. The process of grinding tyres is also called "tyre derived aggregate" which produces different variations of crumb rubber by different methods as water jet grinding, cryogenic grinding, and ambient temperature grinding. The size of crumb rubber resulting from these operations depends on the requested market or sequential operations and processes needed. The lower size limit of rubber dust is smaller than 0.3 mm and has a very rough surface [12]. Table 4.4 shows some different type shapes and sizes of rubber aggregate and their applications.

Rubber crumb can be used as a low-cost raw material in various types of applications, such as agricultural purposes, drainage ability, polymer compositions or blends with other materials like concrete or asphalt in civil engineering works [27]. Another application that uses crumb rubber is reclamation, also known as "devulcanization". In this process, a fine powder of crumb rubber is mixed with chemical additives powder to produce rubber sheets, rubber tubing, and rubber mats or new rubber products. Selected examples on the usages of shredded or crumbs are:

- mixture with concrete as gravel substitute to improve tensile capacity in the buildings.
- with asphalt paved roads for better traction.
- incorporate waste tyres in road construction as a layer of crumbs rubber mixed with bitumen to reduce the noise that is associated with road traffic.
- Soft surfaces for playgrounds, day-care centers, schools, private homes, recreation, parks, running, walking paths and racetracks, with advantages of absorbing impact, reducing risk of injury and being of low-cost materials.
- reclaim rubber or/and to extract oil, gas, and char by thermochemical conversion processers.

TABLE 4.4

Different Type Shapes and Sizes of Rubber Aggregate

Shape or name	Size	Application	Reference
Fine powder	<500µm	a replacement for cement or binding materials	[10]
Buffing	0–40 mm	assemble construction machines and transport technology	[10]
Granulated rubber	<0.425 to12 mm	Civil engineering applications, pyrolysis using, Size reduced rubber for insulation products: lumber and other construction product, as a replacement for cement	[9, 10, 23]
Ground rubber	<0.425 mm.	cement replacement, lightweight self-compacting concrete, As a replacement for cement	[10, 24]
Crumb rubber	0.425 to 4.75 mm	sand replacement, Road construction, explosives, adhesives, disposal, and flooring	[24]
Rubber dust	<0.8 mm	Replacement of natural fine, bituminous mixture	[12, 21, 25]
Rubber granulates	0.8 to 20 mm	Road construction, explosives, adhesives, disposal and flooring, civil construction, size reduced rubber for insulation products: lumber and other construction product	[10, 12, 21, 26]
Rubber chips	10 to 50 mm	Road construction, explosives, adhesives, disposal, and flooring; Civil construction, Soil Moisturizer, fuel	[10,12, 21, 22, 26]
Coarse rubber	13 to 76 mm	Civil construction, replacement for natural gravel, Soil Moisturizer, fuel	[10, 24, 26]
Shreds	20–400 mm	Civil construction, concrete/cement composites, rubber modified concrete	[10, 12, 26]
Cut tyres	>300 mm	Energy recovery (e.g., pyrolysis), Playground, Footpath, Animal farm as a slip resistance	[10, 12, 26]

III) Separation of waste tyres contents into different materials

Waste tyres contents that are discussed in the above (Tyres components) presented opportunities for the recovery of materials and energy, such as metal (bead wires and steel cords), latex (natural) rubber, styrene-butadiene (synthetic) rubber, carbon black and various additives [13]. These materials can be reused as raw material for different applications.

The statistical data related to recycling and reuse of waste tyres material indicates 47% of materials recovery, 31% landfilled, stockpiled, and other unknown or not recovered, 20% energy recovery, 2% civil engineering construction and backfilling [28]. A summary of the waste tyre recycling process is presented in Figure 4.2.

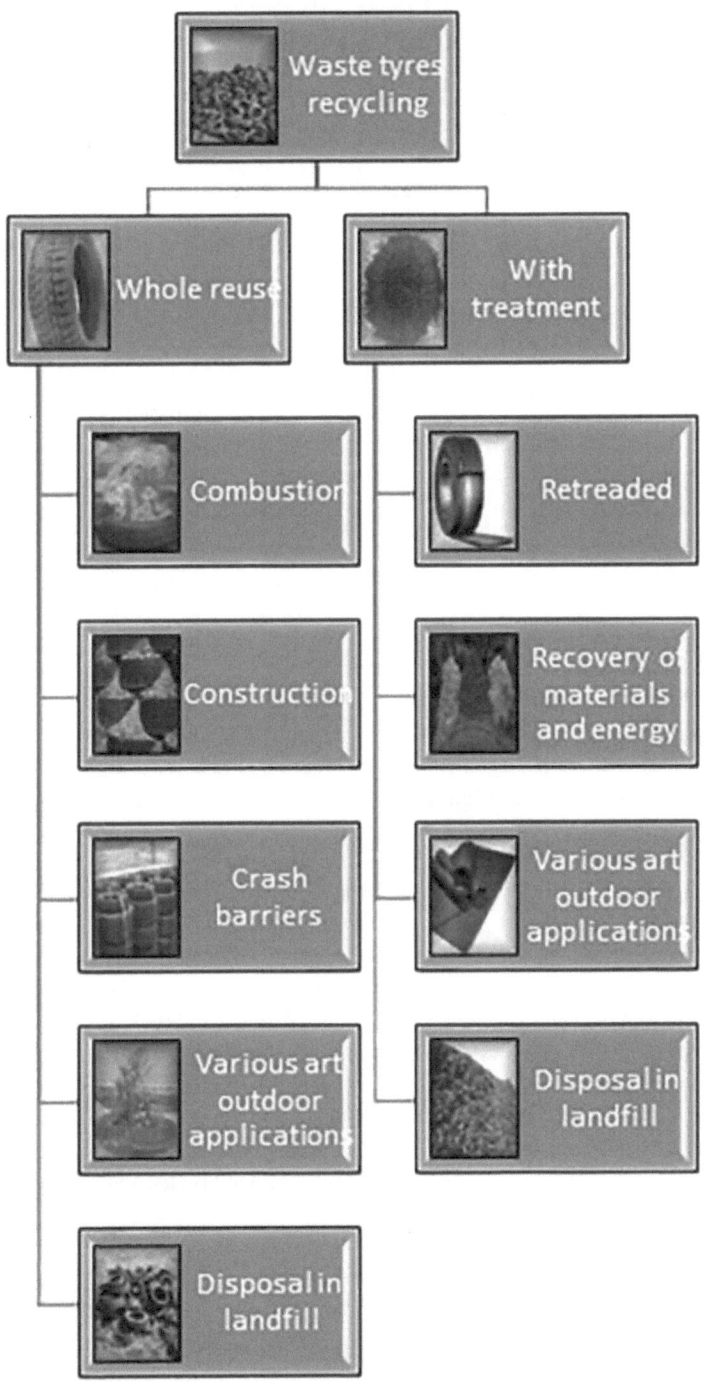

FIGURE 4.2 Waste tyres recycling processes.

4.6 THERMOCHEMICAL CONVERSION OF WASTE TYRES INTO USEFUL ENERGY

Energy recovering from waste provides an added value for the waste materials. Thermal conversion technologies can be applied in waste tyres treatment as individual processes or in an integrated way. Each conversion technology provides a different set of products and combines different input requirements, uses different equipment configurations, and operates in different modes. Furthermore, each technology has its own advantages and disadvantages.

Incineration is the most widely applied thermal conversion process. Pyrolysis and gasification differ from incineration in that they can be used to recover chemical value from waste, rather than its active value. In some cases, chemical derivatives can be used as feedstock for other processes or as a secondary fuel. However, when applied to wastes, pyrolysis, gasification and combustion-based processes are often combined, usually on the same site as part of an integrated process.

The application of plasma gasification to treat waste tyres is scarcely found because the plasma gasification applied to hazardous waste and waste tyres are not considered hazardous waste.

4.6.1 BURNING PROCESS

The tyre combustion as an alternative energy recovery is more attractive than material recycling [12]. When tyres are burned in an open area, heat and thick black smoke are produced, and quickly spread over the whole burning zone (disposal area). The smoke causes air pollution, and deposits oily and ash residues that are plummeted on the soil and contaminate it.

An alternative solution, waste tyres are directly used as an alternative fuel for combustion processes, like in cement kilns, industrial facilities, as utilities in paper mills and boilers or in dedicated electricity generators where the smoke is under control. Waste tyres have a high average calorific value (35 MJ/kg), which makes them competitive compared to other types of fuel (Table 4.2). This application has an advantage of reducing the cost of fuels and eliminates the waste tyres due to the high temperature that is used in cement kilns (range of 1455 to 1510 °C), which ensures the complete combustion of all tyre components. In addition, the chimney and its filtering component guarantee a reduced air pollution. As an example of cost reduction, it was found that the electricity needed to produce one tonne of new tyres is 1019 kWh, while the electricity consumption needed for waste tyre treatment is 770.5–800 kWh. Thus, energy consumption can be significantly reduced by using recycled tyres [3].

4.6.2 PYROLYSIS TECHNOLOGY

Thermochemical conversion processes have focused on the production of fuels. Pyrolysis process is one of thermochemical conversion processes that is adequate for converting waste tyres into fuels in gaseous, liquid (condensate) and solid states. All of these are considered as alternative fuels. One study found that waste tyre processing by pyrolysis produced 40% of liquid, 42% of char, and 18% of gas

[16]. While another study found that the range of production of pyrolysis process of a prototype plant were 35–45 wt% of char, 25–30 wt% of liquid, 18–25 wt% of gas and 8–12 wt% of steel cord when 1500 kg of whole tyres of 150 tyres are treated [15].

In general, the yield of the pyrolysis process depends on the operation conditions and the purpose of treatment. For example, tyre pyrolysis oil (TPO) extracted from vacuum pyrolysis can produce fuel oil similar to diesel fuel, but it needs other treatment to be used in engines [29]. The optimal temperature of the production of TPO in the pyrolysis plant was found at 500°C. The estimation of TPO was around 230 to 280 thousand m³ per year that is about 2% of the onshore petroleum and fuel oil (FO) produced in Brazil [30].

Therefore, besides the environmental and sustainability benefits, tyre pyrolysis contributes economically to the generation of cheaper fuels due to the high calorific value of waste tyres. This confirms the possibility of using waste tyres as alternative fuels. However, numerous research studies show that the characteristics of the products obtained by the tyre pyrolysis process differ based on the operating conditions such as temperature, pressure and residence time [15].

There are several studies on production of oil produced from waste tyres by pyrolysis process that has similar properties to diesel [30–32]. In general, fuel production from waste tyres could form part of the solution to this global environmental problem as it has high energy: the HHV is about 43 MJ/kg [31]. The HHV of volatile and solid fractions of tyre pyrolysis are around 42 and 31 MJ/kg respectively [15, 33].

In comparison with other thermochemical processes, the pyrolysis seems to be more attractive to solve the waste tyres disposal problem. In fact, waste tyre materials used as a feedstock in pyrolysis reactors need to have pre-treatment, such as cutting the material into small pieces, and separate steel wires [33]. Figure 4.3 provides the steps of the pre-treatment process.

The pyrolysis process breaks down the organic part of a tyre into low molecular weight compounds in the absence of oxygen condition. The products are found in three states; these are gas, liquid and solid (Figure 4.4). In the first step, the waste is converted into a secondary energy carrier (a combustible liquid, gas, or solid product), while in a second step this secondary energy carrier is burned (in a steam turbine, gas turbine or gas engine) in order to produce heat and/or electricity. The conversion of solid wastes to secondary energy carriers is a cleaner and more efficient process.

FIGURE 4.3   The materials pre-treatment steps.

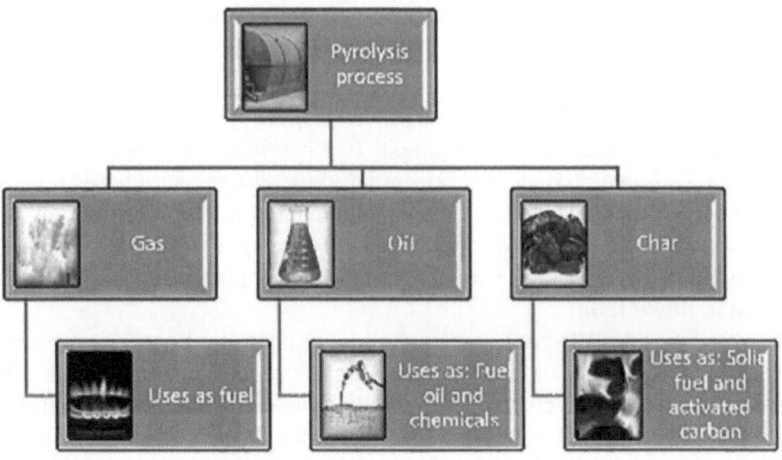

FIGURE 4.4   A diagram showing pyrolysis products and their usages.

4.6.3 Gasification Technology

Like pyrolysis, gasification is the thermochemical reduction of a waste based upon partial oxidation (limited oxygen supply) without direct combustion. The waste materials react with agents, usually air and/or steam under the application of heat. Process temperatures are in the range of 500–1500°C. The main product is typically a gas that is called "Syngas" and solid known as char. Syngas is a mixture of gases; these contain methane (CH_4), carbon monoxide (CO), carbon dioxide (CO_2), hydrogen (H_2), water vapor (H_2O), nitrogen (N_2), few amounts of higher hydrocarbons as (C_nH_m), hydrogen sulphide (H_2S), and particulates as inorganic impurities.

There are several utilizations of the syngas product; it can be burned to create heat and steam to produce electricity. When syngas is directly used in internal combustion engines, it needs higher cleaning, while in other cases it needs less treatment. It also can be converted into methanol, ethanol, and other chemicals or liquid fuels. The solid residues (char) can be burned to provide heat for the gasifier reactor itself or to produce steam.

Plasma based pyrolysis and gasification techniques were applied on waste tyres treatment. The char yield produced was 31% and 27 % for pyrolysis and gasification, respectively, while the syngas yield was 69% and 73% for pyrolysis and gasification, respectively. The syngas produced has 47.45 and 54.50 MJ/Kg of high calorific value for plasma pyrolysis and gasification process, respectively [34]. Gas heating value produced from 11 m³/h of waste tyres is 6 MJ/m³ [8].

4.7 TYRES DISPOSAL IN LANDFILLS

Disposal of waste tyres through illegal dumping, open dumping, open burning, or landfilling is still practiced in some countries [4, 8, 21]. The disposal of tyres (as whole or shredded) in landfill is permitted under the regulations of environmental protection. Disposal of tyres in landfills requires particular attention due to their resistance to high temperatures, biodegradation, chemical and/or photochemical decomposition and physical compaction [19]. The natural degradation of waste tyres needs around 80–90 years [10]. Some countries banned the disposal of waste tyres in landfill such as European countries and Brazil [35]. In the European Union, a ban on stockpiling whole tyres in landfills has been in effect since 2003, and a ban on stockpiling shredded tyres since 2006 [12].

4.8 CONCLUSION

In this chapter, we have provided an overview of waste tyre recycling in terms of its significance, volume produced, regulations and basic recycling technologies. In fact, the expected volume of waste tyres produced is several billion tonnes by the end of 2030. Waste tyre recycling, thus, is vital in terms of safeguarding the environment and as an alternative source of energy.

Waste tyre management is governed by the well-known 4Rs rule: reduce, reuse, recycle, and recover. Waste tyre management programs aim to increase recycling and recovery at local, regional, and national levels. Furthermore, specific direct and indirect responsibilities, and roles of stakeholders are set and well defined in the adapted environmental regulations.

Waste tyre technologies treat the components of tyres for the sake of providing an alternative energy resource and reduce pollution. The tyres are composed of about 40% natural rubber and about 60% synthetic rubbers.

Energy recovering out of these materials provides an added value for the waste tyre. Several recycling technologies were discussed. Among these are thermal conversion technologies that provide a different set of products and combine different input requirements. Incineration technology is a widely used thermal conversion process. Pyrolysis and gasification differ from incineration in that they can be used to recover chemical value from waste, rather than its active value.

The advantages of waste tyre recycling are numerous, among which we mention the reduction of the amount of waste, preventing pollution by reducing the need to collect new raw materials, and saving energy. Furthermore, recycling of waste tyres will contribute directly and indirectly to the reduction of greenhouse gases emissions, thus mitigating the effects of global warming and climate changes. Economically, recycling could create new job opportunities in waste recycling and manufacturing industries.

REFERENCES

[1] Connor K., Cortesa S., Issagaliyeva S., Meunier A., Bijaisoradat O., Kongkatigumjorn N., Wattanavit K., Seth Tuler S., Stanley Selkow S., Tantayanon S. (2013) Report. Developing a sustainable waste tyre management strategy for Thailand. National science and technology development agency. An interactive qualifying project report. B.S. thesis, Faculty of Worcester Polytechnic Institute, Chulalongkorn University, Bangkok, Thailand.

[2] Turer A. (2012) *Recycling of scrap tyres, chapter in material recycling – trends and perspectives*, pp. 195–212. Available from: www.intechopen.com/books/materialrecycling-trends-and-perspectives/use-of-scrap-tyres-in-industry

[3] Dong Y., Zhao Y., Hossain M.U., He Y., Liu P. (2021) Life cycle assessment of vehicle tyres: A systematic review. *Cleaner Environmental Systems* 2: 100033. https://doi.org/10.1016/j.cesys.2021.100033

[4] Yaqoob H., Teoh Y.H., Jamil M.A., Gulzar M. (2021) Potential of tyre pyrolysis oil as an alternate fuel for diesel engines: A review. Journal of the Energy Institute 96: 205–221. https://doi.org/10.1016/j.joei.2021.03.002

[5] Yakub A., Ansari T, Khalid Iqbal M.J., Khan, A. (2021) Use of waste tier in construction of bitumen road–A review. *IOSR Journal of Engineering (IOSR JEN).* Available from: www.iosrjen.org

[6] Aylón E., Fernández-Colino, A., Murillo, R., Navarro, M.V., García, T., Mastral, A.M. (2010) Valorisation of waste tyre by pyrolysis in a moving bed reactor. *Waste Management* 30: 1220–1224. http://dx.doi.org/10.1016/j.wasman.2009.10.001

[7] Jain A. (2016) *Compendium of technologies for the recovery of materials/energy from End of Life (EoL) tyres* (Final Report). Nueng: International Environmental Technology Centre of United Nations Environment Program, Asian Institute of Technology Regional Resource Centre for Asia and the Pacific.

[8] Rowhani A., Rainey T.J. (2016) Scrap tyre management pathways and their use as a fuel—A review. *Energies* 9: 888. http://doi:10.3390/en9110888

[9] Liu L.L., Cai G.J., Zhang J., Liu X.Y., Liu K. (2020) Evaluation of engineering properties and environmental effect of recycled waste tyre-sand/soil in geotechnical engineering: A compressive review. *Renewable and Sustainable Energy Reviews* 126. https://doi.org/10.1016/j.rser.2020.109831.

[10] Afrin H., Huda N., Abbasi R. (2021) Study on End-of-Life Tyres (ELTs) recycling strategy and applications. *IOP Conference Series: Materials Science and Engineering* 1200: 012009. http://doi:10.1088/1757-899X/1200/1/012009

[11] Al Arni S., Elwaheidi, M. (2021) *Concise handbook of waste treatment technologies.* London: CRC Press and Taylor & Francis Group.

[12] Sienkiewicz M., Kucinska-Lipka J., Janik H., Balas A. (2012) Progress in used tyres management in the European Union: A review. *Waste Management* 32: 1742–1751. http://dx.doi.org/10.1016/j.wasman.2012.05.010

[13] Westall B. (1968) The molecular weight distribution of natural rubber latex. *Polymer* 9: 243–248. https://doi.org/10.1016/0032-3861(68)90035-9

[14] Roff W.J., Scott J.R. (1971) *Fibers, films, plastics and rubbers.* 1st edition. London: Butterworths.

[15] Martinez J.D., Puy N., Murillo R., Garcia T., Navarro M.V., Mastral A.M. (2013) Waste tyre pyrolysis—A review. *Renewable and Sustainable Energy Reviews* 23: 179–213. http://dx.doi.org/10.1016/j.rser.2013.02.038

[16] Ayanoglu A., Yumrutas R. (2016) Production of gasoline and diesel like fuels from waste tyre oil by using catalytic pyrolysis. *Energy* 103: 456–468. http://dx.doi.org/10.1016/j.energy.2016.02.155

[17] Wang F., Gao N., Quan C., López G. (2020) Investigation of hot char catalytic role in the pyrolysis of waste tyres in a two-step process. *Journal of Analytical and Applied Pyrolysis* 146: 104770. https://doi.org/10.1016/j.jaap.2019.104770

[18] Mousavi H. Z., Hosseynifar A., Jahed V., Dehghani S.A.M. (2010) Removal of lead from aqueous solution using waste tyre rubber ash as an adsorbent. *Brazilian Journal of Chemical Engineering* 27(1): 79–87.

[19] Mmereki D, Machola B., Mokokwe K. (2019) Status of waste tyres and management practice in Botswana. *Journal of the Air & Waste Management Association* 69: 10, 1230–1246, https://doi.org/10.1080/10962247.2017.1279696

[20] Ouyang S., Xiong D., Li Y., Zou L., Chen J. (2018) Pyrolysis of scrap tyres pretreated by waste coal tar. *Carbon Resources Conversion* 1: 218–227. https://doi.org/10.1016/j.crcon.2018.07.003

[21] Torretta V., Rada E.C., Ragazzi M., Trulli E., Istrate I. A., Cioca L.I. (2015) Treatment and disposal of tyres: Two EU approaches. A review. *Waste Management* 45: 152–160. http://dx.doi.org/10.1016/j.wasman.2015.04.018

[22] Pehlken A., Essadiqi E. (2005) Report, scrap tyre recycling in Canada. *Canmet Materials Technology Laboratory Report MTL* 2005–2008 (CF).

[23] Martínez J.D., Campuzano F., Cardona-Uribe N., Arenas C.N., Muñoz-Lopera D. (2020) Waste tire valorization by intermediate pyrolysis using a continuous twin-auger reactor: Operational features. *Waste Management* 113: 404–412. https://doi.org/10.1016/j.wasman.2020.06.019

[24] Bušic R., Milicevic I., Šipoš T.J., Strukar K. (2018) Recycled rubber as an aggregate replacement in self-compacting concrete—literature overview. *Materials* 11: 1729. http://dx.doi.org/10.3390/ma11091729

[25] Mavridou S. D. and Kaisidou E. N. (2016) Examination of bituminous mixtures made of conventional aggregates and recycled materials. *Cyprus 2016 4th International Conference on Sustainable Solid Waste Management*, 23–25 June 2016, Atlantica Miramare Beach Hotel, Limassol, Cyprus.

[26] Verma P., Zare A., Jafari M., Bodisco T.A., Rainey T., Ristovski Z.D., Brown R.J. (2018) Diesel engine performance and emissions with fuels derived from waste tyres. *Scientific Reports* 1–13. https://doi.org/10.1038/s41598-018-19330-0.

[27] Tyrecycle (2022) Products and uses. Available from: https://tyrecycle.com.au/products-uses/.

[28] Hamdi A., Abdelaziz, G., Farhan, K.Z. (2020) Scope of reusing waste shredded tyres in concrete and cementitious composite materials: A review. *Journal of Building Engineering*. Available from: https://doi.org/10.1016/j.jobe.2020.102014

[29] Islam and Nahian (2016) Improvement of waste tyre pyrolysis oil and performance test with diesel in CI engine. *Journal of Renewable Energy* 2016: 5137247. http://dx.doi.org/10.1155/2016/5137247

[30] Gamboa A.R., Ana M.A., Rochab, L.R., dos Santosc, J.A., de Carvalho, J. (2020) Tyre pyrolysis oil in Brazil: Potential production and quality of fuel. *Renewable and Sustainable Energy Reviews* 120: 109614. https://doi.org/10.1016/j.rser.2019.109614

[31] Abdul Aziz M., Rahman M.A., Molla, H. (2018) Design, fabrication and performance test of a fixed bed batch type pyrolysis plant with scrap tyre in Bangladesh. *Journal of Radiation Research and Applied Sciences* 11: 311–316. https://doi.org/10.1016/j.jrras.2018.05.001

[32] Xu S., Dengguo L., Zeng X., Zhang L., Han Z., Cheng J., Wu R., Mašek O., Xu G. (2018) Pyrolysis characteristics of waste tyre particles in fixed-bed reactor with internals. *Carbon Resources Conversion* 1: 228–237. https://doi.org/10.1016/j.crcon.2018.10.001

[33] Williams P.T. (2013) Pyrolysis of waste tyres: A review. *Waste Management* 33: 1714–1728. http://dx.doi.org/10.1016/j.wasman.2013.05.003.

[34] James G., Nema S.K., Anantha Singh T. S., Murugan P. V. (2019) A comparative analysis of pyrolysis and gasification of tyre waste by thermal plasma technology for environmentally sound waste disposal. Available from: http://uest.ntua.gr/heraklion2019/proceedings/pdf/HERAKLION2019_James_etal.pdf

[35] Gomes T.S, Neto G.R, de Salles A.C.N., Visconte L.L.Y., Pacheco E.B.A.V. (2019) End-of-life tyre destination from a life cycle assessment perspective. *New Frontiers on Life Cycle Assessment-Theory and Application IntechOpen*. http://dx.doi.org/10.5772/intechopen.82702

5 Electrochemical Removal of Organic Compounds from Municipal Wastewater

R. Elkacmi

CONTENTS

5.1 Introduction 73
5.2 Electro-Fenton 77
5.3 Electrocoagulation 82
5.4 Anodic Oxidation 85
5.5 Combined Processes 86
5.6 Conclusion 88
References 89

5.1 INTRODUCTION

Since the start and the emergence of industrialization, waste management has always been and remains a major concern for human beings. In general, waste in its three forms (liquid, solid and gaseous) comes mainly from the municipal, industrial and medical sectors. Over 80% of wastewater generated by public or private enterprises including households, trade and commerce, small businesses, office buildings and institutions (schools, hospitals and government structures) possesses a major threat to the biosphere and natural resources [1, 2].

Municipal/domestic wastewater mainly comes from households or a mixture of wastewater from households and industrial sources as well as precipitation water. It comes mainly from toilets, kitchens, baths, sinks, washing machines and industrial processes. The main sources of this liquid waste are summarized in Figure 5.1.

According to the United Nations reports, the global demand for water will be increased by 22–34% by 2050 as the world population will reach 9.4–10.2 billion [3]. Industries have become one of the main sources of hazardous wastewater due to their increasing consumption and dependency on freshwater at various stages of production. In addition, rapid growth of the human population contributes to intensive production of water pollution. Therefore, the purification of liquid wastes is required to reduce the concentration of contaminants before discharge.

DOI: 10.1201/9781003260738-5

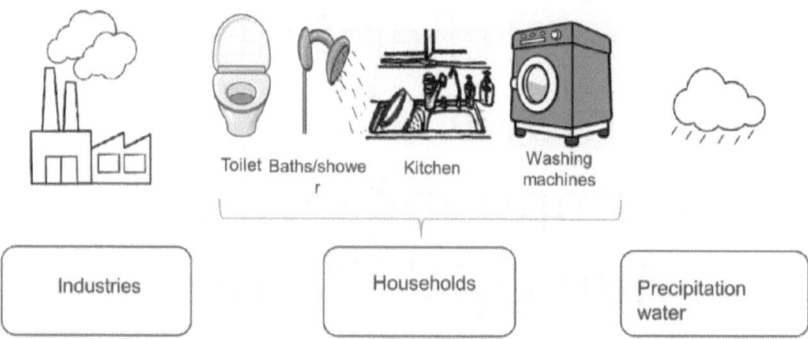

FIGURE 5.1   Main sources of municipal wastewater.

Typically, pH, chemical oxygen demand (COD), biochemical oxygen demand (BOD), total dissolved solid (TDS), total suspended solid (TSS), dissolved oxygen (DO), total nitrogen (TN), total phosphate (TP), and potassium are the main parameters of municipal wastewater (MWW) [4, 5]. Table 5.1 illustrates the overall composition of MWW given by several researchers.

In order to emphasize the better and suitable wastewater treatment method to provide clean water and reduce the negative impact on the environment. Numerous processes nowadays have been employed for the treatment of MWW, which include: biological methods, such as sand filters [11], aerobic, and anaerobic digestion [12]; physical processes, such as filtration/reverse osmosis [13], distillation [14], and adsorption [15]; chemical processes, such as coagulation–flocculation [16], chlorination [17], ion exchange [18], and advanced oxidation [19].

The techniques mentioned above suffer from a significant disadvantage, such as large space requirements, low process efficiencies, high energy consumption, large sludge formation and high operating cost. In this context, Table 5.2 represents the main advantages and disadvantages of each process.

TABLE 5.1

Physicochemical Characteristics of Raw Municipal Wastewater

	Units	Hernández-Flores et al. 2017 [6]	Singh et al. 2017 [7]	Shah et al 2014 [8]	Paing, & Voisin 2005 [9]	Kumar & Chopra 2012 [10]
pH		6.7 ± 0.03	8.11	–	–	8.39 ± 0.19
TSS	mg/l	106 ± 11	–	–	330	1824.42 ± 8.46
COD	mg/l	219 ± 21	211.59 ± 86.32	236	744	1420.54 ± 8.16
BOD$_5$	mg/l	145 ± 15	98.49 ± 11.57	132	324	620.27 ± 6,82
Conductivity	(mS/ cm)	0.31 ± 0.01	–	–	–	284. ± 21
Nitrogen	mg/l	12.2 ± 2.0	–	2.65	76	84.99 ± 10.92
Sulphates	mg/l	0.41 ± 0.03	–	–	–	–
Phosphorus	mg/l	–	–	2.1	11	124.42 ± 5.52
Alkalinity	mg/l	200 ± 10	232.22 ± 10.54	–	–	254.33 ± 8,85

TSS: total suspended solid; COD: chemical oxygen demand; BOD$_5$: biological oxygen demand.

TABLE 5.2
Advantages and Disadvantages of Various Techniques for Municipal Wastewater Treatment

Techniques	Advantages	Disadvantages
Biological methods		
Aerobic and anaerobic digestion	High operational stability, Low capital cost, high removal efficiency, no need for chemicals.	High operating cost, require large operational area, release of bad odors, high time consuming, formation of toxic by-products, ineffective for some effluents containing toxic elements
Physical methods		
Filtration/reverse osmosine	Simple, high efficiency, less space requirement, no need for chemical reagents	High operational cost, lower productivity with time, membrane fouling, sludge formation
Adsorption	Simple, effective of a wide range of organic compounds	High cost of activated carbon, inefficient method using some natural or low-cost adsorbents, high cost of regeneration of adsorbents
Membrane distillation	High efficiency, no requirement of high maintenance, no need for chemicals.	High capital cost, low productivity
Chemical methods		
Coagulation/ flocculation	Reduced precipitation time, removal of fine particles	High operating cost, slow technique, high sludge production, high pH requirement, complex dosing
Ion exchange	Environment friendly, no sludge production, no requirement of high maintenance, insensitive to flow variations	High capital cost, ineffective for high metal concentrations, inapplicable for large scales, regeneration produces secondary pollution
Advanced oxidation	Rapid reaction, total mineralisation of organic matter, no sludge production	High operating cost, high energy consumption, high maintenance cost, presence of residual hydrogen peroxide

Due to its environmental adaptability, wide applicability, stability, high efficiency and ease of operation, electrochemical technologies have drawn great attention as one of the most effective processes for removing a wide range of impurities from water and wastewater, including colloidal particles and dissolved organic substances [20].

Among the available electrochemical technologies for organic pollutants removal suitable for the treatment of these hazardous effluents are electro-Fenton (EF) [21], anodic oxidation (AO) [22] and electrocoagulation (EC) [20] processes which have received significant attention in recent decades. The process efficiency has been experimentally confirmed under different conditions by different authors for the removal of various organic and inorganic pollutants.

Compared to the conventional processes, the electrochemical way has the advantage of allowing better process control since the main species involved in the oxidation process, known as oxidizing agents, are generated in situ. Furthermore, electricity used as an energy source offers a clean solution without producing secondary pollutants, since the end products of oxidation are CO_2, H_2O, and some inorganic ions.

There are two main forms of electrochemical oxidation: (i) Direct: by direct action of electric current during electrolysis (AO), (ii) Indirect: based on the electrochemical generation of an oxidant that will subsequently react in solution with the pollutants (EF, EC).

The direct electrochemical oxidation induced by the electric current consists in carrying out an electron transfer directly on the organic compound; it is an exchange of electrons between organic species and the electrode surface. In the case of indirect oxidation, a strong oxidant is generated at the anode such as hydroxyl radical ($^{\bullet}$OH), active chlorine or others and reacts chemically with the organic substances, different reactive species generated during electrolysis are oxidizied in the oxidation process, which can be carried out either by electron transfer or by oxygen atom transfer. The electrogeneration of strong oxidants can be carried out from salts (chlorides, sulphates, and phosphates) or water. These salts are present in wastewater or added as supporting electrolytes to increase the ionic conductivity of the solution to reduce the energy cost of the process.

There are several parameters influencing the efficiency of the electrochemical processes in wastewater treatment such as current density (CD), electrolysis time, electrode position and agitation speed, as well as the physico-chemical properties of the liquid waste before treatment.

Research on MWW and its electrochemical treatment tools were studied as well as published in retrospective and prospective studies. Based on the Clarivate Web of Science database, general research was conducted using a combination of keywords: "municipal wastewater", "domestic wastewater", "electro-Fenton", "anodic oxidation" and "electrocoagulation". A total of 40625 papers were published between 2010/2021. Figure 5.2 shows the tendency of the electrochemical technologies for

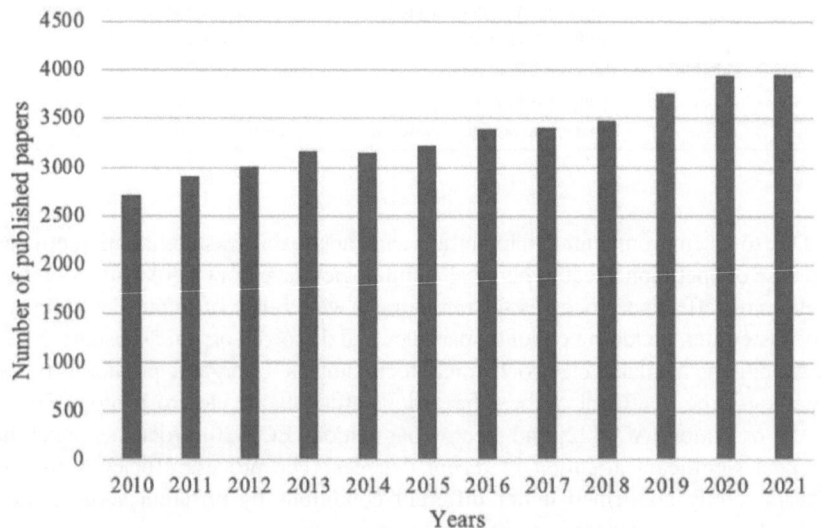

FIGURE 5.2 Distribution of scientific papers on municipal wastewater over the last ten years (2010–2021).

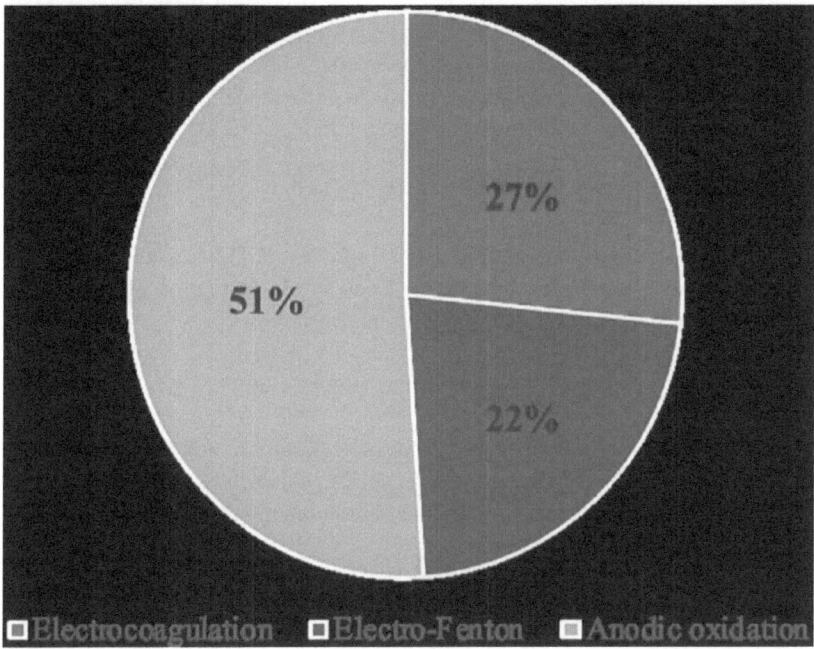

FIGURE 5.3 Researches on application of different electrochemical processes for municipal wastewater treatment from 2010–2021.

treating MWW wastewater. Among these electrochemical techniques, AO is widely applied for the removal of pollutants from MWW, as shown in Figure 5.3.

This chapter will focus mainly on the MWW treatment by electrochemical processes which involve EF, AO and EC, and summarize the current status of this liquid waste treatment based on these three electrochemical methods. Meanwhile, this work gives a general up to date view on the mechanism, process parameters as well as their effects on pollutant removal. Moreover, the latest applications of these technologies as well as the combinations with other processes, will be described aiming to solve most of the issues related to toxic substances migration into the environment.

5.2   ELECTRO-FENTON

The history of Fenton's chemistry began in 1894 with the publication of Henry J. Fenton [23] concerning the oxidation of tartaric acid by hydrogen peroxide (H_2O_2) in the presence of ferrous ions. Several papers have been published on the application of this chemical technique to reduce pollutant load in various types of toxic liquid discharges. However, the Fenton process, as any other treatment, has drawbacks that limit its application. (i) a high concentration of Fe^{2+} generates a large amount of iron sludge; (ii) operation in acidic conditions; and (iii) the difficulty of transporting and storing of H_2O_2.

The electro-Fenton process was developed to overcome these drawbacks while ensuring high pollutant removal efficiency and offering a competitive investment and treatment cost. This combination of Fenton's process with electrochemistry is based on the continuous in situ generation of hydrogen peroxide H_2O_2, in an acidic medium, via the electrochemical reduction of O_2 at the cathode.

$$O_2 + 2H^+ + 2e^- \rightarrow H_2O_2 \tag{5.1}$$

In this acidic condition, adding Fe^{2+} improves H_2O_2's oxidizing power significantly, because the H_2O_2/Fe^{2+} system leads to the formation of strong oxidant hydroxyl radical $^{\cdot}OH$ according to Fenton reaction:

$$Fe^{2+} + H_2O_2 \rightarrow Fe^{3+} + {}^{\cdot}OH + OH^- \tag{5.2}$$

Fenton's reagents can be produced electrochemically in solution, with the simultaneous generation of H_2O_2 by cathodic reduction of dissolved molecular oxygen and regeneration of iron ions consumed by the Fenton reaction.

$$Fe^{3+} + e^- \rightarrow Fe^{2+} \tag{5.3}$$

Benefits of the EF process have been identified by several authors as follows: (i) high removal rate at relatively low cost by optimizing the operating parameters; (ii) in-situ production of H_2O_2 avoiding the risks of its transport and storage; (iii) no use of chemical reagents and sludge formation; and (iv) easy to use and easy control of the degradation kinetics [24–26]. These advantages have prompted the rapid development of the EF process in the field of wastewater treatment.

It is worth mentioning that the efficiency of this electrochemical process depends on several operational parameters such as initial pH solution, CD, hydrogen peroxide concentration, molar ratio H_2O_2/Fe^{2+}, pollutants concentration, electrode materials, inter-electrode distance, operating temperature, feeding mode, oxygen sparging rate, and reaction time [27].

The choice of electrode material is of major importance for the performance of the EF process. In recent years, much effort has been devoted to finding the most suitable material for the treatment of wastewater polluted by organic compounds, and the most effective materials are those with high oxygen overvoltage. Due to its excellent conductivity and high stability, the platinum (Pt) electrode was the most widely used for the degradation of pollutants in the EF process [27–29]. However, from a price point of view, this material is rarely used in the practical wastewater's treatment. Therefore, different electrodes were tested on the detoxification of MWW, including: IrO_2 [30] and RuO_2 [31] as materials with low oxygen evolution overpotential, while PbO_2 [32] and SnO_2 [33] as materials with high oxygen evolution overpotential. A new material has emerged: Boron-doped diamond (BDD), which has a higher oxygen overvoltage than previous materials. Despite its high price, this material is widely used as a sacrificial anode for the electrocatalytic oxidation of organic pollutants. The production of H_2O_2 in solution is achieved by the electronic reduction of oxygen on different cathodes constituted mainly by carbon felt, copper, titanium or stainless steel [34–36].

Several research studies indicate that the EF process has been successfully applied for municipal/domestic wastewater treatment. It must be noted that the performance evaluation of the EF process in terms of removal of organic pollutants is typically measured as COD, total organic carbon (TOC) or coloration. Table 5.3 presents a compilation of these reports, with a summary of the main operating conditions and the COD, TOC or coloration removals obtained.

As can be seen from Table 5.3, the EF process has been widely used for the removal of various pollutants present in MWW. As reviewed by Meijide et al. [45], many publications have considered pharmaceutical products to be the main toxic organic load of MWW. The EF technique has been successfully used for the oxidation of levofloxacin [46], enoxacin [47], amoxicillin [48], chloroquine [49], norfloxacin [50] paracetamol [51], and other drugs [52–55]. In all these works, good removal efficiencies have been recorded.

As mentioned above, the BDD anode has a higher oxidation efficiency than the metal oxide anode in terms of hydroxyl radical production. This is mainly established by the large number of publications confirming its use for the detoxification of wastewater. Titchou et al. [56] investigated the reduction of dye Direct Red 23. At optimum conditions, around 90% TOC removal efficiency was achieved after 6 h treatment in sulfate medium while around 83% TOC removal efficiency in chloride medium using BDD/carbon graphite electrodes. More recently, Mbaye et al. [57] concluded that fungicide thiram oxidation over BDD anode was more effective in the degradation of 92% TOC than the Pt anode. Another study by Kulaksız et al. [58] showed that the EF process achieved the highest degradation efficiency of anticancer drug 5-FU and the highest TOC removal rate compared to TiO_2/UV photocatalysis and H_2O_2 subcritical water methods. Al-Zubaidi and Pak [59] tested the application of EF for the treatment of an aqueous solution containing 50 mg/L of parachlorophenol. This study yielded 86.4% and 4.2 kW h/kg$_{COD}$ of COD removal efficiency and energy consumption, respectively.

Iron is another key electrode material for wastewater treatment, Davarnejad et al. [60] have made a comparison between aluminum and iron plate electrodes on COD and color removal from Petrochemical wastewaters. The results showed that the iron electrode had the highest efficiency, with yields of up to 67.3% COD and 71.58% color. Soltani et al. [61] proposed for the first time mineral iron-based natural catalysts. In this study, four natural catalysts were tested as sources of ferrous iron (Fe^{2+}) ions: ilmenite ($FeTiO_3$), pyrite (FeS_2), chromite ($FeCr_2O_4$), and chalcopyrite ($CuFeS_2$). Experimental results exhibited that $CuFeS_2$ was suitable for total cefazolin degradation.

Despite its low energy consumption compared to other techniques, few studies have been found in the literature regarding the treatment of MWW by the EF process, this is mainly due to the narrow working pH (acidic medium), the difficult management of the metal residues formed, the addition of H_2O_2 and Fe^{2+} in some cases and also the generation of additional pollution by soluble salts in the treated wastewater.

TABLE 5.3
Summary of Some Relevant Studies on the Application of the EF Process for Industrial/Municipal Wastewater Treatment

Type of effluent	Initial parameter	Operating condition	Electrodes	Removal efficiency (%)	Location	References
Alcoholic wastewater	pH = 4.5–4.8 $[COD]_0$ = 60–70 g/l	pH = 2.84, t = 69.29 min, CD = 51.54 mA, $H_2O_2/$ Fe^{2+} = 3.7	Fe_2O_3-graphite	66.15% COD	Iran	Davarnejad and Azizi [37]
Pharmaceutical wastewater	$[TOC]_0$ = 8.86± 0.45 mg/L $[CBZ]_0$ = 60–70 µg/L	pH = 3.0, t = 120 min, CD = 0.2 A	BDD anode/carbon felt cathode	52 % TOC 73 % CBZ	Canada	Komtchou et al. [38]
Municipal wastewater	pH = 7.0 $[TOC]_0$ = 7.7 mg/L	t = 3 h CD = 40 mA/cm² $[Fe^{2+}]$ = 0.3 mmol/L	BDD/BDD	92 % TOC	Mexico	Villanueva-Rodríguez et al. [39]
Coking wastewater	pH = 9.14 $[COD]_0$ = 2.97 g/L $[NH_3-N]_0$ = 80.4 mg/L	pH = 3.88, t = 69.29 min, ED = 1 cm, AV = 10 V	DSA anode/Fe/AC/ Ni cathode	88.91–96.65% COD 100% NH_3-N	China	Meng, G et al. [40]
Pulp and paper mill wastewater	pH = 7 $[COD]_0$ = 1.2 g/L	pH = 2, t = 60 min, CD = 20 mA/cm² $[H_2O_2]$ = 0.2 mol/L	Fe/Fe	80 % COD	Turkey	Un et al. [41]
Textile wastewater	pH = 9.84 $[COD]_0$ = 0.544 g/L	pH = 3, t = 90 min, I = 32 A $[Fe^{2+}]$ = 0.53 mmol/L	Ti/RuO₂ anode/ Al cathode	100 % COD 90.30 % Color	India	Kaur et al. [42]

Wastewater	Initial conditions	Operating conditions	Electrodes	Removal	Country	Reference
Pesticides wastewaters	pH = 6.6–7.2, $[TOC]_0$ = 1.236–1.358 g/L, $[TC]_0$ = 23.7–25.1 mg/L	pH = 3, t = 180 min, CD = 2.22 mA/cm², $[Fe^{2+}]$ = 0.2 mmol/L, $[Na_2SO_4]$ = 0.99 g/L	Fe anode/ carbon plate cathode	75.4% TOC 93.7% TC	Viet Nam	Pham et al. [43]
Tannery wastewaters	$[TOC]_0$ = 0.2 mg/L	pH = 3, CD = 30 mA/cm², F = 12 L/min, $[Fe^{2+}]$ = 0.3 mmol/L, $[Na_2SO_4]$ = 50 mmol/L	BDD/BDD	100% TOC	Mexico	Villaseñor-Basulto et al. [44]

COD: chemical oxygen demand; TOC: total organic carbon; CD: current density; t: reaction time; ED: electrode distance; CBZ: carbamazepine; AV: applied voltage; DSA: dimensionally stable anodes; TC: tricyclazole; F: flow rate.

5.3 ELECTROCOAGULATION

Electrocoagulation (EC) technique is one of the most popular applications of electrochemical technology for MWW treatment. This process can effectively remove a wide range of dissolved organic matter present in the wastewater [62]. This process is considered to be the cost-effective process with advantages of COD and BOD$_5$ reduction, high color removal efficiency, sludge reduction and effective breakdown of recalcitrant pollutants.

The EC process is based on the principle of soluble anodes. It consists of imposing a current (or potential) between two electrodes (iron or aluminum) immersed in an electrolyte contained in a reactor to generate, in situ (Fe^{2+}, Fe^{3+}, Al^{3+}) ions able to produce a coagulant in solution that will cause coagulation of the pollutants. Figure 5.4 shows the principle of the process with aluminum electrodes.

The main reactions that take place with the electrodes (in the case of aluminum electrodes) are:

At the anode:

$$Al \rightarrow Al^{3+} + 3e^- \tag{5.4}$$

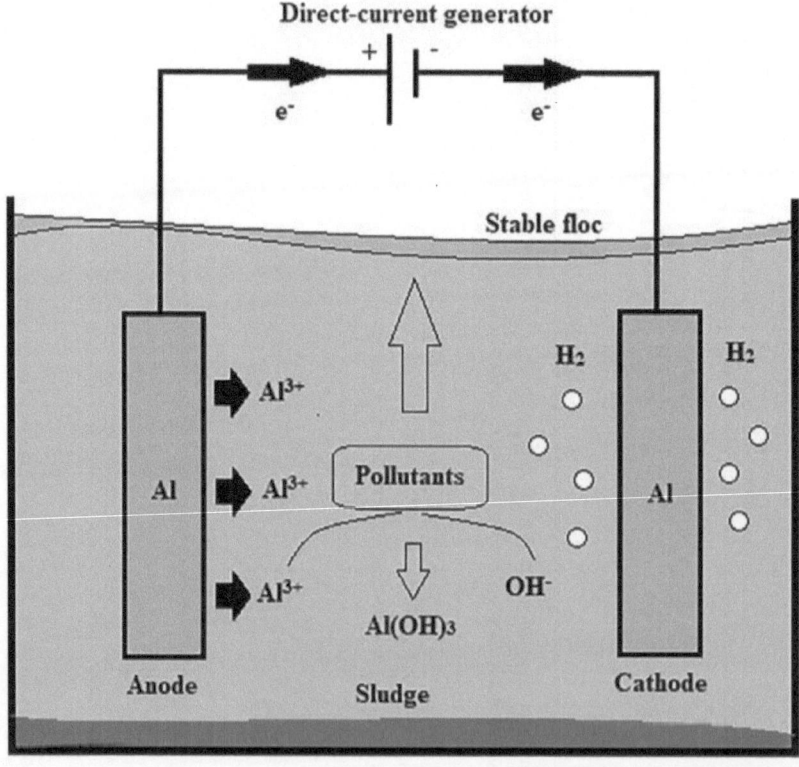

FIGURE 5.4   General Schematic diagram of electrocoagulation technique with two aluminum electrodes.

At the cathode:

$$2NO_2 + CH_4 \rightarrow 2H_2O + CO_2 + N_2 \tag{5.5}$$

$$H_2O + e^- \rightarrow 1/2H_2 + OH^- \tag{5.6}$$

It is worth noting that, among several metals used as soluble anode, aluminum and iron remain the most widely used due to their low price and high valence ionic form. Several studies were investigated in order to evaluate EC's efficiency in treating wastewater and especially removing toxic compounds.

Nguyen et al. [63] studied the EC process of local MWW. The authors examined the effect of a nanosecond pulse power supply (NPS) in comparison with a direct current. The yielded maximum COD removal efficiency was found to be 72% and 82% using DC and NPS, respectively. Follmann et al. [64] optimized the reduction of nutrients and organic matter present in MWW and improved the filterability during the EC process in a Submerged Membrane Electro-Bioreactor. Under optimum conditions, removal efficiencies for 99.51% PO_4^{3-}-P, 98.56% NH_4^+-N, and 99.04% COD were observed after 48 hours of electrolysis. Yang and his research group [65] addressed EC for the removal of phosphate using aluminum (Al-EC) and iron (Fe-EC) as electrodes for comparison, under low and high dissolved oxygen (DO) concentrations. They concluded that the removal rate of PO_4^{3-}-P in Fe-EC can reach 98% with low and high DO concentration. Also, it should be noted that a slight lower removal rate during Al-EC was observed compared to Fe-EC.

In another study, Qi et al. [66] examined the influence of horizontal bipolar electrodes (BPEs) on EC performances for treating MWW using various combinations of BPEs (iron or aluminum). The main results obtained revealed that a considerable reduction of the TOC and turbidity was attained throughout Al-BPEs. Moreover, a decrease of the energy consumption of around 75.2% for total phosphorus (TP) removal with Fe-BPEs and 81.5% with Al-BPEs.

Smoczynski et al. [67] investigated the treatment of natural MWW using EC in a pilot-scale reactor. They were able to reduce 86–99.5% of color, 60.8–63.5% of COD, 94.5–96% of TP and 100% of TSS after 2 h of treatment. Moreno et al. [68] published data on COD, coliforms and colony forming units (CFU) reduction from Mexican MWW using the EC process. The results obtained after a reaction time of 30 s showed 77–94% of COD, 80% of coliforms and 99.98% of CFU removal efficiencies. Koyuncu and Arıman [69] also examined the treatment of MWW by the EC process in a real-scale plant. EC successfully removed up to 67–80% COD, 69–81% BOD, 72–83% TSS, 21–47% TN and 27–46% for TP concentration of MWW.

To avoid the most serious drawback of this electrochemical approach, which is the high cost of electrical energy. Several researchers have proposed combining this electrochemical process with a new form of renewable energy such as solar photovoltaic system. Recently, Nawarkar and Salkar [70] conducted a study on MWW collected from sewers in Maharashtra City, India. Six aluminum electrodes were connected in a parallel configuration coupled with a solar photovoltaic module.

Global removals of 92.01%, 93.97%, and 49.78% were achieved for COD, turbidity, and TDS, respectively.

As is widely known, all types of industrial liquid discharges are considered to be a significant part of the pollutant load of MWW. Table 5.4 elucidates the application of EC technique on several types of industrial and domestic wastewater. In general, researchers obtained higher and faster pollutant removals with the EC process.

TABLE 5.4

Electrocoagulation Processes Applied for the Treatment of Domestic and Industrial Wastewaters

Type of effluent	Initial parameter	Operating condition	Electrodes	Removal efficiency (%)	References
Olive oil mill wastewater	pH = 5.6 $[COD]_0 = 31.4$ g/L $[TPh]_0 = 6.54$ g/L $A_{395} = 10.12$	pH = 5.6, t = 40 min, CD = 32.14 mA/cm²	Aluminium	79.24% COD, 94.82% TPh, 97.87% color	Elkacmi et al. [71]
Pulp and paper mill wastewaters	pH = 7.65 $[COD]_0 = 0.534$ g/L $[TOC]_0 = 0.209$ g/L $[TDS]_0 = 1.858$ g/L	pH = 7, t = 40 min, CD = 24.8 mA/cm² [NaCl] = 1.0 g/L	Stainless steel	82% COD, 99.4% color, 79% TOC, 43% TDS	Kumar and Sharma [72]
Pesticides wastewater	–	pH = 7, t = 150 min, CD = 7.5 mA/cm² [NaCl] = 2.5 mg/L	Iron	79% COD, 78% Cl2, 77% Br	Ramyaa et al. [73]
Textile wastewater	pH = 7.5–8.5 $[COD]_0 =$ 0.28–0.295 $[TOC]_0 =$ 0.26–0.22g/L $[TSS]_0 = 0.034$g/L	pH = 5, t = 120 min, CD = 25 mA/cm²	Aluminum	42.5% TOC, 18.6 % COD, 83.5% turbidity, 64.7% TSS, 90.3–94.9% color	Bener et al. [74]
Tannery wastewater	pH = 8.34 $[COD]_0 = 5.634$ g/L $[TSS]_0 = 0.91$ g/L $[Cr]_0 = 0.075$ g/L	pH = 7.2, t = 15 min, CD = 68 mA/cm²	Iron anode /Stainless steel cathode	64 % COD, 96% TSS, 99% Cr	Aboulhassan et al. [75]
Domestic wastewater	pH = 8.4 ± 0.6 $[COD]_0 = 0.420 ±$ 0.040 g/L $[PO_4-P]_0 = 0.0153$ ± 0.003 g/L $[E. coli]_0 = 392 ±$ 100 × 10⁶ MPN/ml	pH = 7.8, t = 20 min, CD = 100 A/m²	Aluminum anode/ carbon felt cathode	90.9 % COD, 97% PO_4-P, 96% turbidity 90.6% E. coli	Özyonar and Korkmaz. [76]

TPh: total phenolic compounds; COD: chemical oxygen demand; TOC: total organic carbon; TDS: Total dissolved solid; Cl_2: chloride; Br: bromide; TSS: total suspended solid; Cr: chromium; E. coli: Escherichia coli

Despite of these good results, scaling EC to industrial level remains a major challenge because of the generation of a huge volume of sludge, the high capital cost, and the high energy consumption [77].

5.4 ANODIC OXIDATION

Anodic oxidation (AO) has been widely used in water and wastewater treatment for the removal of organic and inorganic contaminants. A considerable number of studies have been undertaken on the use of this electrochemical technique for a complete mineralization and fractionation of MWW.

In AO, a highly and very powerful reactive oxidizing agents such as $^{\bullet}OH$ are produced in situ, which will then react with organic matter to give hydroxylated derivatives and finally be transformed into CO_2, inorganic ions, and water. The general mechanism can be represented as follows:

$$M + H_2O \rightarrow M(^{\bullet}OH) + H^+ + e^-$$ (5.7)

$$M(^{\bullet}OH) + R \rightarrow M + H_2O + CO_2$$ (5.8)

where M denotes the anode material, R is the organic molecule and $M(^{\bullet}OH)$ is the hydroxyl radical adsorbed on the anode surface.

As previously described in the EF process section, several insoluble electrode materials with high oxygen overvoltage are commonly tested, knowing that BDD anode has been used widely due to its high stability, strong corrosion resistance and high efficiency [77]. Up to date, various comprehensive reviews on the AO treatment of real and synthetic wastewater have been published. For example, in a recent study on the case of landfill leachate wastewater as a part of MWW, under optimum conditions, Wilk et al. [78] investigated the effectiveness of this electrochemical oxidation in removing over 70 % of COD using BDD anodes. Klidi et al. [79] studied the efficiency of AO for the degradation of the organic matter load present in paper mill wastewater through BDD and $TiRuSnO_2$ anodes and a stainless-steel cathode. Results revealed that the BDD allowed maximum and rapid rate COD abatement rate with low energy consumption compared to $TiRuSnO_2$ anode. Nevertheless, the latter is particularly effective in removing ammonium. Other studies related to AO of organic pollutants using different anode materials are summarized in Table 5.5. It was shown that BDD electrodes, which have a much higher overvoltage, allow total mineralization of different pollutants due to a higher production of hydroxyl radicals.

However, although the AO process has been studied widely and has demonstrated its efficiency for the treatment of various wastewater, few studies have been found in the literature regarding the detoxification of fresh and raw MWW by this technique alone. For this reason, several researchers have studied the combination of AO with other oxidation processes or combined with other technologies, in order to improve the efficiency of pollutant degradation, reduce both the energy consumption and the treatment time.

TABLE 5.5

Summary of Some Relevant Studies in AO Wastewater Treatment

Type of effluent	Initial parameter	Operating condition	Electrodes	Removal efficiency (%)	References
Pesticides wastewater	pH = 6.84, $[COD]_0$ = 1.81 g/L	pH = 6, t = 6 h, CD = 5 A/dm²	Ti/MMO anode /Stainless steel cathode	76% COD	Babu et al. [80]
Winery wastewater	pH = 5.6, $[COD]_0$ = 3.49 g/L, $[TOC]_0$ = 1.32 g/L	pH = 7.5 ± 0.2, t = 420 min, CD = 60 mA/cm², $[Na_2SO_4]$ = 50 mmol/L	BDD anode/ Stainless steel cathode	99.81% COD, 99.89% TOC	Candia-Onfray et al. [81]
Oilfield wastewater	pH = 6.5, $[TOC]_0$ = 0.22 g/L	t = 50 min, AC = 1 A, [NaCl] = 10 g/L, [NaOH] = 10 g/L, $[Fe(NO_3)_3]$ = 10 g/L	Ti/Ru-Ir anode/ graphite cathode	89.4% TOC	Liu et al. [82]
Petroleum refinery wastewater	pH = 8, $[COD]_0$ = 0.5–0.51 g/L, $[TOC]_0$ = 0.22 g/L	pH = 4, t = 120 min, CD = 50 mA/cm², $[Na_2SO_4]$ = 0.035 mol/L	PbO_2 anode / stainless steel cathode	84.8% COD	Ghanim and Hamza [83]
		pH = 7, t = 120 min, CD = 50 mA/cm², $[Na_2SO_4]$ = 0.035 mol/L	Carbon felt anode / stainless steel cathode	86.3% COD	
Coke oven wastewater	pH = 8, $[COD]_0$ = 1.52 g/L, $[CN^-]_0$ = 0.28 g/L, $[TPh]_0$ = 0.9 g/L	pH = 3.95, t = 10 h, CD = 6.7 mA/cm², [NaCl] =1 g/L	PbO_2 anode/ graphite	85,2% COD, 99.1% CN⁻, 99.7% TPh	Pillai and Gupta [84]
Slaughterhouse wastewater	pH = 6.4, $[COD]_0$ = 1.368 g/L, $[Turbidity]_0$ = 320 NTU	pH = 5, t = 100 min, CD = 3.83 mA/cm², [NaCl] = 1%	BDD/ BDD	100% COD, 80% turbidity, 90% color	Abdelhay et al. [85]
Pharmaceutical wastewater	pH = 7.80 ± 0.02, $[COD]_0$ = 0.0355 g/L, $[TOC]_0$ = 0.07262 g/L	pH = 7.66, t = 300 min, CD = 6 mA/cm²,	BDD anode/ stainless steel cathode	88% TOC	Calzadilla et al. [86]

Ti/MMO: oxide coated titanium; AC: applied current; CN⁻: cyanide; TPh: total phenolic compounds

5.5 COMBINED PROCESSES

As already pointed out, these three electrochemical processes have proven to be effective methods for the complete degradation and mineralization of synthetic and real wastewater contaminated with different kind of organic pollutants present in MWW such as landfill leachates, pesticides, industrial effluents, agricultural runoff, pharmaceuticals products, and domestic sewage.

To improve the performance of electrochemical techniques and to overcome the drawback of high energy consumption and treatment cost, several researchers tried a combination of these processes with other biological or physico-chemical technologies.

Most of the combined processes proposed up to the date for the treatment of MWW are based on biological tools. Makwana and Ahammed [87] applied continuous EC for the treatment of the anaerobically digested MWW. The results suggested that over 99.8% of both total coliform and fecal coliform removal efficiencies could be achieved at optimum operating conditions. Nguyen et al. [88] proposed a combined treatment system consisting of a rotating hanging media bioreactor (RHMBR) and submerged membrane bioreactor (SMBR) with a post-EC process to remove organic and nutrient pollutants from MWW. Under optimal conditions, final average of COD, BOD_5, TN, TP and NH_4^+-N close to 11.46, 0.26, 3.81, 0.03, and 0 mg/L were respectively achieved. Moreover, they found that the effluent quality was achieved with 5 mg/L of TSS and 30 MPN/L of total coliform.

Sharma and Chopra [89] examined a coupled treatment process for MWW comprising biological treatment and then EC processing using iron electrodes. The optimum removal of COD (92.35%) was achieved at a pH value of 7.5, CD of 2.82 A/m^2 upon 40 min residence time, as well as 84.88% BOD_5 reduction after 30 min settling time. Moreover, the mean operating cost, electrode consumption and energy consumption, estimated by the authors were 1.08 $US\$/m^3$, 78.48 × 10^5 kg Al/m^3 and 108.48 kWh/m^3, respectively.

Several publications proposed to treat different types of industrial wastewater by an electrochemical process coupled with biological treatments, a selection of which is presented in Table 5.6. It is worthy to denote here that electrochemical processes combined with biological treatments such as aerobic and anaerobic digestion technologies have shown effective removal of COD >90% for various types of liquid discharges.

In some cases, electrochemical processes were coupled with physico-chemical treatments whenever the aim is to improve the degradation of organic compounds and reduce the toxicity of MWW. Park et al. [97] proposed a treatment process for MWW based on combining microfiltration technology with electrochemical

TABLE 5.6
Literature Review of Some Industrial Wastewater Treatment by Electrochemical-Biological Processes Combination

Type of effluent	Processes	Removal efficiency (%)	References
Slaughterhouse wastewater	BP + EF	97% COD	Vidal et al. [90]
Landfill leachate wastewater	BP + EF	97% COD	Baiju et al. [91]
Agricultural wastewater	BP + EC	78.5% COD, 91.6% TPh	Hanafi et al. [92]
Pharmaceutical wastewater	BP + EF	94% COD	Popat et al. [93]
Petroleum refinery wastewater	BP + EF	95% COD, 94% BOD	Dehboudeh et al. [94]
Dairy wastewater	BP + AO	100 % COD	Katsoni et al. [95]
Pistachio processing wastewater	BP + EC	86.41 % COD, 91.09 % TPh	Ozay et al. [96]

BP: biological process

oxidation. The results showed that it was possible to reduce 80% TOC and >99% turbidity with a considerable reduction in membrane fouling. Natarajan et al. [98] studied the treatment of MWW by using a novel natural coagulant prepared from *Pisum sativum* in an EC reactor. The experimental results reveal that 94.6% COD abatement was achieved upon 20 V voltage, 1.5 A current intensity, 4 g/L coagulant dose, and an optimal pH of 5.0.

Abdel-Fatah et al. [99] examined the treatment of MWW by means of EC followed by ultra-filtration membrane; the final values of COD, BOD_5 and TP at the outlet of the pilot plant were around 95%, 93% and 100% respectively. Finally, the high-quality produced water can be used for irrigation or fish farming. As well as, the final sludge was apt to use as fertilizer due to its quality.

The applicability of this combination for the treatment of industrial effluents has also been analyzed in several studies. Roa-Morales et al. [100] have combined three processes, namely, coagulation, EC and ozonation for the treatment of industrial wastewater containing offset printing dyes. In this combined system, 99.99% of color and 99.35% of COD were removed at 4 A for 50 min with a dose of 20 mg/L aluminum hydroxychloride and 15 min of ozonation. In another research paper, Gunawan and his co-workers [101] proposed a yarn dyed wastewater treatment process comprising EC followed by Fenton oxidation in a continuous system. The combined system yielded 80 % of COD abatement under optimal conditions. Moreover, they found that the chemical cost of the continuous mode (160 IDR/L) is lower than the batch system with 256 IDR/L. Zakeri et al. [102] proposed new combination system between chemical coagulation and EF, the research group were able to remove 90.3%, 87.25%, and 87% of COD, BOD_5 and TSS, respectively, under the optimum working conditions of pH 3, voltage of 20 V, poly aluminum chloride (PAC) concentration of 100 mg/L, H_2O_2 concentration of 1.5 g/L, reaction time of 60 min and electrode distance of 2 cm. Torres et al. [103] used a combination of coagulation-flocculation and electrochemical oxidation to remove TOC and turbidity and reduce the toxicity of textile wastewater. According to the authors, 82% of TOC were removed at 3 mA/cm^2 with a good toxicity reduction.

In conclusion of this section, we would like to highlight that this work deals with the application of electrochemical processes as individual or combined systems for the detoxification of MWW with the focus on some new applications on industrial liquid effluents belonging to the MWW family.

5.6 CONCLUSION

This chapter emphasizes on treating MWW by employing various electrochemical technologies. Details of EC, EF and AO processes are presented in this chapter including performance, mechanism, application, limitations, and cost-effectiveness. The main objective of these tools is to minimize the chemical reagents amount by producing the oxidants directly in the toxic effluent by electrochemistry. Based on technical, environmental and economic reasons confirmed by the number of publications, the AO has demonstrated its efficiency for the treatment of MWW. Several insoluble electrodes with high oxygen overvoltage are tested for a high removal efficiency. However, the main limitation of this process is the high operating cost due to the high energy consumption. The design of EC technique is revived to fulfil the needs of this method for the removal of pollutants from MWW. This method can be used to optimize the parameters of the electrochemical process to ensure good treatment performance at acceptable costs. Up to this moment, many alternative strategies of MWW treatment by EC are practiced in order to reduce the polluting load. However, further investigation is needed to move to a real industrial scale with full economic analysis.

REFERENCES

[1] Sas E, Hennequin L, Frémont A, Jerbi A, Legault N, Lamontagne J, et al. Biorefinery potential of sustainable municipal wastewater treatment using fast-growing willow. Sci Total Environ 2021; 792:148146.

[2] Renuka N, Ratha S, Kader F, Rawat I, Bux F. Insights into the potential impact of algae-mediated wastewater beneficiation for the circular bioeconomy: A global perspective. J Environ Manage 2021; 297:113257.

[3] Boretti A, Rosa L. Reassessing the projections of the world water development report. NPJ Clean Water 2019; 2(1):1–6.

[4] Wijaya I, Soedjono E. Physicochemical characteristic of municipal wastewater in tropical area: Case study of Surabaya City, Indonesia. IOP Conference Series: Earth Environ Sci 2018; 135:012018.

[5] Shah M, Hashmi H, Ali A, Ghumman A. Performance assessment of aquatic macrophytes for treatment of municipal wastewater. J Environ Health Sci 2014; 12(1):1–12.

[6] Hernández-Flores G, Solorza-Feria O, Poggi-Varaldo H. Bioelectricity generation from wastewater and actual landfill leachates: A multivariate analysis using principal component analysis. Int J Hydrog Energy 2017; 42(32):20772–20782.

[7] Singh K, Vaishya C R, Gupta A. Evaluation of Consortia performance under continuous process treating municipal wastewater with low concentration of heavy metals, antibiotic (gentamicin) and diesel oil. Int J Eng Technol 2017; 9(5):3448–3457.

[8] Shah M, Hashmi H, Ali A, Ghumman A. Performance assessment of aquatic macrophytes for treatment of municipal wastewater. J Environ Health Sci 2014; 12(1).

[9] Paing J, Voisin J. Vertical flow constructed wetlands for municipal wastewater and septage treatment in French rural area. Water Sci Technol 2005; 51(9):145–155.

[10] Kumar V, Chopra A. Monitoring of physico-chemical and microbiological characteristics of municipal wastewater at treatment plant, Haridwar City (Uttarakhand) India. J Environ Sci Technol 2012; 5(2):109–118.

[11] Mulugeta S, Helmreich B, Drewes J, Nigussie A. Consequences of fluctuating depth of filter media on coliform removal performance and effluent reuse opportunities of a biosand filter in municipal wastewater treatment. J Environ Chem Eng 2020; 8(5):104135.

[12] Wang K, Martin Garcia N, Soares A, Jefferson B, McAdam E. Comparison of fouling between aerobic and anaerobic MBR treating municipal wastewater. H2 Open J 2018; 1(2):131–159.

[13] Arola K, Van der Bruggen B, Mänttäri M, Kallioinen M. Treatment options for nanofiltration and reverse osmosis concentrates from municipal wastewater treatment: A review. Crit Rev Environ Sci Technol 2019; 49(22):2049–2116.

[14] Kim H, Shin J, Won S, Lee J, Maeng S, Song K. Membrane distillation combined with an anaerobic moving bed biofilm reactor for treating municipal wastewater. Water Res 2015; 71:97–106.

[15] Coimbra R, Calisto V, Ferreira C, Esteves V, Otero M. Removal of pharmaceuticals from municipal wastewater by adsorption onto pyrolyzed pulp mill sludge. Arab J Chem 2019; 12(8):3611–3620.

[16] Abomohra AE-F, Jin W, Sagar V, Ismail GA. Optimization of chemical flocculation of scenedesmus obliquus grown on municipal wastewater for improved biodiesel recovery. Renew Energy 2018; 115:880–886.

[17] Zhang, Y, Zhuang Y, Geng J, Ren H, Zhang Y, Ding L, Xu K. Inactivation of antibiotic resistance genes in municipal wastewater effluent by chlorination and sequential UV/chlorination disinfection. Sci Total Environ 2015; 512–513:125–32.

[18] Huang X, Guida S, Jefferson B, Soares A. Economic evaluation of ion-exchange processes for nutrient removal and recovery from municipal wastewater. NPJ Clean Water 2020; 3(1).

[19] Serna-Galvis EA, Botero-Coy AM, Martínez-Pachón D, Moncayo-Lasso A, Ibáñez M, Hernández F, et al. Degradation of seventeen contaminants of emerging concern in municipal wastewater effluents by sonochemical advanced oxidation processes. Water Res 2019; 154:349–60.

[20] Elkacmi R., Bennajah M. New techniques for treatment and recovery of valuable products from olive mill wastewater. Handbook of Environmental Materials Management, Springer International Publishing., 2019, p. 1–20.

[21] Nidheesh PV, Gandhimathi R. Trends in electro-fenton process for water and wastewater treatment: An overview. Desalination 2012; 299:1–15.

[22] Hu Z, Cai J, Song G, Tian Y, Zhou M. Anodic oxidation of organic pollutants: Anode fabrication, process hybrid and environmental applications. Curr Opin Electrochem 2021; 26:100659.

[23] Fenton HJ. LXXIII.—oxidation of tartaric acid in presence of iron. J Chem Soc, Trans 1894; 65:899–910.

[24] Oturan N, Oturan MA. Electro-Fenton process: Background, new developments, and applications. Electrochemical Water and Wastewater Treatment 2018; 193–221.

[25] Trellu C, Olvera Vargas H, Mousset E, Oturan N, Oturan MA. Electrochemical Technologies for the treatment of pesticides. Curr Opin Electrochem 2021; 26:100677.

[26] Sirés I, Brillas E, Oturan MA, Rodrigo MA, Panizza M. Electrochemical advanced oxidation processes: Today and tomorrow. A Review. Environ Sci Pollut Res 2014; 21(14):8336–67.

[27] He H, Zhou Z. Electro-fenton process for water and wastewater treatment. Crit Rev Environ Sci Technol 2017; 47(21):2100–31.

[28] Çelebi MS, Oturan N, Zazou H, Hamdani M, Oturan MA. Electrochemical oxidation of carbaryl on platinum and boron-doped diamond anodes using electro-fenton technology. Sep Purif Technol 2015; 156:996–1002.

[29] Zazou H, Oturan N, Sönmez-Çelebi M, Hamdani M, Oturan M. A. Study of degradation of the fungicide imazalil by electro-Fenton process using platinum and boron-doped diamond electrodes. J Mater Environ Sci 2015; 6(1):107–13.

[30] Wang H, Wang J, Bo G, Wu S, Luo L. Degradation of pollutants in polluted river water using ti/iro2–TA2O5 coating electrode and evaluation of electrode characteristics. J Clean Prod 2020; 273:123019.

[31] Zhi D, Zhang J, Wang J, Luo L, Zhou Y, Zhou Y. Electrochemical treatments of coking wastewater and coal gasification wastewater with Ti/TI4O7 and TI/ruo2–iro2 anodes. J Environ Manage 2020; 265:110571.

[32] He Y, Huang W, Chen R, Zhang W, Lin H, Li H. Anodic oxidation of aspirin on PbO 2, BDD and porous TI/BDD electrodes: Mechanism, kinetics and utilization rate. Sep Purif Technol 2015; 156:124–31.

[33] Yang C, Fan Y, Li P, Gu Q, Li X-yan. Freestanding 3-dimensional macro-porous SNO_2 electrodes for efficient electrochemical degradation of antibiotics in wastewater. Chem Eng J 2021; 422:130032.

[34] Yuan S, Gou N, Alshawabkeh AN, Gu AZ. Efficient degradation of contaminants of emerging concerns by a new electro-fenton process with Ti/MMO Cathode. Chemosphere 2013; 93(11):2796–804.

[35] Sopaj F, Oturan N, Pinson J, Podvorica FI, Oturan MA. Effect of cathode material on electro-fenton process efficiency for electrocatalytic mineralization of the antibiotic sulfamethazine. Chem Eng J 2020; 384:123249.

[36] Qi H, Sun X, Sun Z. Cu-doped fe2o3 nanoparticles/etched graphite felt as bifunctional cathode for efficient degradation of sulfamethoxazole in the heterogeneous electro-fenton process. Chem Eng J 2022; 427:131695.

[37] Davarnejad R, Azizi J. Alcoholic wastewater treatment using electro-fenton technique modified by FE2O3 nanoparticles. J Environ Chem Eng 2016; 4(2):2342–9.

[38] Komtchou S, Dirany A, Drogui P, Bermond A. Removal of carbamazepine from spiked municipal wastewater using electro-fenton process. Environ Sci Pollut Res 2015; 22(15):11513–25.

[39] Villanueva-Rodríguez M, Bello-Mendoza R, Hernández-Ramírez A, Ruiz-Ruiz EJ. Degradation of anti-inflammatory drugs in municipal wastewater by heterogeneous photocatalysis and electro-fenton process. Environ Technol 2018; 40(18):2436–45.

[40] Meng G, Jiang N, Wang Y, Zhang H, Tang Y, Lv Y, et al. Treatment of coking wastewater in a heterogeneous electro-fenton system: Optimization of treatment parameters, characterization, and removal mechanism. J Water Process Eng 2022; 45:102482.

[41] Un UT, Topal S, Oduncu E, Ogutveren UB. Treatment of tissue paper wastewater: Application of electro-fenton method. Int J Environ Sci Dev 2015; 6(6):415–8.

[42] Kaur P, Sangal VK, Kushwaha JP. Parametric Study of electro-fenton treatment for real textile wastewater, disposal study and its cost analysis. Int J Environ Sci Technol 2018; 16(2):801–10.

[43] Pham TL, Boujelbane F, Bui HN, Nguyen HT, Bui X-T, Nguyen DN, et al. Pesticide production wastewater treatment by electro-fenton using taguchi experimental design. Water Sci Technol 2021; 84(10–11):3155–71.

[44] Villaseñor-Basulto D, Picos-Benítez A, Bravo-Yumi N, Perez-Segura T, Bandala ER, Peralta-Hernández JM. Electro-Fenton mineralization of diazo dye black NT2 using a pre-pilot flow plant. J Electroanal Chem 2021; 895:115492.

[45] Meijide J, Dunlop PS, Pazos M, Sanromán MA. Heterogeneous electro-fenton as "green" technology for pharmaceutical removal: A Review. Catal 2021; 11(1):85.

[46] Barhoumi N, Labiadh L, Oturan MA, Oturan N, Gadri A, Ammar S, et al. Electrochemical mineralization of the antibiotic levofloxacin by electro-fenton-pyrite process. Chemosphere 2015; 141:250–257.

[47] Özcan A, Atılır Özcan A, Demirci Y, Şener E. Preparation of FE2O3 modified kaolin and application in heterogeneous electro-catalytic oxidation of enoxacin. Appl Catal B: Environ 2017; 200:361–371.

[48] Zhang J, Zheng C, Dai Y, He C, Liu H, Chai S. Efficient degradation of amoxicillin by scaled-up electro-fenton process: Attenuation of toxicity and decomposition mechanism. Electrochimi Acta 2021; 381:138274.

[49] Midassi S, Bedoui A, Bensalah N. Efficient degradation of chloroquine drug by electro-fenton oxidation: Effects of operating conditions and degradation mechanism. Chemosphere 2020; 260:127558.

[50] Özcan A, Atılır Özcan A, Demirci Y. Evaluation of mineralization kinetics and pathway of norfloxacin removal from water by electro-fenton treatment. Chem Eng J 2016; 304:518–526.

[51] Le TX, Charmette C, Bechelany M, Cretin M. Facile preparation of porous carbon cathode to eliminate paracetamol in aqueous medium using electro-fenton system. Electrochimi Acta 2016; 188:378–384.

[52] Droguett C, Salazar R, Brillas E, Sirés I, Carlesi C, Marco JF, et al. Treatment of antibiotic cephalexin by heterogeneous electrochemical Fenton-based processes using chalcopyrite as sustainable catalyst. Sci Total Environ 2020; 740:140154.

[53] Görmez F, Görmez Ö, Gözmen B, Kalderis D. Degradation of chloramphenicol and metronidazole by electro-fenton process using graphene oxide-fe3o4 as heterogeneous catalyst. J Environ Chem Eng 2019; 7(2):102990.

[54] Sopaj F, Oturan N, Pinson J, Podvorica F, Oturan MA. Effect of the anode materials on the efficiency of the electro-fenton process for the mineralization of the antibiotic sulfamethazine. Appl Catal B: Environ 2016; 199:331–341.

[55] Shueai Y, Ghizlan K, Mariam K, Miloud E, Azeem A, Abdelkader Z, Kacem E. Mineralization of Ofloxcacin Antibiotic in Aqueous Medium by Electro-Fenton Process Using a Carbon Felt Cathode: Influencing Factors Anal Bioanal Electrochem 2020; 12(4): 425–36.

[56] Titchou FE, Zazou H, Afanga H, El Gaayda J, Ait Akbour R, Hamdani M, et al. Electro-fenton process for the removal of direct red 23 using BDD anode in chloride and sulfate media. J Electroanal Chem 2021; 897:115560.

[57] Mbaye M, Diaw PA, Mbaye OM, Oturan N, Gaye Seye MD, Trellu C, et al. Rapid removal of fungicide thiram in aqueous medium by electro-fenton process with PT and BDD Anodes. Sep Purif Technol 2022; 281:119837.

[58] Kulaksız E, Kayan B, Gözmen B, Kalderis D, Oturan N, Oturan MA. Comparative degradation of 5-fluorouracil in aqueous solution by using H_2O_2-modified subcritical water, photocatalytic oxidation and electro-fenton processes. Environ Res 2022; 204:111898.

[59] Al-Zubaidi DK, Pak KS. Degradation of parachlorophenol in synthetic wastewater using batch electro-fenton process. Mater Today: Proc 2020; 20:414–9.

[60] Davarnejad R, Mohammadi M, Ismail AF. Petrochemical wastewater treatment by electro-fenton process using aluminum and iron electrodes: Statistical comparison. J Water Process Eng 2014; 3:18–25.

[61] Soltani F, Navidjouy N, Khorsandi H, Rahimnejad M, Alizadeh S. A novel bio-electro-fenton system with dual application for the catalytic degradation of tetracycline antibiotic in wastewater and bioelectricity generation. RSC Adv 2021; 11(44):27160–73.

[62] Elkacmi R, Kamil N, Bennajah M. Upgrading of Moroccan olive mill wastewater using electrocoagulation: Kinetic study and process performance evaluation. J Urban Environ Eng 2017; 30–41.

[63] H. Nguyen Q. Electrocoagulation with a nanosecond pulse power supply to remove cod from municipal wastewater using iron electrodes. Int J Electrochem Sci 2020; 493–504.

[64] Follmann HV, Souza E, Aguiar Battistelli A, Rubens Lapolli F, Lobo-Recio MÁ. Determination of the optimal electrocoagulation operational conditions for pollutant removal and filterability improvement during the treatment of municipal wastewater. J Water Process Eng 2020; 36:101295.

[65] Yang Y, Li Y, Mao R, Shi Y, Lin S, Qiao M, et al. Removal of phosphate in secondary effluent from municipal wastewater treatment plant by iron and aluminum electrocoagulation: Efficiency and mechanism. Sep Purif Technol 2022; 286:120439.

[66] Qi Z, You S, Liu R, Chuah CJ. Performance and mechanistic study on electrocoagulation process for municipal wastewater treatment based on horizontal bipolar electrodes. Front Environ Sci En 2020; 14(3).

[67] Smoczyński L, Kalinowski S, Ratnaweera H, Kosobucka M, Trifescu M, Pieczulis-Smoczyńska K. Electrocoagulation of municipal wastewater—A pilot-scale test. Desalination Water Treat 2017; 72:162–8.

[68] Moreno H, Parga JR, Gomes AJ, Rodríguez M. Electrocoagulation treatment of municipal wastewater in Torreon Mexico. Desalination Water Treat 2013; 51(13–15):2710–7.

[69] Koyuncu S, Arıman S. Domestic wastewater treatment by real-scale electrocoagulation process. Water Sci Technol 2020; 81(4):656–67.

[70] Nawarkar CJ, Salkar VD. Solar powered electrocoagulation system for municipal wastewater treatment. Fuel 2019; 237:222–6.

[71] Elkacmi R, Boudouch O, Hasib A, Bouzaid M, Bennajah M. Photovoltaic electrocoagulation treatment of Olive Mill wastewater using an external-loop airlift reactor. Sustain Chem Pharm 2020; 17:100274.

[72] Kumar D, Sharma C. Remediation of pulp and paper industry effluent using electrocoagulation process. J Water Resour Prot 2019; 11(3):296–310.

[73] Ramya T, Premkumar P, Thanarasu A, Velayutham K, Dhanasekaran A, Sivanesan S. Degradation of pesticide-contaminated wastewater (coragen) using electrocoagulation process with iron electrodes. Desalination Water Treat 2019; 165:103–10.

[74] Bener S, Bulca Ö, Palas B, Tekin G, Atalay S, Ersöz G. Electrocoagulation process for the treatment of real textile wastewater: Effect of operative conditions on the organic carbon removal and kinetic study. Process Saf Environ Prot 2019; 129:47–54.

[75] Aboulhassan MA, El Ouarghi H, Ait Benichou S, Ait Boughrous A, Khalil F. Influence of experimental parameters in the treatment of tannery wastewater by electrocoagulation. Sep Sci Techno 2018; 53(17):2717–26.

[76] Özyonar F, Korkmaz MU. Sequential use of the electrocoagulation-electrooxidation processes for domestic wastewater treatment. Chemosphere 2022; 290:133172.

[77] Elkacmi R, Bennajah M. Advanced oxidation technologies for the treatment and detoxification of olive mill wastewater: A general review. J Water Reuse Desalin 2019; 9(4):463–505.

[78] Wilk BK, Szopińska M, Luczkiewicz A, Sobaszek M, Siedlecka E, Fudala-Ksiazek S. Kinetics of the organic compounds and ammonium nitrogen electrochemical oxidation in landfill leachates at boron-doped diamond anodes. Mater 2021; 14(17):4971.

[79] Klidi N, Clematis D, Delucchi M, Gadri A, Ammar S, Panizza M. Applicability of electrochemical methods to paper mill wastewater for reuse. Anodic oxidation with BDD and Tirusno2 anodes. J Electroanal Chem 2018; 815:16–23.

[80] Babu BR, Kanimozhi R, Venkatesan P, Meera KS. Electrochemical degradation of methyl parathion. Int J Environ Eng 2013; 5(3):311.

[81] Candia-Onfray C, Espinoza N, Sabino da Silva EB, Toledo-Neira C, Espinoza LC, Santander R, et al. Treatment of winery wastewater by anodic oxidation using BDD electrode. Chemosphere 2018; 206:709–17.

[82] Liu K, Yi Y, Zhang N. Anodic oxidation produces active chlorine to treat oilfield wastewater and prepare ferrate(vi). Journal of Water Process Engineering. 2021; 41:101998.

[83] Ghanim AN, Hamza AS. Evaluation of direct anodic oxidation process for the treatment of petroleum refinery wastewater. J Environ Eng 2018; 144(7):04018047.

[84] Sasidharan Pillai IM, Gupta AK. Anodic oxidation of coke oven wastewater: Multiparameter optimization for simultaneous removal of cyanide, cod and phenol. J Environ Manage 2016; 176:45–53.

[85] Abdelhay A, Jum'h I, Abdulhay E, Al-Kazwini A, Alzubi M. Anodic oxidation of slaughterhouse wastewater on boron-doped diamond: Process variables effect. Water Sci Technol 2017; 76(12):3227–35.

[86] Calzadilla W, Espinoza LC, Diaz-Cruz MS, Sunyer A, Aranda M, Peña-Farfal C, et al. Simultaneous degradation of 30 pharmaceuticals by anodic oxidation: Main intermediaries and by-products. Chemosphere 2021; 269:128753.

[87] Makwana AR, Ahammed MM. Continuous electrocoagulation process for the post-treatment of anaerobically treated municipal wastewater. Process Saf Environ Prot 2016; 102:724–33.

[88] Nguyen DD, Ngo HH, Yoon YS. A new hybrid treatment system of bioreactors and electrocoagulation for superior removal of organic and nutrient pollutants from municipal wastewater. Bioresour Technol 2014; 153:116–25.

[89] Sharma AK, Chopra AK. Removal of cod and Bod from biologically treated municipal wastewater by electrochemical treatment. J Appl Nat Sci 2013; 5(2):475–81.

[90] Vidal J, Huiliñir C, Salazar R. Removal of organic matter contained in slaughterhouse wastewater using a combination of anaerobic digestion and solar photoelectro-Fenton Processes. Electrochim Acta 2016; 210:163–70.

[91] Baiju A, Gandhimathi R, Ramesh ST, Nidheesh PV. Combined heterogeneous electro-fenton and biological process for the treatment of stabilized landfill leachate. J Environ Manage 2018; 210:328–37.

[92] Hanafi F, Belaoufi A, Mountadar M, Assobhei O. Augmentation of biodegradability of olive mill wastewater by electrochemical pre-treatment: Effect on phytotoxicity and operating cost. J Hazard Mater 2011; 190(1–3):94–9.

[93] Popat A, Nidheesh PV, Anantha Singh TS, Suresh Kumar M. Mixed industrial wastewater treatment by combined electrochemical advanced oxidation and biological processes. Chemosphere 2019; 237:124419.

[94] Dehboudeh M, Dehghan P, Azari A, Abbasi M. Experimental investigation of petrochemical industrial wastewater treatment by a combination of integrated fixed-film activated sludge (IFAS) and electro-fenton methods. J Environ Chem Eng 2020; 8(6):104537.

[95] Katsoni A, Mantzavinos D, Diamadopoulos E. Coupling digestion in a pilot-scale UASB reactor and electrochemical oxidation over BDD anode to treat diluted cheese whey. Environ Sci Pollut Res 2014; 21(21):12170–81.

[96] Ozay Y, Ünşar EK, Işık Z, Yılmaz F, Dizge N, Perendeci NA, et al. Optimization of electrocoagulation process and combination of anaerobic digestion for the treatment of pistachio processing wastewater. J Clean Prod 2018; 196:42–50.

[97] Park H, Choo K-H, Park H-S, Choi J, Hoffmann MR. Electrochemical oxidation and microfiltration of municipal wastewater with simultaneous hydrogen production: Influence of organic and particulate matter. Chem Eng J 2013; 215–216:802–10.

[98] Natarajan R, Al Fazari F, Al Saadi A. Municipal waste water treatment by natural coagulant assisted electrochemical technique—parametric effects. Environ Technol Innov 2018; 10:71–7.

[99] Abdel-Fatah MA, Shaarawy HH, Hawash SI. Integrated treatment of municipal wastewater using advanced electro-membrane filtration system. SN Appl Sci 2019; 1(10).

[100] Roa-Morales G, Barrera-Díaz C, Balderas-Hernández P, Zaldumbide-Ortiz F, Reyes Perez H, Bilyeu B. Removal of color and chemical oxygen demand using a coupled coagulation-electrocoagulation-ozone treatment of industrial wastewater that contains offset printing dyes. J Mex Chem Soc 2017; 58(3). 362–68.

[101] Gunawan D, Kuswadi VB, Sapei L, Riadi L. Yarn dyed wastewater treatment using hybrid electrocoagulation-fenton method in a continuous system: Technical and Economical Viewpoint. Environ Eng Res 2017; 23(1):114–9.

[102] Zakeri HR, Yousefi M, Mohammadi AA, Baziar M, Mojiri SA, Salehnia S, et al. Chemical coagulation-electro fenton as a superior combination process for treatment of dairy wastewater: Performance and modelling. Int J Environ Sci Technol 2021; 18(12):3929–42.

[103] Torres NH, Souza BS, Ferreira LF, Lima ÁS, dos Santos GN, Cavalcanti EB. Real textile effluents treatment using coagulation/flocculation followed by electrochemical oxidation process and ecotoxicological assessment. Chemosphere 2019; 236:124309.

6 Photocatalytic Membrane for Emerging Pollutants Treatment

Nur Hashimah Aliasa, Nur Hidayati Othman, Woei Jye Lau, Fauziah Marpani, Muhammad Shafiq Shayuti, Zul Adlan Mohd Hir, Juhana Jaafar, and Mohd Haiqal Abd Aziz

CONTENTS

6.1 Overview of Emerging Pollutants in Wastewater 95
6.2 Photocatalysis for Emerging Pollutants Treatment......................................98
6.3 Photocatalytic Membrane for Emerging Pollutants Treatment....................100
    6.3.1 Fabrication of Photocatalytic Membranes 101
    6.3.2 Performance of Photocatalytic Membrane on Emerging
        Pollutants Treatment 103
6.4 Challenges and the Future Ways Forward 107
6.5 Conclusion 108
References..109

6.1 OVERVIEW OF EMERGING POLLUTANTS IN WASTEWATER

Environmental pollution and insufficient clean and natural energy resources are among today's most serious global issues. The rampant unregulated industrial growth and the increase in world population have accelerated energy consumption. Thus, these uncontrolled unabated releases of hazardous waste into waterways and air have increased pollution-related diseases, global warming, and abnormal climate changes [1]. As a result, approximately 1.8 billion people, as estimated by the United Nations Water (UN-Water), will undergo water paucity. On top of that, among this population, over 70% of the people lack sanitation [2]. It is also expected that two-thirds of the world population in 2025 will have to live under water stress conditions [3]. These claims also agreed with the World Health Organization (WHO) report that more than 1 billion people could not access clean water [4]. Therefore, the deterioration of global water quality and diseases caused by polluted water could be directly related to the death of 3900 children per day and millions of people annually, as reported by the health monitoring authorities [4, 5]. Furthermore, if efficient

DOI: 10.1201/9781003260738-6

treatment methodologies are not instantly developed or implemented, there will be more chronic health issues soon due to water contamination due to increased discharge of pollutants. According to the US Environmental Protection Agency, toxic contaminants (e.g., aliphatic series (C_1–C_{50})), polycyclic aromatic hydrocarbons, phenols, volatile organic compounds, and heavy metals) also have harmed ecosystems, including water bodies.

On the other hand, the increase in pollution levels due to rapid industrialization is considered one of the leading global challenges of the 21st century. Emerging pollutants (EPs) are a new concern because their release can affect the natural environment and drinking water resources and potentially pose risks when present at low concentrations. EPs are identified as natural or synthetic chemicals that possibly penetrate the environment and cause harm to ecological and human health. According to Teodosiu et al. [6], "Emerging" refers to either new pollutants identified in aquatic media and organisms or to new characteristics and impacts of compounds that already exist in the environment. Meanwhile, the European Environmental Agency says that EPs (referred to also as "hazardous substances and chemicals") need to be closely monitored as concentrations and effects since they are increasingly being found in water bodies across the EU [6].

Unfortunately, they are not regularly monitored [7]. In other serious cases, the EPs that were released for a long time are unable to be detected until a new method of detection is introduced [2]. In addition, due to improper treatment, the accumulation of these EPs is creating a novel source of EPs that become more toxic and hazardous. Therefore, most of the recent literature has classified EPs as a new group of chemicals that are still not regulated or not registered by discharged regulations authority, and their adverse health impacts are not entirely discovered [8–10]. In addition, Tang et al. [11] reported that EPs, also known as contaminants of emerging concern (CECs), are yet to be regulated in many countries. Ever since EPs attracted the global attention, countless attempts have been made to shed light on the concern of EPs released into the environment and further encourage policymakers to take related measures to prevent the ecological risks.

More than 20 categories have been identified as the source of origin for EPs. The major categories are pesticides (agriculture), pharmaceuticals (urban, stock farming), disinfection by-products (urban, industry), wood preservation and industrial chemicals (industry) [2]. They have high polarity to boost bioaccumulation and are highly resistant to biodegradation. In general, there are several classifications of EPs which commonly found, including heavy metals (arsenic, lead, mercury, cadmium, chromium, copper, nickel, silver and zinc), nanocompounds (silver nanoparticles, iron nanoparticles and zinc nanoparticles), pharmaceuticals (triclosan, triclocarban, atenolol, ibuprofen, and oxybenzone, erythromycin, estrone hormone, ibuprofen and diclofenac, personal care products (perfumes, sunscreen, surfactants, disinfectants, insect repellents, and UV-filters), pesticides (insecticides, herbicides and fungicides) and industrial chemicals (dibutyl phthalate, bisphenol A, diethyl hexyl and phthalate [8, 10]. On the other hand, Teodosiu et al. [6] mentioned that more than 1,000 substances had been gathered in 16 different classes of EPs. They are algal toxins, antifoaming and complexing agents, antioxidants, detergents, disinfection by-products, plasticizers, flame retardants, fragrances, gasoline additives, nanoparticles, perfluoroalkylated substances, personal care products, pharmaceuticals, pesticides and anticorrosive.

Overall, EPs are causative of teratogenic and reproductive effects [12]. In addition, some of the EPs are highly toxic to microorganisms and have strong antibacterial effects against many microorganisms, chronically toxic, unrelenting, hardly decomposed, and often soluble in water. However, there is still limited knowledge regarding their behaviour in the environment and the deficiency in analytical and sampling techniques. Thus, multiple levels urgently require serious action to resolve the issues. Figure 6.1 shows the representative sources and routes of micropollutants in the environment while Figure 6.2 illustrates the EPs' pathways and related impacts related to water uses.

In the last decades, various treatment technologies have been used to treat EPs which include coagulation/flocculation [14, 15], aerobic and anaerobic biological treatment [16, 17], catalytic ozonation [18–21], photo-oxidation [22–24], photocatalytic degradation [25–27], ion-exchange [28–30], activated sludge process [31–33], adsorption [34, 35] and membrane filtration [36–38]. Among these treatment methods, photocatalysis via the semiconductor-based has attracted myriad attention as reliable remediations treatment of EPs [8, 39–42] owing to its fascinating features; commercially available with attractive cost, sustainable, ease the operating procedure, promotes higher degradation potential under light irradiation or sunlight irradiation, and could ideally complete the mineralization of pollutants and turn to the safer final products which are carbon dioxide (CO_2) and water.

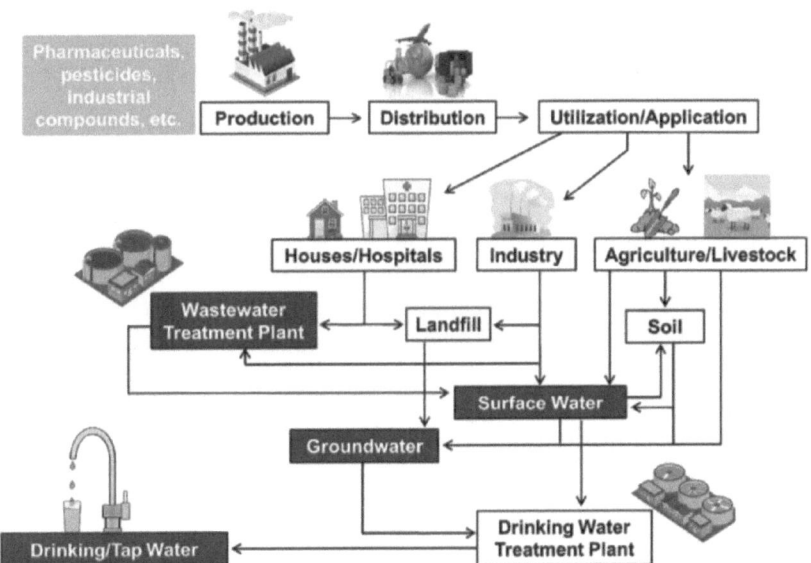

FIGURE 6.1 Representative sources and routes of micropollutants in the environment.

Source: Reproduced with permission from [13], Copyright Water Research, 2016.

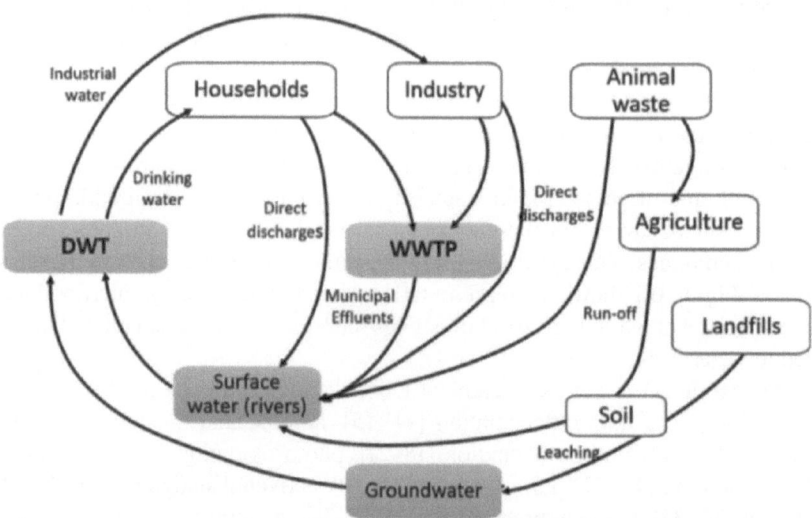

FIGURE 6.2 EPs' pathways and related impacts related to water use (DWT: drinking water treatment; WWTP: wastewater treatment plant).

Source: Reproduced with permission from [6], Copyright Journal of Cleaner Production, 2018.

Herein, this chapter aims to highlight the recent progress on the photocatalytic membrane for EPs treatment. We have reviewed the fundamentals of the photocatalysis process, photocatalysis membranes, their fabrication and performance on EPs treatment. Lastly, the challenges and the future ways forward on photocatalytic membranes for EPs treatment were also discussed and concluded.

6.2 PHOTOCATALYSIS FOR EMERGING POLLUTANTS TREATMENT

According to the International Union of Pure and Applied Chemistry (IUPAC), photocatalysis act as a change in the rate of a chemical reaction or its initiation under the action of ultraviolet, visible, or infrared radiation in the presence of a substance the photocatalyst that absorbs light and is involved in the chemical transformation of the reaction partners [43]. The term "photocatalysis" was used as early as 1921 by a group of researchers [44] to describe a process of photochemical synthesis in the living plant. Later, Doodeve and Kitchener in 1938 discovered that titanium dioxide (TiO_2) could act as a photosensitizer for dye bleaching in the presence of oxygen [45]. In 1972, Fujishima and Honda [46] proudly introduced remarkable findings on semiconductor electrochemical photolysis, classifying their research work as semiconductor photocatalysis. The research in the semiconductor photocatalysis area continues rapidly to the development of heterogeneous photocatalytic oxidation, photo electrocatalytic reduction and many more. Until today these significant discoveries have become a solid reference for future development of photocatalysis research. Thanks to photocatalysis, studies on photocatalysis have been extensively carried out in the past decades not only for photodegradation on EPs but also for organic synthesis, water splitting, nitrogen fixation and CO_2 reduction.

In general, the whole photocatalytic process can be summarized in four steps: (i) absorption of light followed by the separation of the electron-hole couple; (ii) adsorption of the reagents; (iii) redox (reduction and oxidation) reaction and (iv) desorption of the products [47]. Theoretically, the photocatalysis mechanism can be scrutinized as an accelerated photochemical reaction at a solid surface semiconductor with light at a particular wavelength [48, 49]. At least two reactions must occur simultaneously: oxidation from photogenerated holes and reduction from photogenerated electrons. When the light photon irradiates by the source of light (visible light or ultraviolet light), and the band gap energy (E_g) of the semiconductor excites electrons from the valence band (VB) to the conduction band (CB), the electron-hole pair will be generated as shown in Figure 6.3.

The electron promoted to the CB will recombine with the electron-hole in the valence band, whether on the semiconductor surface or inside the particle. Or else, the electron combination can be entrapped on the surface that will react with the donor (D) or acceptor (A) of the adsorbed species. However, since the electron-generated charge carriers in excited states (CB) are not stable and can recombine easily, it will result in low photocatalysis conversion efficiency [51]. A band gap is a gap in a perfect crystal where no electron state can exist between the highest occupied energy band (VB) and the lowest empty band (CB). Each photocatalyst semiconductor material possesses its respective band gap.

The semiconductors' electrons can act as better oxidizing agents or reducing agents when the reaction is conducted in water as a medium with highly reactive

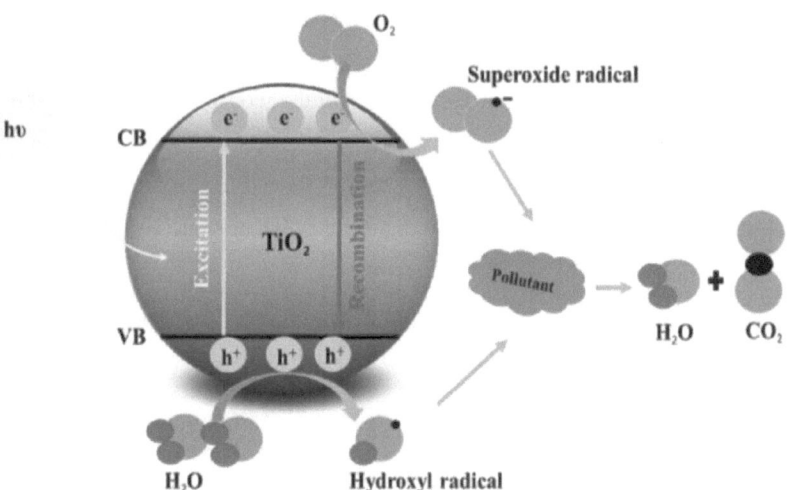

FIGURE 6.3   Conceptual illustration of the degradation of organic compounds, microorganisms, or pollutants by forming photo-induced charge carriers on the surface of TiO_2 photocatalyst.

Source: Reproduced with permission from [50], Copyright Environmental Science and Pollution Research, 2021.

hydroxyl radical (•OH) and superoxide oxygen radical ($\cdot O_2^-$) be generated. Hence, the organic pollutants molecule can be entrapped by the reactive hydroxyl radical (•OH), further degrade, and produce harmless water and CO_2. The hydroxyl radicals (•OH) are considered the pivotal oxidizing species compared to oxygen radicals ($\cdot O_2^-$) in photocatalytic activity. The high oxidizing power of free radicals (•OH) can destroy organics such as carboxylic acids and phenolic derivatives of chlorinated aromatics, transforming them into harmless compounds such as CO_2, mineral salts and water [52]. In addition, other than free radicals (•OH), reactive oxygen species are also vital in the oxidative pathways to perform the degradation of pollutants.

Despite the excellent properties of semiconductor photocatalysts, which typically exist in powder form with a high specific surface area, these photocatalysts must be recovered, regenerated, and removed from treated water before they can be reused or discharged into the environment. It is very difficult to separate and reuse nanosized photocatalysts from treated water, resulting in a low utilization rate and a chance of secondary pollution [53]. Besides, these powder photocatalysts also suffer severe agglomeration in aqueous solution [54, 55], which limits their photocatalytic ability due to low solar light utilization due to the limited penetration of sunlight in deep water [42]. Therefore, considering these drawbacks, many approaches have been developed to immobilize the powder photocatalyst into or onto the membrane matrix organic, inorganic or a combination of organic-inorganic membranes and form photocatalytic membranes.

6.3 PHOTOCATALYTIC MEMBRANE FOR EMERGING POLLUTANTS TREATMENT

Photocatalytic membranes own a smart principle that can provide a synergistic effect of photocatalysis and membrane separation simultaneously with control of the products and by-products when the light is irradiated on the membrane during the process. Therefore, most scientists have come out with the idea and further developed hybrid or modified photocatalysts to obtain more visible light sensibility, high photodegradation efficiency, high interaction with polymer matrix, and hydrophilicity/hydrophobicity of membrane surface [55].

In general, membrane separation technology offers attractive advantages of small footprint, ease of maintenance, excellent separation efficiency, high purity of water quality, low chemical consumption and low environmental impact [53]. Membrane separation processes efficiently remove a wide-ranging of organic, inorganic and solid particles from seawater, surface water or wastewater all through semipermeable materials, which permit the separation of water (flux or permeate) and the concentrate (retentate on membrane surface). Membranes are characterized by their structure, physicochemical and thermal properties through the type of materials used, surface charge, pore size, thickness, wettability, etc., that determine what type of EPs to be separated. A wide variety of thermally and chemically stable polymers or polymer blends and other materials, such as ceramics, metals, glasses or mixed matrix membranes, may be used [6].

As a result, a vast number of membranes have been well developed, such as polymer membranes (polysulfone, polyethersulfone, polyacrylonitrile, polyvinylidene fluoride, polyester and polyamide), inorganic membranes (aluminium oxide (Al_2O_3), TiO_2, silicon dioxide (SiO_2), and perovskite) or combination of both organic and inorganic membranes. Nevertheless, conventional membrane technology faces severe inherent shortcomings, majorly on membrane fouling caused by deposition of the pollutant contaminants on the membrane surface, resulting in a great decrease in the permeation flux, low selectivity, a short membrane life span, an increase in treatment cost and energy consumption. On top of that, membrane processes also pose a disadvantage compared to advanced oxidation processes, where the pollutants are transferred into the concentrate streams and not degraded in the membrane process. Thus, the concentrate needs further treatment and proper disposal [56]. Therefore, the development of hybrid photocatalytic membranes can kill two birds with one stone by overcoming the shortcomings of powder photocatalyst and membranes: photocatalysts powder agglomeration and membrane fouling. Furthermore, these membrane modifications also produce a membrane with superior characteristics, high fouling resistance, and in-situ membrane self-cleaning, thus reducing operational costs. On top of that, photocatalytic membrane reactors (PMRs) possess advantages over traditional photoreactors [57].

Upon light irradiation on a photocatalytic membrane, it can degrade the EPs in the feed solution by the reactive oxygen species and electron-hole generated by photocatalysts. Thus, this prevents cake layer formation that is absorbed on the membrane surface during filtration, reducing pore blocking, hindering membrane fouling to some extent and further reducing pump consumption to collect permeate and reduce the frequency of membrane cleaning. On the other hand, the membrane substrate simultaneously acts as a selective barrier for the EPs species to be removed and degraded into intermediate and final products.

6.3.1 Fabrication of Photocatalytic Membranes

The fabrication of photocatalytic membranes commonly involves two different steps: fabrication of membrane using different techniques such as dry-wet phase inversion, thermally induced phase separation (TIPS), non-solvent induced phase separation (NIPS), electrospinning, stretching, sintering, template leaching and others, followed by immobilization of the photocatalyst in the membrane. The photocatalyst immobilization into or onto the membrane matrix can be done in two ways: 1) direct deposition of the photocatalyst onto the membrane matrix surface and 2) entrapment of the photocatalyst into the membrane as a mixed-matrix membrane. Figure 6.4 illustrates a polymeric photocatalytic membrane fabrication.

The mixed matrix membrane is fabricated before direct deposition of a coated photocatalyst onto the membrane surface is established. Generally, in a mixed matrix membrane, photocatalyst leaching is not as severe as photocatalyst-coated membrane, and the fouling can be prevented by controlling the hydrophilicity of the membrane. Thus, it has better reusability and low surface energy. However, direct deposition of a photocatalyst into the membrane matrix could cause the photocatalyst

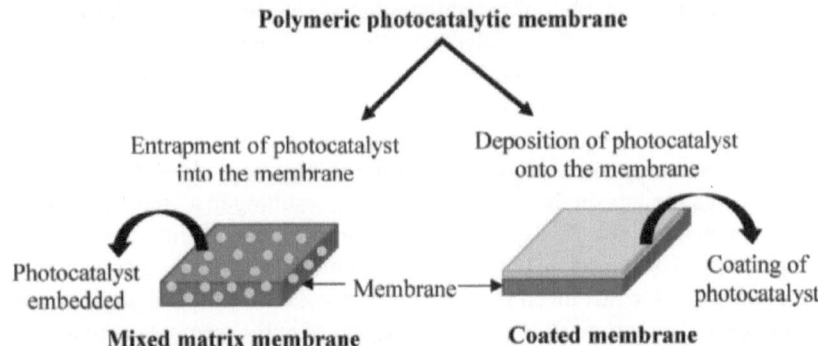

FIGURE 6.4   Schematic illustration of the fabrication of polymeric photocatalytic membrane.

Source: Reproduced with permission from [58], Copyright Chemical Engineering Journal, 2022.

agglomeration, thus reducing the surface-active sites to absorb irradiated light for photocatalysis. Moreover, this agglomeration could contribute to membrane pore blockage and limit the flux during filtration [38]. Thus, a photocatalyst-coated membrane has the advantage over a mixed matrix membrane in terms of the photocatalytic activity and filtration performance.

Various established methods have been used to fabricate mixed matrix membranes. The most common one is the dry-wet phase inversion method. In the dry-wet phase inversion method, the polymer dope solution is prepared by mixing a suitable ratio of photocatalyst with compatible polymer precursors and dispersed solvent. Prior to the preparation, a better understanding of solvent miscibility with photocatalyst and polymer used is crucial to be investigated as homogeneity of dope polymer solution is the utmost priority for membrane fabrication either for flat sheet or hollow fiber membranes. On the other hand, dry-wet co-spinning and dry-wet spinning are also a type of dry-wet phase inversion processes that are attractive for the fabrication of dual layer and single layer photocatalytic hollow fiber membranes. However, as the photocatalyst needs to be accessible to the irradiated light for the photocatalysis process and to minimize agglomeration, the dual-layer photocatalytic hollow fiber membrane is preferred over the single-layer membrane. Furthermore, the dual-layer photocatalytic hollow fiber membrane through co-extruding spinning can save photocatalyst costs as it eliminates the lamination process.

On the other hand, recent approaches have seen many invented methods to coat photocatalyst onto the membrane surface efficiently. Examples are dip-coating, electrospinning, electrospraying, sputtering, layer-by-layer self-assembly, atmospheric plasma spraying (APS) and chemical vapor deposition (CVD) [58]. Therefore, selecting these methods should be carefully done to optimize the deposition of photocatalysts and strengthen it onto the membrane surface to avoid leaching during filtration. The dip-coating method coats the photocatalyst on the membrane surface. It is done by dipping the membrane support in the precursor solution of the photocatalyst or suspension of the photocatalyst, followed by the withdrawal of the membrane support from the solution at a constant speed. Prior to the dip-coating process, the membrane support should be clean and dried properly to avoid contamination. Several studies treated the membranes with specific chemicals such as hydroxide aqueous solution to develop a stronger attachment with photocatalysts or nanoparticles [59]. However, dip-coating on the membrane using a suspension solution of photocatalyst powder without proper control of the photocatalyst size resulted in poor photocatalyst dispersion on the membrane surface due to powder agglomeration. Therefore, it is necessary to sonicate the photocatalyst solution before dip-coating to maximize its dispersion to better the membrane's photocatalytic ability. On the other hand, a comprehensive study by Saflashkar et al. [60] mentioned that the highest rate of cephalexin removal is reported for a membrane coated with a photocatalyst using the spin coating method, followed by interfacial polymerization method, graft polymerization method and blending method, respectively. They also suggested that the blending membrane could obtain a high flux, the interfacial polymerization membrane has high fouling resistance, high photocatalytic degradation via graft polymerization membrane and high separation via spin coating membrane.

Another better idea to disperse the photocatalyst on the membrane surface homogenously is introduced through electrospinning. By electrospinning, the size of the photocatalyst can be controlled efficiently. During the process, the photocatalyst was mixed with solvent and polymer to obtain a dope solution and further electrospun directly onto the membrane surface using an electrospinning machine at high voltage and controlled dope solution flow rate at the desired thickness [36, 38]. Figure 6.4 shows the schematic illustration of photocatalyst nanofiber coating on flat sheets and hollow fiber membranes.

6.3.2 Performance of Photocatalytic Membrane on Emerging Pollutants Treatment

Up until today, many efforts have been devoted to developing photocatalytic membranes to optimize and obtain the best photocatalytic membrane performance in treating EPs. Overall, the photocatalysis process is highly dependent on the concentration of pollutants, type, and amount of the photocatalyst, the temperature of the medium, the surface area of the photocatalyst, the pH of the solution, doping of metal ions and nonmetal of photocatalyst and substrate, and light intensity and irradiation time. In contrast, the membrane performances depend on numerous factors, including the design of the PMR and membrane properties such as pore structure, roughness, polarity, mechanical characteristics, and hydrophilicity. Thus, the photocatalytic membrane's performance in terms of flux and rejection of the targeted contaminants could be affected by several conditions, properties of the photocatalyst embedded and coated on the membrane surface, photocatalyst dispersion and its accessibility to light absorption and intensity and power of the light source. On top of that, it is also found that photocatalytic degradation and filtration are not stand-alone processes in the photocatalytic membrane as the removal of EPs might as well due to in situ or synergistic effects with other processes such as adsorption, repulsion, ionic interaction, etc.

In 2021, Wang et al. [61] developed a 2D-based photocatalytic membrane by incorporating graphitic carbon nitride (g-C_3N_4) nanosheets into an ultrathin metal-organic framework (MOF-2) via a vacuum-assisted self-assembly method for treatment of pesticides and antibiotics pollutants. This 2D heterostructure successfully enforces the aqueous pollutant confinement and steric-hindrance effect, accelerating the radical attack and the capture of pollutants, respectively, on atrazine (ATZ), malathion (MLT), tetracycline (TC) and sulfamethazine (SMT) as the source of pollutants. The results exhibited the enhanced removal of contaminants at 98% for ATZ, 95% for TC, 89% for SMT, and 92% for SMX with a permeation flux of 23.6 $Lm^{-2}h^{-1}bar^{-1}$. Furthermore, a removal mechanism of initial enrichment of pollutants, intensified photocatalytic degradation and electrostatic repulsion of by-products was proposed in this work. On the other hand, work on the treatment of pesticide pollutants was conducted by Krishnan et al. [62], where they fabricated bismuth tungsten (BWO)@ MIL-100(Fe) grafted polyvinylidene fluoride (PVDF) membrane for the treatment of pirimicarb pesticide. The fabricated membrane showed higher permeate flux (25.99 $Lm^{-2}h^{-1}$ at 4 bar) and promising pirimicarb photodegradation (81% at pH 5) compared to the neat PVDF membrane prepared (16.2 L $m^{-2}h^{-1}$ at 4 bar and 8% at pH 5). They also concluded that adding BWO@MIL-100(Fe) nanofiller was able to improve membrane hydrophilicity, antifouling properties and surface energy, which led to performance enhancement.

Holda, Perera and Emanuelsson Patterson (2022) developed a polysulfone-based nanofiltration membrane containing homo-coupled conjugated microporous poly(phenylene butadiynylene) (HCMP) via phase inversion and used the membrane to degrade Rhodamine B (RhB). The removal of RhB was higher for the membrane containing 0.5-M HCMP at 99% rejection with 6.91 L $m^{-2}h^{-1}bar^{-1}$ of permeance. Furthermore, they concluded that the HCMP-containing membrane exhibited good stability in prolonged operation after 8 cycles. In another work on the treatment of dye wastewater, Pang et al. [64] successfully developed an antifouling photocatalytic membrane by immobilising Ti_3C_2/WO_3 onto PVDF substrate. They reported that the 5 wt% Ti_3C_2/WO_3-PVDF membrane exhibited water permeability at 480 L m^2 h^{-1} at 1 bar, with its flux recovery rate maintained at 94% after the antifouling test. Besides, high rejection of anionic dyes and heavy metal ions (Congo red <98% and Cr^{6+}~85.4%) were attained by using this membrane. The findings concluded that efficient charge transfer between the photocatalytic support layer and the Ti_3C_2/WO_3 has contributed to strong photocatalytic ability on targeted EPs. Figure 6.5 shows the band edge positions for 2D/2D heterojunction of Ti_3C_2/WO_3 of the fabricated photocatalytic membrane.

The findings from the study by Pang et al. [64] were also in agreement with a later study by Zhao et al. [65] that mentioned the efficient removal of dyes was attributed to the effective charge separation of the heterojunction photocatalysts. In their study, they fabricated hybrid MXene/tungsten oxide ($Ti_3C_2/W_{18}O_{49}$) photocatalytic membranes by coating Ti_3C_2 and $W_{18}O_{49}$ on the surface of mixed cellulose ester (MCE) support using a vacuum filter method. They reported that the hybrid $Ti_3C_2/W_{18}O_{49}$ photocatalytic membranes obtained high methylene blue (MB) adsorption at q_{max} under the Langmuir model at 35.12 mg/g. This value was

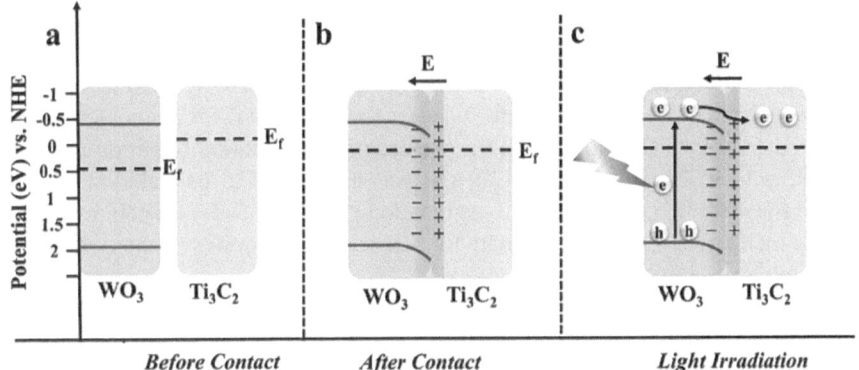

FIGURE 6.5 The band edge positions for 2D/2D heterojunction of Ti_3C_2/WO_3: (a) before contact, (b) after contact and (c) after contact under light irradiation and the charge migration pathway.

Source: Reproduced with permission from [64], Copyright Ceramic International, 2022.

3.9 times higher than the Ti_3C_2 membrane. On top of that, the removal of MB was enhanced under visible light irradiation at a rate constant of 2.32×10^{-2} min^{-1} in 120 min and showed stable reuse performance across ten cycles. They concluded that, during the photocatalysis process, $W_{18}O_{49}$ provided absorption sites and photocatalytic ability while Ti_3C_2 served as an electron trap to form a Schottky junction. The photo-induced electrons in the conduction band of $W_{18}O_{49}$ were transferred to Ti_3C_2 to form $•O_2^-$, while its electron holes remained in the valence band to generate $•OH$, and h^+. $•O_2^-$ and $•OH$ are mainly in charge of degrading dyes. Thus, this efficient membrane has the potential as a photocatalytic membrane for dye removal from wastewater.

Prado et al. [66] utilised recycled membranes to develop a photocatalytic membrane coated with greener TiO_2 and graphene oxide (GO) in different modification routes. Interestingly, the membranes were reproduced by oxidizing end-of-life reverse osmosis (RO) membranes. This process was conducted by immersing RO membrane in a sodium hypochlorite at room temperature, and a contact time with an intensity of 300,000 ppm.h was applied. Based on the results obtained, the photocatalyst coating via dopamine generated better adhesion of the photocatalyst on the membrane substrate than self-assembly deposition of the photocatalyst. This coating method formed a stable photocatalyst coating with better permeability and rejection efficiency. On top of that, the addition of TiO_2 and GO increased the membrane hydrophilicity and formed a hydrated layer that repelled the organic contaminants and reduced surface fouling. Upon irradiation under UV light, the resulting membrane achieved 86.6% removal rate against betamethasone (BET), a pharmaceutically active compound. The study concluded that membrane rejection, adsorption (contribution: ~10%), and photocatalysis (contribution: ~20%) were the possible mechanisms that contributed to the pollutant's removal efficiency.

In another recent study, Zhang et al. [67] developed a glass hollow fiber membrane coated with TiO_2. The glass hollow fiber membranes were prepared using the phase inversion-calcination method, and a TiO_2 sol was then coated on the membrane surface via the dip-coating method. The results showed that the TiO_2-coated membrane, calcined at 550°C, could form a uniform surface without a fall-off phenomenon and cracks. Furthermore, they found that the presence of TiO_2 coating has improved MB removal up to 97.2%. The removal of MB was also found to be stable at around 92.3–93.6% after five photocatalytic degradation cycles. Thus, this excellent photocatalytic membrane is a great candidate for MB removal from wastewater with good long-term stability.

As novel 3D printed materials are receiving much attention in the 21st century, Sreedhar et al. [68] developed a novel β–FeOOH nanorods-coated 3D printed photocatalytic feed spacer via 3D laser sintering (SLS) technology and used it with polyethersulfone membrane for organic foulant treatment. In this study, a PDA/PEI layer was deposited on the printed spacer to create active sites for Fe^{3+} ions binding, forming a strong attachment of the iron nanorods layer onto the spacer surface. In addition to the membrane's basic role as a matrix for support, the spacer also provided two new functions, i.e., degradation of membrane-permeating pollutants in the feed and membrane cleaning. Based on the photocatalytic performance, it was found that the addition of photocatalytic spacer exhibited the ability to clean the membrane surface of three organic foulants (humic acid (HA), sodium alginate (SA) and bovine serum albumin (BSA)), and the flux recover was recorded at 92%, 60% and 54% for SA, HA and BSA, respectively. The obtained findings marked another remarkable achievement in developing a photocatalytic membrane for treating pollutants under UV light irradiation.

Ning et al. [69] proved that photocatalytic membranes are applicable not only for pressure-driven processes such as ultrafiltration (UF) and RO but also for thermal-based processes such as membrane distillation (MD). In the MD process, semi-volatile organic compounds (s-VOCs) contamination could happen as it may transfer into the membrane and impact the permeate quality. In this study, an innovative silver chloride/Materials of Institut Lavoisier: MILs-100(Fe)/ polytetrafluoroethylene (AgCl/MIL-100(Fe)/PTFE) photocatalytic membrane was developed to remove nitrobenzene (NB) – an EP that can cause carcinogenic and genetic toxicity to human beings. The resulting membrane successfully removed NB up to 87.84% and maintained over five removal cycles experiment with a higher than 84.84% removal rate. They also proposed a plausible mechanism of NB removal where the volatile NB molecules were adsorbed on the surface of photocatalytic membranes, and the photoelectrons (e^-) were generated by the immobilized photocatalyst and further effectively removed pollutants under visible light irradiation. On top of that, the combination of adsorption and photocatalysis of this photocatalytic membrane provided an attractive avenue for water reclamation from s-VOCs contaminated wastewater in real applications.

Rathna et al. [70] treated Cr(VI) in a PMR using titanium dioxide-tungsten oxide-polyaniline (TiO_2-WO_3) incorporated PANI membranes under visible light irradiation. Besides being high surface area (173.57 m_2/g), the mesoporous TiO_2-WO_3

nanoparticle was able to promote visible light activity and has high electron acceptance for an effective treatment process. In addition, the presence of optimum nanoparticle loading (5 wt%) could enhance the properties of the membranes in terms of hydrophilicity, flux, permeate quality and Cr(VI) reduction capacity. Thus, the TiO_2-WO_3-PANI membrane was effective to reduce 98.50% of Cr(VI) to Cr(III) in the PMR through filtration. This membrane also showed an improved self-cleaning characteristic. Recently, Kumari et al. [71] developed a Schottky-like photocatalytic-electrocatalytic macro-porous membrane using TiO_2 and zinc oxide (ZnO) based on the atomic layer deposition (ALD). In this study, a batch cross-flow PMR filtration was used to evaluate the degradation of MB, phenol, paracetamol and atrazine. They found out that the photo-electrocatalytic efficiency of the TiO_2 and ZnO coated membranes were able to achieve greater pollutant degradation, offering increased degradation kinetic factors by 2.9 and 2.3 compared to the photocatalysis and electrocatalysis, respectively. These synergistic effects between photoelectrocatalysis and membrane separation in cross-flow filtration mode offer a unique approach to the design of stimuli-responsive and self-cleaning membrane materials, with lower operating costs and greater performance outputs.

6.4   CHALLENGES AND THE FUTURE WAYS FORWARD

Overall, the performance of photocatalytic membranes for EPs treatment may differ based on several factors. For example, the fabrication method could play an important role in determining the membranes' overall stability, integrity and lifespan. Besides, the photocatalytic ability of the membrane is majorly influenced by the state of distribution of the photocatalyst coated or embedded into the membrane matrix. Therefore, it is important to ensure that the effective loading of the photocatalyst in or onto the membrane matrix could absorb maximum light irradiation for excellent results of photocatalysis reaction. On the other hand, higher loading of the photocatalyst without proper distribution control may cause photocatalyst agglomeration, lumps, and reduction of photocatalyst surface area, thus hindering the photocatalyst from light penetration, limiting the photocatalytic activity and EPs removal. Furthermore, photocatalyst agglomeration could also cause membrane pore blockage and reduce the flux of the membrane.

Kundu and Karak (2022) mentioned that incorporating photocatalyst onto the polymeric membrane using the sputtering technique is the most promising among various methods for photocatalyst coating. However, this technique suffers from photocatalyst leaching as this method is unable to develop a strong chemical bond between sputtered photocatalysts and membrane support. Thus, in this case, the properties of the photocatalyst, such as precursor selection and ratio, synthesis conditions, size of the photocatalyst, type of solvent used for photocatalyst delamination, etc., are required to be optimized. On top of that, the photocatalytic ability also needs to be enhanced by narrowing the photocatalyst's band gap energy and minimizing the electron-hole pairs' recombination rate. Photocatalysts with narrow band gap energy can be activated under solar light irradiation, which is favorable for energy-saving system applications. However, this photocatalyst could also suffer from high electron-hole pairs' recombination rate when irradiated under highly intensive light. Thus, extensive research has been conducted to nanoarchitecture the heterojunction photocatalysts that possess higher photocatalytic activity due to the spatial separation of photogenerated electron-hole pairs.

Among all the challenges faced in developing photocatalytic membranes, altering membrane wettability and enhancing the strength of photocatalyst attachment to the membranes are the most crucial and difficult to be optimized. Membrane wettability will determine the threshold between the flux performance and membrane fouling rate. Therefore, the developed photocatalytic membrane properties should be appropriate to the characteristics of EPs to be treated. On the other hand, the strength of photocatalyst attachment in/onto the membrane matrix should become the utmost priority in the development of photocatalytic membrane as the leached photocatalyst from the membrane could go into the water body, thus become a secondary pollutant and cause harm to human health and environment. Furthermore, as many photocatalysts are nanometer size, their presence or detachment from the membrane could only be detected via a high-tech analytical instrument. Therefore, it is important to ensure the photocatalyst stays intact throughout the filtration operation. Thus, an appropriate selection of membrane materials and photocatalysts, as well as the fabrication methods, are essential in this matter.

As the future ways forward in the development of photocatalytic membrane for treatment of EPs, future research should focus on developing naturally occurring polymers, bio-based abundantly available, renewable and green membrane materials. Although much work has been exploring bio-based membranes such as chitosan, cellulose and biomass, their application as a photocatalytic membrane matrix for huge-scale and industrial applications has not been reported much. Besides, the usage of conventional solvents and organic solvents have still championed the choice of solvent used for membrane fabrication over the green solvent due to the low cost and availability in mass production. However, these conventional solvents can negatively impact operational safety and costs, the environment, and human health. Therefore, an urgent need is required to mass-produce the green solvent to replace the conventional solvents.

On the other hand, photocatalyst synthesis should opt for facile and green synthesis routes and eco-friendly green photocatalyst precursors. Nevertheless, according to Suresh et al. [72], green synthesized photocatalysts suffer several drawbacks such as the recovery of the photocatalyst during their synthesis, limitation of light penetration due to its absorption by surface adsorbed organic contents on photocatalyst, the opacity of the photocatalyst slurry, and requirement of large biomass quantities. Therefore, multiple works should be done to make these green photocatalysts as competitive as conventional photocatalysts become a reality. Therefore, environmentally benign materials are essential for wastewater treatment and environmental remediation to align with the United Nations' Sustainable Development Goals (SDGs) for 2030.

6.5 CONCLUSION

This chapter reviews recent approaches in the photocatalytic membrane for EPs treatment. Although extensive research has been conducted recently, a big loophole still needs to be overcome to minimize and control the impact of these released

EPs on human health and the environment. This chapter highlights the overview of EPs in wastewater, the photocatalysis process to treat EPs, the advancement of the photocatalytic membrane as a potential method to treat EPs efficiently, and their fabrication and performances. Besides, several challenges and future ways forward in applying photocatalytic membrane for EPs treatment were also discussed. Most of the research concluded that photocatalytic membrane provides an excellent treatment of EPs wastewater through synergistic adsorption, photocatalytic degradation, and filtration under light irradiation. These processes could be directly involved as a mechanism step during the treatment. Although ideally, mineralization of pollutants produces harmless pollutants (CO_2 and water) in the photocatalytic membrane reactor, the persistent pollutants or intermediate products from the degradation process will still be left in the treated wastewater discharge, creating secondary pollution that may be more harmful than initial pollutants. Therefore, the rapid research studies on photocatalytic membranes should be intensively continued to seek further the ideal photocatalytic membrane that could completely mineralize pollutants and turn them into safer final products for the huge application of EPs of industrial wastewater.

REFERENCES

[1] Schneider J, Matsuoka M, Takeuchi M, Zhang J, Horiuchi Y, Anpo M, et al. Understanding TiO_2 photocatalysis mechanisms and materials. Chem Rev. 2014;114(9):9919–9986.

[2] Geissen V, Mol H, Klumpp E, Umlauf G, Nadal M, van der Ploeg M, et al. Emerging pollutants in the environment: A challenge for water resource management. Int Soil Water Conserv Res [Internet]. 2015; 3(1):57–65. Available from: http://dx.doi.org/10.1016/j.iswcr.2015.03.002

[3] Navarro-Ortega A, Acuña V, Bellin A, Burek P, Cassiani G, Choukr-Allah R, et al. Managing the effects of multiple stressors on aquatic ecosystems under water scarcity. The GLOBAQUA project. Sci Total Environ. 2015; 503–504:3–9.

[4] Dharupaneedi SP, Nataraj SK, Nadagouda M. Membrane-based separation of potential emerging pollutants. Sep Purif Technol. 2019; 210:850–866.

[5] Shannon MA, Bohn PW, Elimelech M, Georgiadis JG, Mariñas BJ, Mayes AM. Science and technology for water purification in the coming decades. Nature. 2008; 452(7185):301–310.

[6] Teodosiu C, Gilca AF, Barjoveanu G, Fiore S. Emerging pollutants removal through advanced drinking water treatment: A review on processes and environmental performances assessment. J Clean Prod. 2018; 197:1210–1221.

[7] Salazar H, Martins PM, Santos B, Fernandes MM, Reizabal A, Sebastián V, et al. Photocatalytic and antimicrobial multifunctional nanocomposite membranes for emerging pollutants water treatment applications. Chemosphere. 2020; 250:126299

[8] Ahmed S, Khan FSA, Mubarak NM, Khalid M, Tan YH, Mazari SA, et al. Emerging pollutants and their removal using visible-light responsive photocatalysis – A comprehensive review. J Environ Chem Eng. 2021; 9(6):106643.

[9] Taoufik N, Sadiq M, Abdennouri M, Qourzal S. Recent advances in the synthesis and environmental catalytic applications of layered double hydroxides-based materials for degradation of emerging pollutants through advanced oxidation processes. Mater Res Bull. 2022; 154:111924.

[10] Gondi R, Kavitha S, Yukesh Kannah R, Parthiba Karthikeyan O, Kumar G, Kumar Tyagi V, et al. Algal-based system for removal of emerging pollutants from wastewater: A review. Bioresour Technol. 2022; 344:126245.

[11] Tang Y, Yin M, Yang W, Li H, Zhong Y, Mo L, et al. Emerging pollutants in water environment: Occurrence, monitoring, fate, and risk assessment. Water Environ Res. 2019; 91(10):984–991.

[12] Vimalkumar K, Sangeetha S, Felix L, Kay P, Pugazhendhi A. A systematic review on toxicity assessment of persistent emerging pollutants (EPs) and associated microplastics (MPs) in the environment using the Hydra animal model. Comp Biochem Physiol Part—C Toxicol Pharmacol. 2022; 256:109320.

[13] Barbosa MO, Moreira NFF, Ribeiro AR, Pereira MFR, Silva AMT. Occurrence and removal of organic micropollutants: An overview of the watch list of EU Decision 2015/495. Water Res. 2016; 94:257–279.

[14] Zhao Y, Zhang Q, Yuan W, Hu H, Li Z, Ai Z, et al. High efficient coagulant simply by mechanochemically activating kaolinite with sulfuric acid to enhance removal efficiency of various pollutants for wastewater treatment. Appl Clay Sci. 2019; 180:105187

[15] Hu X, Hu P, Yang H. Influences of charge properties and hydrophobicity on the coagulation of inorganic and organic matters from water associated with starch-based coagulants. Chemosphere. 2022; 298:134346.

[16] Yang L, Xu X, Wang H, Yan J, Zhou X, Ren N, et al. Biological treatment of refractory pollutants in industrial wastewaters under aerobic or anaerobic condition: Batch tests and associated microbial community analysis. Bioresour Technol Reports. 2022; 17:100927.

[17] Ahmad HA, Ahmad S, Cui Q, Wang Z, Wei H, Chen X, et al. The environmental distribution and removal of emerging pollutants, highlighting the importance of using microbes as a potential degrader: A review. Sci Total Environ. 2022; 809:151926.

[18] Araújo A, Soares OSGP, Orge CA, Gonçalves AG, Rombi E, Cutrufello MG, et al. Metal-zeolite catalysts for the removal of pharmaceutical pollutants in water by catalytic ozonation. J Environ Chem Eng. 2021; 9:106458.

[19] Li P, Miao R, Wang P, Sun F, Li X yan. Bi-metal oxide-modified flat-sheet ceramic membranes for catalytic ozonation of organic pollutants in wastewater treatment. Chem Eng J. 2021; 426:131263.

[20] Guo Y, Long J, Huang J, Yu G, Wang Y. Can the commonly used quenching method really evaluate the role of reactive oxygen species in pollutant abatement during catalytic ozonation? Water Res. 2022; 215:118275.

[21] Issaka E, AMU-Darko JNO, Yakubu S, Fapohunda FO, Ali N, Bilal M. Advanced catalytic ozonation for degradation of pharmaceutical pollutants—A review. Chemosphere. 2022; 289:133208.

[22] Fatma Ece S, Karatas O, Gengec E, Khataee A. Treatment of real printing and packaging wastewater by combination of coagulation with Fenton and photo-Fenton processes. Chemosphere. 2022; 306:135539.

[23] Mandal S, Adhikari S, Choi S, Lee Y, Kim DH. Fabrication of a novel Z-scheme Bi2MoO6/GQDs/MoS2 hierarchical nanocomposite for the photo-oxidation of ofloxacin and photoreduction of Cr(VI) as aqueous pollutants. Chem Eng J. 2022; 444:136609.

[24] Yang JZ, Ie IR, Lin ZB, Yuan CS, Shen H, Shih CH. Oxidation efficiency and reaction mechanisms of gaseous elemental mercury by using CeO_2/TiO_2 and CuO/TiO_2 photo-oxidation catalysts at low temperatures in multi-pollutant environments. J Taiwan Inst Chem Eng. 2021; 125:413–423.

[25] Alias NH, Jaafar J, Samitsu S, Yusof N, Hafiz M, Othman D, et al. Photocatalytic degradation of oilfield produced water using graphitic carbon nitride embedded in electrospun polyacrylonitrile nanofibers. Chemosphere. 2018; 204:79–86.

[26] Azuwa M, Jaafar J, Zain MFM, Jeffery L, Kassim MB, Saufi M, et al. In-depth understanding of core-shell nanoarchitecture evolution of g-C3N4@C,N co-doped anatase/rutile: Efficient charge separation and enhanced visible-light photocatalytic performance. Appl Surf Sci. 2018; 436:302–318.

[27] Alias NH, Jaafar J, Samitsu S, Ismail AF, Othman MHD, Rahman MA, et al. Efficient removal of partially hydrolysed polyacrylamide in polymer-flooding produced water using photocatalytic graphitic carbon nitride nanofibres. Arab J Chem. 2020; 13(2):4341–4349.

[28] Feng Y, Yang S, Xia L, Wang Z, Suo N, Chen H, et al. In-situ ion exchange electro-catalysis biological coupling (i-IEEBC) for simultaneously enhanced degradation of organic pollutants and heavy metals in electroplating wastewater. J Hazard Mater. 2019; 364:562–570.

[29] Kodispathi T, Jacinth Mispa K. Fabrication, Characterization, Ion-Exchange studies and binary separation of Polyaniline/Ti(IV) iodotungstate composite Ion-Exchanger for the treatment of water pollutants. Environ Nanotechnol Monit Manag. 2021; 100555.

[30] Charles J, Bradu C, Morin-Crini N, Sancey B, Winterton P, Torri G, et al. Pollutant removal from industrial discharge water using individual and combined effects of adsorption and ion-exchange processes: Chemical abatement. J Saudi Chem Soc. 2016; 20:185–194.

[31] Li M, Xia D, Xu H, Guan Z, Li D. Iron oxychloride composite sludge-derived biochar for efficient activation of peroxymonosulfate to degrade organic pollutants in wastewater. J Clean Prod. 2021; 329:129656.

[32] Chen J, Bai X, Yuan Y, Zhang Y, Sun J. Printing and dyeing sludge derived biochar for activation of peroxymonosulfate to remove aqueous organic pollutants: Activation mechanisms and environmental safety assessment. Chem Eng J. 2022; 446:136942.

[33] Fan Z, Yang S, Zhu Q, Zhu X. Effects of different oxygen conditions on pollutants removal and the abundances of tetracycline resistance genes in activated sludge systems. Chemosphere. 2022; 291:132681.

[34] Fuzil NS, Othman NH, Jamal NASRA, Mustapa AN, Alias NH, Dollah 'Aqilah, et al. Bisphenol A Adsorption from Aqueous Solution Using Graphene Oxide-Alginate Beads. J Polym Environ. 2022; 30(2):597–612.

[35] Abdullah N, Othman FEC, Yusof N, Matsuura T, Lau WJ, Jaafar J, et al. Preparation of nanocomposite activated carbon nanofiber/manganese oxide and its adsorptive performance toward leads (II) from aqueous solution. J Water Process Eng. 2020; 37:101430.

[36] Nor NAM, Jaafar J, Ismail AF, Mohamed MA, Rahman MA, Othman MHD, et al. Preparation and performance of PVDF-based nanocomposite membrane consisting of TiO_2 nanofibers for organic pollutant decomposition in wastewater under UV irradiation. Desalination. 2016; 391:89–97.

[37] Abdullah N, Yusof N, Lau WJ, Jaafar J, Ismail AF. Recent trends of heavy metal removal from water/wastewater by membrane technologies. J Ind Eng Chem. 2019; 76:17–38.

[38] Alias NH, Jaafar J, Samitsu S, Matsuura T, Ismail AF, Huda S, et al. Photocatalytic nanofiber-coated alumina hollow fiber membranes for highly efficient oilfield produced water treatment. Chem Eng J. 2019; 360:1437–1446.

[39] Jiang D, Fang D, Zhou Y, Wang Z, Yang ZH, Zhu J, et al. Strategies for improving the catalytic activity of metal-organic frameworks and derivatives in SR-AOPs: Facing emerging environmental pollutants. Environ Pollut. 2022; 306:119386.

[40] Murgolo S, Franz S, Arab H, Bestetti M, Falletta E, Mascolo G. Degradation of emerging organic pollutants in wastewater effluents by electrochemical photocatalysis on nanostructured TiO_2 meshes. Water Res. 2019; 164:114920.

[41] Bernabeu A, Vercher RF, Santos-Juanes L, Simón PJ, Lardín C, Martínez MA, et al. Solar photocatalysis as a tertiary treatment to remove emerging pollutants from wastewater treatment plant effluents. Catal Today. 2011; 161:235–240.

[42] Motora KG, Wu C-M, Naseem S. Magnetic recyclable self-floating solar light-driven WO2.72/Fe3O4 nanocomposites immobilized by Janus membrane for photocatalysis of inorganic. J Ind Eng Chem. 2021; 102:25–34.

[43] Braslavsky SE, Braun AM, Cassano AE, Emeline A V., Litter MI, Palmisano L, et al. Glossary of terms used in photocatalysis and radiation catalysis (IUPAC recommendations 2011). Pure Appl Chem. 2011; 83(4):931–1014.

[44] Baly ECC, Heilbron IM, Barker WF. Photocatalysis. Part I. The synthesis of formaldehyde and carbohydrates from carbon dioxide and water. J Chem Soc Trans. 1921; 119:1025–1035.

[45] Wu J, Zheng W, Chen Y. Definition of photocatalysis: Current understanding and perspectives. Curr Opin Green Sustain Chem. 2022; 33:1–6.

[46] Fujishima a, Honda K. Electrochemical photolysis of water at a semiconductor electrode. Nature. 1972; 238:37–38.

[47] Molinari R, Lavorato C, Argurio P. Recent progress of photocatalytic membrane reactors in water treatment and in synthesis of organic compounds. A review. Catal Today. 2017; 281:144–164.

[48] Wang X, Yu J, Sun G, Ding B. Electrospun nanofibrous materials: A versatile medium for effective oil/water separation. Mater Today. 2016; 19(7):403–414.

[49] Mohamed MA, Jaafar J, Zain MFM, Minggu LJ, Kassim MB, Rosmie MS, et al. In-depth understanding of core-shell nanoarchitecture evolution of g-C_3N_4@C,N co-doped anatase/rutile: Efficient charge separation and enhanced visible-light photocatalytic performance. Appl Surf Sci. 2018; 436:302–318.

[50] Chakhtouna H, Benzeid H, Zari N, el Kacem Quaiss A, Bouhfid R. Recent progress on Ag/TiO_2 photocatalysts: photocatalytic and bactericidal behaviors. Environ Sci Pollut Res. 2021; 28:22638–44666.

[51] Wang Y, Bai X, Qin H, Wang F, Li Y, Li X, et al. Facile one step synthesis of hybrid graphitic carbon nitride and carbon composites as high performance catalysts for CO_2 photocatalytic conversion. ACS Appl Mater Interfaces. 2016; 8(27):17212–17219

[52] Ahmad R, Ahmad, Zaki, Khan AU, Mastoi NR, Aslam M, Kim J. Photocatalytic systems as an advanced environmental remediation: Recent developments, limitations and new avenues for applications. J Environ Chem Eng. 2016; 4(4):4143–4164.

[53] Shi Y, Huang J, Zeng G, Cheng W, Hu J. Photocatalytic membrane in water purification: is it stepping closer to be driven by visible light? J Memb Sci. 2019; 584:364–392.

[54] Alias NH, Nor AM, Mohamed MA, Jaafar J, Othman NH. Photocatalytic materials-based membranes for efficient water treatment. In: Hussain CM, Mishra AK, editors. Handbook of smart photocatalytic materials. London: Elsevier; 2020, pp. 209–230.

[55] Kusworo TD, Budiyono, Kumoro AC, Utomo DP. Photocatalytic nanohybrid membranes for highly efficient wastewater treatment: A comprehensive review. J Environ Manage. 2022; 317:115357.

[56] Rodriguez-Mozaz S, Ricart M, Köck-Schulmeyer M, Guasch H, Bonnineau C, Proia L, et al. Pharmaceuticals and pesticides in reclaimed water: Efficiency assessment of a microfiltration-reverse osmosis (MF-RO) pilot plant. J Hazard Mater. 2015; 282:165–173.

[57] Molinari R, Caruso A, Argurio P, Poerio T. Degradation of the drugs Gemfibrozil and Tamoxifen in pressurized and de-pressurized membrane photoreactors using suspended polycrystalline TiO_2 as catalyst. J Memb Sci. 2008; 319:54–63.

[58] Kundu S, Karak N. Polymeric photocatalytic membrane: An emerging solution for environmental remediation. Chem Eng J. 2022; 438:135575.

[59] Li J, Guo S, Xu Z, Li J, Pan Z, Du Z, et al. Preparation of omniphobic PVDF membranes with silica nanoparticles for treating coking wastewater using direct contact membrane distillation: Electrostatic adsorption vs. chemical bonding. J Memb Sci. 2019; 574:349–357.

[60] Amin M, Homayoonfal M, Davar F. Achieving high separation of cephalexin in a photocatalytic membrane reactor: What is the best method for embedding catalyst within the polysulfone membrane structure? Chem Eng J. 2022; 450:138150.

[61] Wang Z, He M, Jiang H, He H, Qi J, Ma J. Photocatalytic MOF membranes with two-dimensional heterostructure for the enhanced removal of agricultural pollutants in water. Chem Eng J. 2022; 435:133870

[62] Krishnan SAG, Sasikumar B, Arthanareeswaran G, László Z, Nascimben Santos E, Veréb G, et al. Surface-initiated polymerization of PVDF membrane using amine and bismuth tungstate (BWO) modified MIL-100(Fe) nanofillers for pesticide photodegradation. Chemosphere. 2022; 304:135286.

[63] Holda AK, Perera S, Emanuelsson Patterson EA. Photocatalytic membranes containing homocoupled conjugated microporous poly(phenylene butadiynylene) for chemical-free degradation of organic micropollutants. Catal Commun. 2022; 168:106463.

[64] Pang X, Xue S, Zhou T, Xu Q, Lei W. 2D/2D nanohybrid of Ti_3C_2 MXene/WO3 photocatalytic membranes for efficient water purification. Ceram Int. 2022; 48:3659–3668.

[65] Zhao X, You Y, Ma Y, Meng C, Zhang Z. Ti_3C_2/W18O49 hybrid membrane with visible-light-driven photocatalytic ability for selective dye separation. Sep Purif Technol. 2022; 282:120145.

[66] Prado C, Oliveira M De, Rezende V, Abner Y, Lebron R, Vasconcelos B De, et al. Converting recycled membranes into photocatalytic membranes using greener TiO_2-GRAPHENE oxide nanomaterials. Chemosphere. 2022; 306:135591.

[67] Zhang Y, Tan H, Wang C, Li B, Yang H, Hou H, et al. TiO_2-coated glass hollow fiber membranes: Preparation and application for photocatalytic methylene blue removal. J Eur Ceram Soc. 2022; 42(5):2496–2504.

[68] Sreedhar N, Kumar M, Al Jitan S, Thomas N, Palmisano G, Arafat HA. 3D printed photocatalytic feed spacers functionalized with β-FeOOH nanorods inducing pollutant degradation and membrane cleaning capabilities in water treatment. Appl Catal B Environ. 2022; 300:120318.

[69] Ning R, Yan Z, Lu Z, Wang Q, Wu Z, Dai W, et al. Photocatalytic membrane for in situ enhanced removal of semi-volatile organic compounds in membrane distillation under visible light. Sep Purif Technol. 2022; 292:121068.

[70] Rathna T, PonnanEttiyappan JB, RubenSudhakar D. Fabrication of visible-light assisted TiO_2-WO_3-PANI membrane for effective reduction of chromium (VI) in photocatalytic membrane reactor. Environ Technol Innov. 2021; 24:102023.

[71] Kumari P, Bahadur N, Conlan XA, Laleh M, Kong L, O'Dell LA, et al. Atomically-thin Schottky-like photo-electrocatalytic cross-flow membrane reactors for ultrafast remediation of persistent organic pollutants. Water Res. 2022; 218:118519.

[72] Suresh R, Rajendran S, Hoang TKA, Vo DVN, Siddiqui MN, Cornejo-Ponce L. Recent progress in green and biopolymer based photocatalysts for the abatement of aquatic pollutants. Environ Res. 2021; 199:111324.

7 Membrane and Advanced Oxidation Processes for Pharmaceuticals and Personal Care Products Removal

Ryosuke Homma and Haruka Takeuchi

CONTENTS

7.1 Introduction...115
7.2 UV Photolysis of PPCPs...116
    7.2.1 UV Irradiation...116
    7.2.2 PPCPs Degradation by UV Irradiation 116
7.3 UV/H$_2$O$_2$, UV/O$_3$ and O$_3$/H$_2$O$_2$ 118
    7.3.1 Characteristics of UV/H$_2$O$_2$, UV/O$_3$, and O$_3$/H$_2$O$_2$ 118
    7.3.2 Degradation of PPCPs by O$_3$/H$_2$O$_2$, UV/H$_2$O$_2$ and O$_3$/UV 118
7.4 Degradation of PPCPs by UV/TiO$_2$..121
    7.4.1 Characteristics and Reaction Mechanism of UV/TiO$_2$....................121
    7.4.2 PPCPs Degradation by UV/TiO$_2$ 122
    7.4.3 Features and Types of Photocatalytic Membrane Reactors..............125
7.5 Prospects for Future Photocatalytic Technologies in Sewage Reclamation 127
7.6 Conclusions..127
References..128

7.1 INTRODUCTION

Advanced oxidation process (AOP) is a technique that efficiently generates hydroxyl radicals (•OH) by using a combination of O$_3$, ultraviolet (UV), H$_2$O$_2$, ultrasound, and photocatalysis. •OH has been reported to have the highest reactivity and oxidizability of all reactive oxygen species, and is nearly non-selective in its reactions with various chemicals [1–3]. However, it has been reported that, in the case of the degradation of persistent pharmaceuticals and personal care products (PPCPs), •OH reacts with

DOI: 10.1201/9781003260738-7

coexisting organic matter and ions other than the target substance due to its short lifetime [4]. Therefore, in order to improve the efficiency of •OH production and reaction with PPCPs, combination techniques such as O_3/H_2O_2, UV/H_2O_2, O_3/UV, and UV/photocatalyst have been attracting attention [5, 6].

Recently, AOP for photocatalytic degradation has attracted attention as a low-cost and environmentally friendly technology for wastewater treatment. Photocatalytic materials, in particular titanium dioxide (TiO_2), are capable of producing •OH under UV radiation, which allows effective decomposition of organic pollutants in water. From around 2000 to 2022, research on photocatalytic membrane reactors (PMRs) has attracted attention as an efficient reactor configuration for photocatalytic AOP. However, it is not fully clear what type of membrane is used and which pore size or membrane material is appropriate. This chapter provides the characteristics of PMRs in terms of reactor configuration.

This chapter also introduces the characteristics of AOPs (UV/O_3, UV/H_2O_2, O_3/H_2O_2, and UV/photocatalyst) and their degradation performance for PPCPs. Rate constants of PPCPs degradation by UV irradiation and AOPs were summarized and compared with each other to provide the removal characteristics of PPCPs frequently detected in wastewater.

7.2　UV PHOTOLYSIS OF PPCPS

7.2.1　UV Irradiation

UV light has been confirmed to be capable of inactivating chlorine-resistant pathogenic protozoa such as *Cryptosporidium* and *Giardia* even with low UV irradiation doses, and is also being highly re-evaluated because of its lower installation cost and space-saving compared to ozone disinfection [7]. In recent years, UV irradiation has attracted attention not only for inactivation of pathogenic microorganisms, but also for photolysis of chemical substances, and research on treatment of organohalogen compounds, nitro compounds, agricultural chemicals, and pharmaceuticals has been reported [8, 9]. Synergistic effects are also expected when used in conjunction with H_2O_2 and O_3 treatment systems [10–13]. In water and wastewater treatment, many studies have reported the use of low-pressure mercury vapor lamps: UV254, which can irradiate UV-C with a main wavelength of 254 nm, and UV Light Emitting Diode (UV-LED) which can select the wavelength of irradiation [14–16].

7.2.2　PPCPs Degradation by UV Irradiation

In their findings on the photodegradation of PPCPs by UV254, Kim et al. [17] reported that 8 out of 18 PPCPs require UV irradiation doses of 924–2769 mJ/cm² for the degradation of 1 log of PPCPs. This means that photodegradation of PPCPs by UV irradiation would require about 50 to 1385 times the amount of UV irradiation compared to inactivation of *E. coli* or *cryptosporidium*. Therefore, in this chapter, we summarized the findings on the rate constants for the reaction of PPCPs with UV irradiation for the most frequently investigated PPCPs reported by Miege *et al.* (Table 7.1). As shown in Table 7.1, the reaction rate constants (or photolysis rate

TABLE 7.1

Reaction Rate Constants of PPCPs in Pure Water Systems Irradiated With UV254

Compound	Kinetics [min⁻¹]	Reference	Compound	Kinetics [min⁻¹]	Reference
2QCA *	7.2×10^{-3}–2.4×10^{-2}	[18,32,63]	indometacin	7.2×10^{-3}–2.4×10^{-2}	[18,32,63]
acetaminophen	2.0×10^{-3}–4.3×10^{-2}	[32,63,64]	isopropylantipyrine	1.4–2.0×10^{-1}	[32,63]
antipyrine	1.9–2.9×10^{-1}	[32,63]	ketoprofen	1.3–3.9	[18,19,32,63]
atenolol	9.3×10^{-3}	[19]	mefenamic acid	1.1–1.7×10^{-2}	[32,63]
azithromycin	1.9–6.9×10^{-1}	[65]	metoprolol	1.4×10^{-3}–2.7×10^{-2}	[19,32,63,66]
bezafibrate	1.9–8.0×10^{-2}	[19,65]	naproxen	1.7–5.3×10^{-2}	[32,63,67]
caffeine	7.6×10^{-3}	[68]	norfloxacin	1.2×10^{-2}	[69]
carbamazepine	1.3×10^{-3}–1.9×10^{-2}	[32,63,67]	oxytetracycline	1.3×10^{-2}–1.2×10^{-1}	[32,63,70]
chlortetracycline	8.6×10^{-3}–6.2×10^{-1}	[32,63,70]	propranolol	1.9–5.5×10^{-2}	[19,32,63]
ciprofloxacin	8.9×10^{-2}–1.3×10^{-1}	[45]	salbutamol	1.9×10^{-2}	[19]
clarithromycin	5.2×10^{-3}–1.7×10^{-2}	[18,32,63]	sulfadimethoxine	8.4–9.5×10^{-2}	[32,63]
clofibric acid	7.0×10^{-2}–1.0×10^{-1}	[19,67]	sulfadimidine	4.9–7.6×10^{-2}	[32,63]
crotamiton	1.9–3.1×10^{-2}	[32,63]	sulfamethoxazole	2.3×10^{-2}–5.3×10^{-1}	[19,32,63,64]
cyclophosphamide	2.0–6.7×10^{-3}	[18,32,63]	sulfamonomethoxine	2.1–3.4×10^{-1}	[32,63]
DEET **	5.5–6.7×10^{-3}	[32,63]	tetracycline	5.0×10^{-3}–2.4×10^{-1}	[32,63,70]
diclofenac	5.0–7.3×10^{-1}	[18,19,32,63]	theophylline	3.6–8.6×10^{-3}	[32,63]
disopyramide	1.3–1.7×10^{-1}	[32,63]	trimethoprim	9.3×10^{-3}	[19]
enrofloxacin	7.6×10^{-2}	[69]	tylosin	6.7×10^{-1}–2.5	[65]
fenoprofen	1.3–2.4×10^{-1}	[32,63]	triclosan	1.8×10^{-2}–3.0×10^{-1}	[64,71,72]
furosemide	4.6×10^{-2}	[19]			

* 2-quinoxaline carboxylic acid
** N,N-Diethyl-3-methylbenzamide

constants) of 39 PPCPs by UV254 were all expressed as first-order reactions. Their rate constants ranged from a minimum value of 1.3×10^{-3} min⁻¹ (carbamazepine) to a maximum value of 3.9 min⁻¹ (ketoprofen).

Degradation characteristics of PPCPs by UV254 were investigated in previous studies. According to the results of Kim *et al.* [18], PPCPs such as ketoprofen ($k = 1.3$ min⁻¹) and diclofenac ($k = 5.1 \times 10^{-1}$ min⁻¹) were easily photodegraded among the selected 30 PPCPs. On the other hand, PPCPs such as 2-quinoxaline carboxylic acid

(2QCA) ($k = 2.4 \times 10^{-3}$ min^{-1}), cyclophosphamide ($k = 2.0 \times 10^{-3}$ min^{-1}), and clarithromycin ($k = 5.2 \times 10^{-3}$ min^{-1}) were not easily photodegraded.

Wols *et al.* [19] reported that ketoprofen ($k = 3.92$ min^{-1}) is a chemical that is easily photodegraded according to the results of UV irradiation experiments in a complex system with 40 different PPCPs. They also reported that diclofenac ($k = 7.32 \times 10^{-1}$ min^{-1}) and sulfamethoxazole ($k = 5.29 \times 10^{-1}$ min^{-1}) were photodegraded moderately while cyclophosphamide ($k = $ less than 2.0×10^{-3} min^{-1}) and the other 30 PPCPs were hardly photodegraded. As for future work, that research work also points out the need to understand the photodegradability of PPCPs from their physical properties and molecular structure.

7.3   UV/H$_2$O$_2$, UV/O$_3$ AND O$_3$/H$_2$O$_2$

7.3.1   CHARACTERISTICS OF UV/H$_2$O$_2$, UV/O$_3$, AND O$_3$/H$_2$O$_2$

Among AOP, UV/H$_2$O$_2$, UV/O$_3$, and O$_3$/H$_2$O$_2$ have many practical examples, and many findings on the degradation of PPCPs have been reported [20, 21]. Although H$_2$O$_2$ and O$_3$ have the characteristic of being able to generate •OH by themselves, the efficiency of •OH generation is poor, so each of these technologies can be used together to generate •OH more efficiently.

The mechanism of •OH production in UV/H$_2$O$_2$ is said to be that H$_2$O$_2$ absorbs UV below 300 nm, producing an •OH concentration of 2 mol/L from an H$_2$O$_2$ concentration of 1 mol/L. Low-pressure mercury lamps (UV185 and UV254) irradiate only 185 and 254 nm. However, 254 nm is widely used for UV lamps. The advantage of H$_2$O$_2$ is that it is inexpensive and readily available. Nevertheless, the drawback of H$_2$O$_2$ is not being able to produce enough •OH when the concentration of H$_2$O$_2$ is underestimated. Since H$_2$O$_2$ also works as a supplement to •OH, there is a concern that H$_2$O$_2$ and •OH may be consumed ineffectively before reacting with the target substance, even if •OH is efficiently generated by H$_2$O$_2$/UV [22].

In O$_3$/UV, as with UV/H$_2$O$_2$, many techniques have been reported using UV185 or UV254 as the light source [23, 24]. UV185 can irradiate at a wavelength of 185 nm, which produces •OH from H$_2$O in water and O$_3$ from O$_2$ in water. The irradiation wavelength of 254 nm can efficiently produce •OH from O$_3$ in water. Therefore, the O$_3$/UV185 combination is considered to be an efficient way to produce •OH, which reacts nonselectively. However, the shorter the wavelength of UV, the shorter the arrival distance due to Raman scattering, making it more difficult to irradiate a wide area.

O$_3$/H$_2$O$_2$ is an AOP that combines O$_3$ and H$_2$O$_2$ to produce •OH more efficiently and to decompose persistent chemicals. Lin *et al.* [25] state that there is an optimal range for the addition of H$_2$O$_2$.

7.3.2   DEGRADATION OF PPCPs BY O$_3$/H$_2$O$_2$, UV/H$_2$O$_2$ AND O$_3$/UV

Reported reaction rate constants of PPCPs by O$_3$/H$_2$O$_2$, UV/H$_2$O$_2$, and O$_3$/UV for the most frequently investigated PPCPs are summarized in Table 7.2 [26–32]. This confirms that the reaction rate constants of PPCPs by AOP are generally

TABLE 7.2
Reaction Rate Constants of PPCPs by O_3/H_2O_2, UV/H_2O_2, and O_3/UV

Compound	Kinetics [min⁻¹]					
	O_3/H_2O_2	Reference	UV/H_2O_2	Reference	O_3/UV	Reference
2QCA*	$1.0-1.3 \times 10^{-1}$	[26,32]	1.2×10^{-1}	[17]	$1.1-1.3 \times 10^{-1}$	[26,32]
acetaminophen	$3.1-7.2 \times 10^{-1}$	[26,32]	1.0×10^{-1}	[17]	$3.5-5.9 \times 10^{-1}$	[26,32]
antipyrine	$1.7-7.2 \times 10^{-1}$	[26,32]	4.3×10^{-1}	[17]	$5.0-7.2 \times 10^{-1}$	[26,32]
carbamazepine	$1.9-4.6 \times 10^{-1}$	[26,32]	$2.2 \times 10^{-2-}$ 2.2×10^{-1}	[17,28]	$2.5 \times 10^{-1-}$ 5.8×10^{-1}	[26,28]
chlortetracycline	–	–	4.3×10^{-1}	[17]	6.4×10^{-1}	[26]
ciprofloxacin	$1.2-1.7 \times 10^{-1}$	[27]	–	–	–	–
clarithromycin	$1.9-3.1 \times 10^{-1}$	[26,32]	$1.9 \times 10^{-2-}$ 1.5×10^{-1}	[17,28]	$3.1 \times 10^{-2-}$ 1.2×10^{-1}	[26,28]
clofibric acid	–	–	3.6×10^{-2}	[28]	$1.1-4.0 \times 10^{-1}$	[28]
crotamiton	$1.8-2.5 \times 10^{-1}$	[26,32]	2.4×10^{-1}	[17]	$2.0-3.6 \times 10^{-1}$	[26,32]
cyclophosphamide	$1.6-8.7 \times 10^{-2}$	[26,27,32]	7.6×10^{-2}	[17]	$8.2-9.0 \times 10^{-2}$	[26,32]
DEET**	$9.7 \times 10^{-2-}$ 1.4×10^{-1}	[26,32]	1.2×10^{-1}	[17]	$1.6-2.5 \times 10^{-1}$	[26,32]
diclofenac	$2.7 \times 10^{-2-}$ 5.4×10^{-1}	[26,32]	$2.5-7.9 \times 10^{-1}$	[26,32]	$1.2-1.8$	[26,28,32]
disopyramide	$1.0-1.5 \times 10^{-1}$	[26,32]	3.8×10^{-1}	[17]	$3.1-5.4 \times 10^{-1}$	[26,32]
fenoprofen	$1.6-3.1 \times 10^{-1}$	[26,32]	$4.2 \times 10^{-2-}$ 4.3×10^{-1}	[17,28]	$8.4 \times 10^{-2-}$ 6.1×10^{-1}	[26,32]
indometacin	$4.5-7.4 \times 10^{-1}$	[26,32]	$5.6 \times 10^{-2-}$ 3.1×10^{-1}	[17,28]	$5.6 \times 10^{-1}-1.6$	[26,32]
isopropylantipyrine	$5.4-7.9 \times 10^{-1}$	[26,32]	$8.2 \times 10^{-2-}$ 3.8×10^{-1}	[17,28]	$7.6 \times 10^{-1}-3.3$	[26,32]
ketoprofen	$1.0-2.5 \times 10^{-1}$	[26,32]	$1.0-2.4$	[17,28]	$1.6-4.8$	[26,28,32]
mefenamic acid	$1.3-1.9$	[26,32]	1.7×10^{-1}	[17]	$1.4-2.1$	[26,32]
metoprolol	$1.4-2.5 \times 10^{-1}$	[26,32]	1.9×10^{-1}	[17]	$1.5-2.3 \times 10^{-1}$	[26,32] [26,32]
naproxen	$2.2-5.1 \times 10^{-1}$	[26,32]	$7.5 \times 10^{-2-}$ 3.4×10^{-1}	[17,28]	$3.7 \times 10^{-2}-1.0$	[26,28,32]
oxytetracycline	$1.2-2.3$	[26,32]	3.4×10^{-1}	[17]	$1.2-1.8$	[26,32]
propranolol	$2.7-4.6 \times 10^{-1}$	[26,32]	2.9×10^{-1}	[17]	$2.7-4.9 \times 10^{-1}$	[26,32]
sulfadimethoxine	$2.5-5.4 \times 10^{-1}$	[26,32]	3.1×10^{-1}	[17]	$4.1-6.1 \times 10^{-1}$	[26,32]
sulfadimidine	$2.3-5.4 \times 10^{-1}$	[26,32]	2.4×10^{-1}	[17]	$3.4-5.6 \times 10^{-1}$	[26,32]
sulfamethoxazole	$2.1-2.8 \times 10^{-1}$	[26]	5.0×10^{-1}	[17]	$6.1-7.9 \times 10^{-1}$	[26,32]
sulfamonomethoxine	$2.7-4.9 \times 10^{-1}$	[26,32]	5.3×10^{-1}	[17]	$5.2-7.2 \times 10^{-1}$	[26,32]
tetracycline	$3.0 \times 10^{-1}-1.3$	[26,32]	3.1×10^{-1}	[17]	$1.2-5.2$	[26,32]
theophylline	$1.4 \times 10^{-1}-1.1$	[26,32]	9.8×10^{-2}	[17]	$2.2-3.6 \times 10^{-1}$	[26,32]
triclosan	–	–	3.2×10^{-2}	[28]	6.5×10^{-2}	[28]

* 2-quinoxaline carboxylic acid
** N,N-Diethyl-3-methylbenzamide

in the range of 1.6×10^{-2} to 5.2 min^{-1}. Kim *et al.* [32] and Giri *et al.* [28] evaluated the reaction rate constants of PPCPs by AOP using a large number of PPCPs. In the report by Kim *et al.* on 30 PPCPs, cyclophosphamide was the least degraded in O_3/H_2O_2, followed by DEET, disopyramide, 2QCA, and others. On the other hand, oxytetracycline, tetracycline, and mefenamic acid were easily degraded. In UV/H_2O_2, cyclophosphamide was the least degraded, followed by theophylline, 2QCA, clarithromycin, acetaminophen, and others. On the other hand, ketoprofen and diclofenac were easily degraded. For UV/O_3, cyclophosphamide was the least degraded, followed by clarithromycin and 2QCA. On the other hand, ketoprofen, oxytetracycline, tetracycline, and mefenamic acid were easily degraded. In a report on 16 PPCPs by Giri *et al.* [28] ibuprofen and carbamazepine were less degraded in UV/H_2O_2. On the other hand, ketoprofen and diclofenac were easily degraded. For UV/O_3, clarithromycin and ibuprofen were less likely to be degraded. On the other hand, isopropylantipyrine and carbamazepine were easily degraded.

For target substances such as PPCPs, a survey was conducted on representative substances that are •OH scavenger. Gerrity *et al.* [33] reported that •OH scavenger (i.e., DOC, HCO_3^-, CO_3^{2-}, NH_4^+, Br$^-$, and NO_2^-) affect the degradation efficiency of atenolol, atrazine, bisphenol A., carbamazepine, DEET, diclofenac, gemfibrozil, ibuprofen, meprobamate, naproxen, phenytoin, primidone, sulfamethoxazole, triclosan, trimethoprim, and NDMA. They also reported higher rate constants for •OH scavenger reactions in treated wastewater from seven sewage treatment plants in the United States, Switzerland, and Australia, in the order of DOC > NO_2^- > HCO_3^- > CO_3^- > Br- > NH_4^+ (Figure 7.1). Since the effect of reaction inhibition by OH scavenger depends on the reaction rate constant of the •OH scavenger and its residual concentration, it is considered necessary to know the components of water quality in advance when evaluating the decomposition efficiency of target substances in actual environmental water.

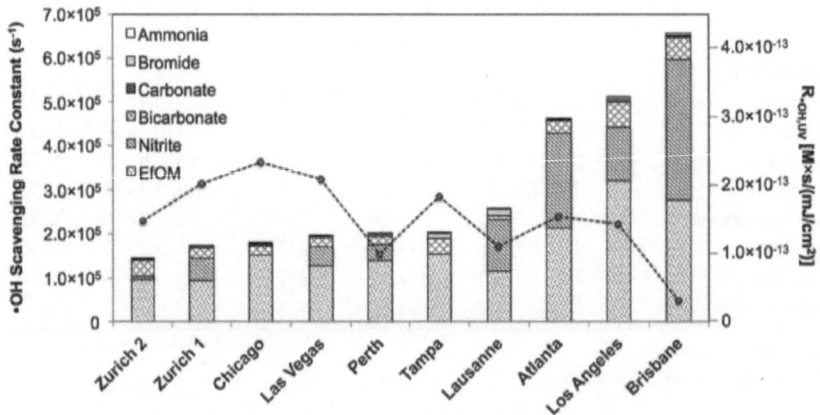

FIGURE 7.1 Contribution of reaction rates of •OH scavengers in treated wastewater from the U.S., Switzerland, and Australia [33].

7.4 DEGRADATION OF PPCPS BY UV/TIO$_2$

7.4.1 CHARACTERISTICS AND REACTION MECHANISM OF UV/TIO$_2$

AOP for photocatalytic degradation has attracted attention as a low-cost and environmentally friendly technology for wastewater treatment. One reason why TiO$_2$ has attracted more attention than other photocatalysts is that it can oxidize not only water to produce •OH but also reduce oxygen to produce •O$_2$– or H$_2$O$_2$ and further produce •OH [34]. The oxidation potential of •OH, i.e., +2.81 eV, is the second highest after that of hydrogen fluoride, and thus it can react with many types of chemicals [35]. Furthermore, since •OH exhibits an extremely high reaction rate, it is believed to be nonselective for a wide variety of chemicals. Two reaction mechanisms are commonly proposed for the photocatalytic reaction of TiO$_2$, and it is often debated whether the end product of the O$_2$ reduction reaction exists as •O$_2$– or as •OH. However, in the photocatalytic reaction of TiO$_2$ in the water treatment field, in addition to the "redox reaction," "reaction in water (pH 4.8~11.6)" and "photoreaction of H$_2$O$_2$" must be taken into account (Figure 7.2).

Focusing on the treatment of target substances with UV/TiO$_2$, since Hirakawa et al. [36] reported that TiO$_2$ photocatalysis in water and wastewater treatment is effective when the concentration range of target substances is ppm and ppb, removal of trace contaminants such as PPCPs should be sufficient. Watanabe et al. [37] reported that photocatalytic inactivation of E. coli, Pseudomonas aeruginosa, Staphylococcus aureus, MRSA, Streptococcus pyogenes, cariogenic bacteria, lactic acid bacteria, Bacillus subtilis, heat-resistant bacterial spores, salmonella, Vibrio parahaemolyticus, E. coli phage QB, and influenza virus was also observed. Since Irie et al. [38] reported inactivation of E. coli by TiO$_2$ photocatalyst even at an irradiation intensity of 1 μW/cm^2 of fluorescent lamp, it is presumed that it can efficiently disinfect even viruses and fungi, which are pathogenic microorganisms. Furthermore, Tanizaki et al. [39] reported that TiO$_2$ photocatalysts can be expected to be effective in removing chloroform, the main component of trihalomethanes, and total organic carbon (TOC) components remaining in tap water.

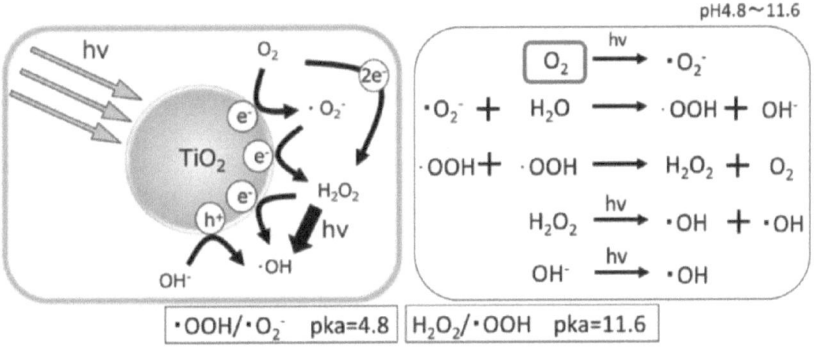

FIGURE 7.2 Reaction mechanism of TiO$_2$ focusing on "redox reaction," "reaction in water," and "photoreaction of products (authors illustrated).

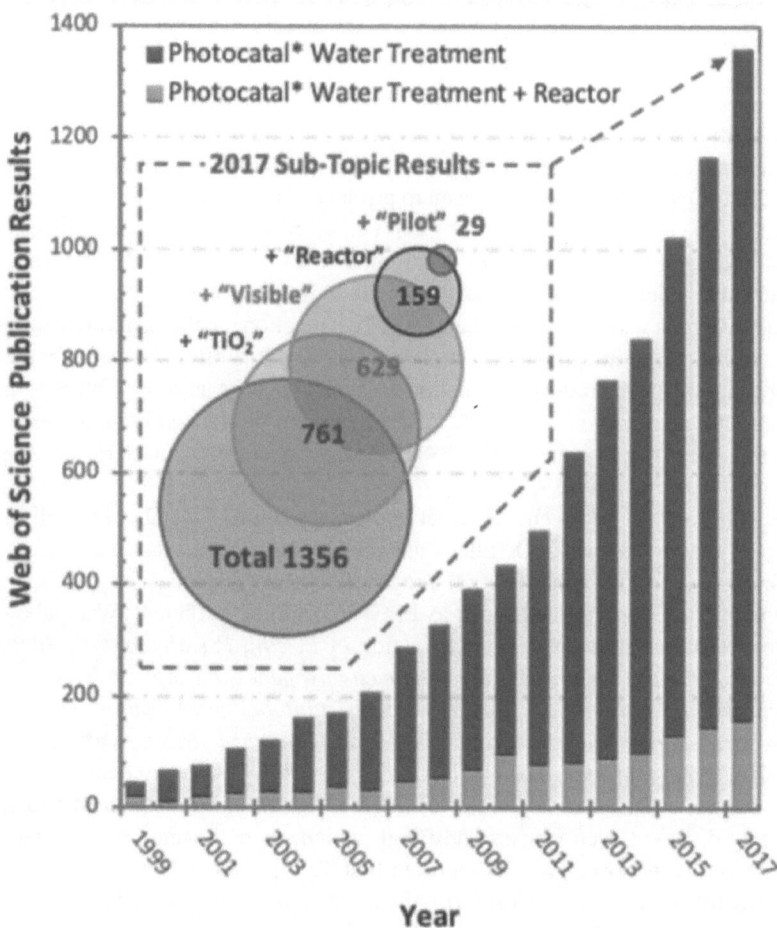

FIGURE 7.3   Number of publications on photocatalytic water treatment from 1999 to 2017 [40].

Loeb *et al.* [40] note that the literature on photocatalytic treatment technologies is growing rapidly, with over 8,000 references since 2000, but the adaptation of photocatalysis in actual treatment systems is extremely limited compared to other promoted oxidation technologies (Figure 7.3). The reason for this, they point out, is that academic research, which focuses on materials design and mechanical evaluation, ignores the consideration of practical applications, which are inherently essential, creating a gap between this research and industrial applications.

7.4.2   PPCPs Degradation by UV/TiO₂

The efficiency of PPCPs degradation by UV/TiO$_2$, as with other AOP, should be evaluated with the understanding that it depends on the experimental source water and experimental conditions. In a review written by Kanakaraju *et al.* [41] on the photocatalytic degradation of PPCPs, the researchers noted that differences in the initial concentration of PPCPs, photocatalyst type, pH, and light source characteristics could affect the efficiency of PPCPs treatment. Awfa *et al.* [42] stated that even with the same target substance, concentration, and irradiation intensity, differences in experimental conditions and other factors could significantly affect the efficiency of treatment to PPCPs. In terms of findings on processing efficiency and light sources for PPCPs, findings using UV-LEDs, black lights, and pseudo-sunlight lamps, which can irradiate UV between 300 and 400 nm wavelengths, were widely presented in both reviews. However, the findings using UV254, which is commonly used for UV disinfection of sewage treatment plants in Europe and the United States, were very few, only 6 out of 114 in these general papers. Mahmoud *et al.* [43] and Tong *et al.* [44] also introduced previous studies that used black lights, UV-LEDs, xenon lamps, and medium-pressure mercury lamps as light sources. Thus, if we restrict the target substance to PPCPs and limit the evaluation of reaction kinetics, we can assume that extremely little is known about the efficiency of degradation of PPCPs by UV/TiO$_2$ using UV254.

Reaction rate constants of 50 PPCPs with UV254/TiO$_2$ are summarized in Table 7.3 [28, 45–53]. The reaction rate constants for PPCPs with UV254/TiO$_2$ ranged from a minimum value of 3.0×10^{-4} min^{-1} (acetaminophen) to a maximum value of 5.1×10^{-1} min^{-1} (ketoprofen). Studies of PPCPs with UV/TiO$_2$ using UV-C light sources irradiating at 254 nm were extremely limited in knowledge compared to UV-A or sunlight light sources, and only 13 PPCPs could be identified. Compared to other AOP, the knowledge on the reaction rate constants of PPCPs by UV254/TiO$_2$ was limited due to the large difference in reactor scale (from 50 mL to 1.2 L), the different reaction systems, and the large difference in experimental test water (about ng/L to mg/L) and experimental conditions, as well as the UV254 /TiO$_2$. Thus, we could not find any studies that elucidated the degradation characteristics of PPCPs by TiO$_2$.

TABLE 7.3
Reaction Rate Constants of PPCPs by UV254/TiO$_2$

Compound	Kinetics [min⁻¹]	Reference
acetaminophen	$3.0 \times 10^{-4} – 2.0 \times 10^{-2}$	[50]
carbamazepine	$2.1 \times 10^{-3} – 1.5 \times 10^{-1}$	[28,48,49]
ciprofloxacin	$8.9 \times 10^{-2} – 1.6 \times 10^{-1}$	[45]
clarithromycin	5.3×10^{-3}	[28]
clofibric acid	8.3×10^{-2}	[28]
cyclophosphamide	$1.0 \times 10^{-3} – 4.3 \times 10^{-2}$	[46,47]
diclofenac	$1.0 \times 10^{-3} – 2.8 \times 10^{-2}$	[28,51,53]
fenoprofen	8.1×10^{-2}	[28]
indometacin	8.5×10^{-2}	[28]
isopropylantipyrine	7.3×10^{-2}	[28]
ketoprofen	5.1×10^{-1}	[28]
naproxen	6.2×10^{-2}	[28]
tetracycline	$2.2 \times 10^{-3} – 2.5 \times 10^{-2}$	[51,52]

The review article by Uyguner-Demirel *et al.* [54] extensively summarized the findings on the reaction rate constants of chemicals mainly by UV-A and sunlight. However, they pointed out that even for those light sources, the reaction rate constants cannot be adequately discussed without the same reactor geometry and under the same irradiation conditions in order to evaluate them. They also pointed out that those papers do not sufficiently clarify whether the degradation is photolysis or photocatalytic degradation, so it is necessary to continue to collect knowledge on the reaction rate constants in light of the above [54].

In terms of the effect of different water quality and experimental conditions on the reaction rate constants of PPCPs, Yang *et al.* [50] and Lalhriatpuia *et al.* [51] evaluated the same reactor geometry and the same light source for acetaminophen and tetracycline, respectively. Yang *et al.* [50] evaluated the effect of acetaminophen on the reaction rate constants for TiO_2 addition concentration, initial concentration of the target substance, DO concentration, pH, and UV irradiation intensity. Experiments were conducted in a TiO_2 suspension system in a semi-batch system. The values of the acetaminophen reaction rate constant increased from 4.9×10^{-3} to 1.43×10^{-2} min^{-1} as the TiO_2 addition concentration increased from 0.04 to 2.0 g/L. However, in the experiment with a TiO_2 addition concentration of 5.0 g/L, the value of the reaction rate constant for acetaminophen (i.e., 1.47×10^{-2} min^{-1}) was not significantly different compared to that with a TiO_2 concentration of 2.0 g/L. This suggests that a certain amount of TiO_2 addition concentration is sufficient for acetaminophen degradation. Second, in experiments with different initial concentrations of acetaminophen, the reaction rate constant decreased significantly from 1.95×10^{-3} to 3.7×10^{-2} min^{-1} as the initial concentration of acetaminophen increased from 2.0 to 10.0 mM. These results suggest that the concentration of the contaminants and the intensity of light irradiation are one of the important factors affecting reaction rate constants in particular when it is evaluated under low-concentration experimental conditions. Regarding DO concentration, the reaction rate constant improved from 1.60×10^{-3} to 1.05×10^{-2} min^{-1} with increased DO concentration from 1.3 to 36.1 mg/L, as is theoretical for the TiO_2 photocatalytic reaction. The improvement of the reaction rate constant was also observed in the supersaturated state, suggesting that the value of the reaction rate constant can be improved by increasing the DO concentration, although the cost of feeding oxygen is high. As for the effect of pH, the highest rate constant was observed at pH 9, followed by a decrease in the rate constant at pH 7.0, 5.5, 3.5, and 11.0. They pointed out that the reason for this is related to the isoelectric point of the photocatalyst and the chargeability of acetaminophen.

In summary, it is clear that the study on the rate constants of PPCPs reaction with UV/TiO_2 using UV254 cannot simply compare the rate constants of PPCPs reaction with other literature values due to the influence of the form of TiO_2 used, experimental conditions, and the difference in water quality. Therefore, in the future, it will be necessary to (1) clarify the degradation characteristics of various PPCPs by UV/TiO_2 in the same reactor configuration and under the same irradiation conditions, (2) distinguish among "adsorption by TiO_2", "photolysis by UV irradiation" and "Oxidation reaction by UV/TiO_2"in UV/TiO_2, and (3) consider what type of reactor design would be appropriate for the evaluation. In addition, there are quite a few research reports on the degradation evaluation of batch-type and pure water systems, even though the AOP using TiO_2 photocatalyst is a continuous system and environmental water treatment technology is required [55, 56]. Hirakawa *et al.* [36] reported that TiO_2 suspension systems were the mainstream in the early stages of their research, but that the method of attaching TiO_2 to some substrate surface is now the mainstream. This leads to an evaluation that is far from practical because TiO_2 suspension systems are not designed for solid-liquid separation of TiO_2. Therefore, it is considered necessary to evaluate the decomposition efficiency of chemicals such as PPCPs using a treatment system that also combines a solid-liquid separation technique for TiO_2.

7.4.3 FEATURES AND TYPES OF PHOTOCATALYTIC MEMBRANE REACTORS

From around 2000 to 2022, research on PMRs has continued to attract attention. Zhang *et al.* [57] reported that approximately 75% or more of the published scientific papers are intended for use in water treatment (Figure 7.4). As the number of publications on photocatalytic membranes is increasing annually, it is expected that many more scientific papers will be reported in the future. It is not fully clear what type of membrane is used and which pore size is appropriate. Regarding membrane materials, UV- and •OH-resistant polytetrafluoroethylene (PTFE) and polyvinylidene difluoride (PVDF) are used as organic membranes, while α-alumina and zeolite are commonly used as inorganic membranes [57, 58]. However, in the case of organic membranes made of other materials, it has been reported that UV and •OH often

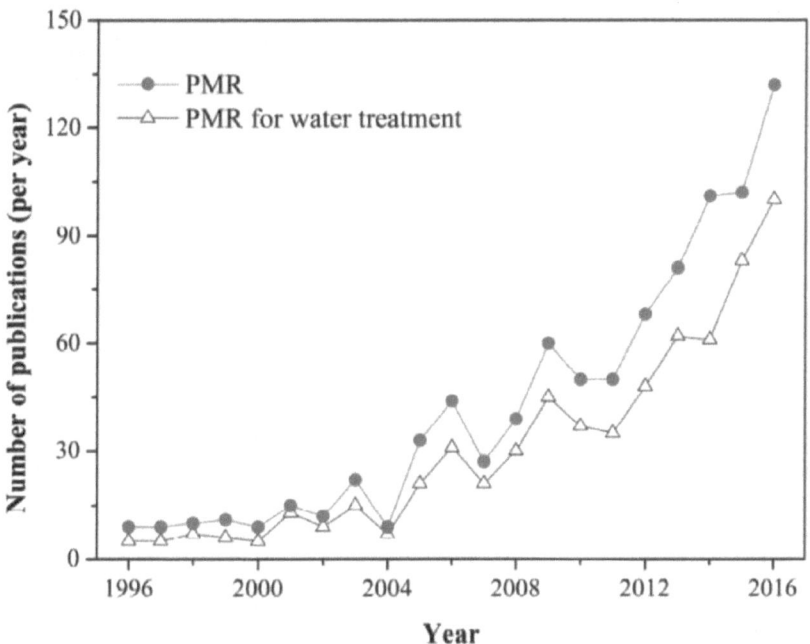

FIGURE 7.4 Annual number of academic publications on photocatalytic membranes [57].

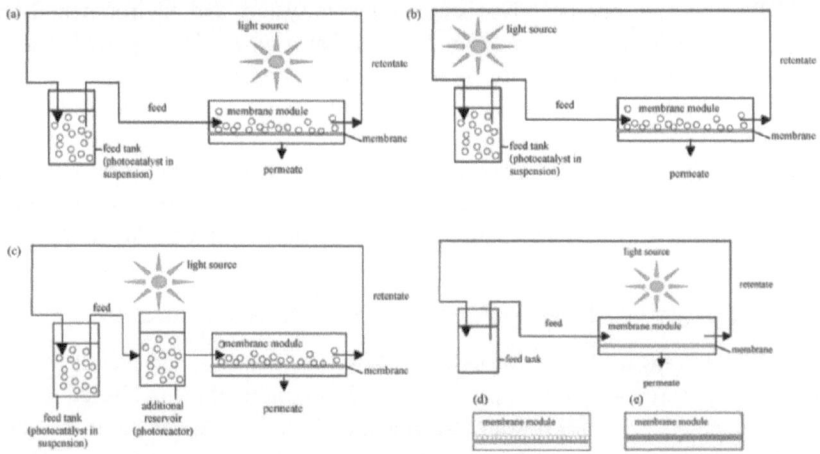

FIGURE 7.5   PMR utilizing photocatalyst in suspension: (a) irradiation of the membrane module; (b) irradiation of the feed tank; (c) irradiation of the additional reservoir (photoreactor) located between the feed tank and membrane module. PMR utilizing photocatalyst immobilized (d) on a membrane and (e) within a membrane structure [58].

destroy the membranes themselves, and there are concerns in the literature about breakage and damage to PTFE and PVDF [58].

Photocatalytic membrane reactors can be classified into two main forms, i.e., slurry batch photoreactor and fixed bed photoreactor (Figure 7.5). In slurry batch photoreactors, TiO_2 is suspended in feed water. These systems have been reported to be further classified into (a) UV irradiation of the feed tank, (b) UV irradiation of the membrane module, or (c) light irradiation of a reactor (Figure 7.5a–c). These systems are characterized by low permeate flux and membrane fouling as disadvantages [58]. Since this is a UV/TiO_2 suspension system, there are concerns about light irradiation interference by TiO_2 particles in terms of reaction efficiency, but if the light source can be UV irradiated at high power in the reactor, large volumes can be processed.

On the other hand, in the fixed bed photoreactors are reactors, TiO_2 is loaded or accumulated in/on the membrane. These systems have been reported to be further classified into (d) TiO_2 immobilization on the membrane and (e) TiO_2 immobilization in the membrane structure (Figure 7.5d, e). For the immobilization of TiO_2 on the membrane, the sintering of TiO_2 and the accumulation of TiO_2 particles are generally known. The sintering method of TiO_2 is characterized by the fact that it is a simple process because it is a method of baking TiO_2 and membranes at high temperatures, but the crystal structure of TiO_2 changes to a rutile form at high temperatures, which is said to reduce photocatalytic activity [57]. In the TiO_2 particle accumulation method, TiO_2 particles accumulated on the membrane on the feed water side may peel off during the crossflow type as shown in Figure 7.5. Therefore, some filtered water is not treated with UV/TiO_2, which raises concerns about treated water quality. The water quality on the feed water side may deteriorate because the reaction by •OH does not occur for chemicals that do not come close to TiO_2 or the membrane.

However, in both methods, the system is easily exposed to light irradiation in water, so it is thought that the ability of the light source can be fully utilized in the decomposition of chemical substances that are easily photodegraded and the inactivation of viruses and *E. coli*. Finally, the characteristics of (e) immobilization of TiO_2 within the membrane structure are similar in many respects to those of (d) immobilization of TiO_2 on the membrane, and since it is formed by the calcination method, there is concern that the photocatalytic activity will be reduced. The ability to take full advantage of the light source is attractive, but there are concerns about the quality of the treated water and membrane fouling due to the inefficient formation of •OH on the membrane.

Thus, although research on photocatalytic membranes is increasing, their use patterns are different, and even the same photocatalytic membrane may have very different reaction mechanisms and treatment efficiencies. Solid-liquid separation of TiO_2 by membrane separation as a treatment technology has shown promise, but it is still unclear what method is appropriate for the best form of use.

7.5   PROSPECTS FOR FUTURE PHOTOCATALYTIC TECHNOLOGIES IN SEWAGE RECLAMATION

Recent academic research has focused on increasing the efficiency of photocatalytic materials, and attention has been paid to the development of visible light responsive photocatalysts by combining them with auxiliary catalysts and transition metals, N-doping, and other techniques [40, 59–61]. Therefore, since the current stage is the development of visible light responsive photocatalysts, the selectivity of photocatalysts is expanding, but this is not fully reflected in practical aspects. In addition, cost evaluations for drinking water applications are often expressed in terms of electricity consumption per m^3 of treated water volume, and UV/TiO_2 suspension systems using UV lamps are estimated to take more than 10 kWh/m^3 [62]. The UV/H_2O_2 and O_3/H_2O_2 are reported to consume less than 1 kWh of electricity per m^3 of treated water, so the actual treatment efficiency of the UV/TiO_2 suspension system is considered poor [62]. This is thought to be due mainly to insufficient progress in optimizing the use of photocatalysts in water and wastewater treatment, including the obstruction of light by photocatalytic particles. In addition, photocatalysts are expected to use sunlight, which will further lower the cost of operation.

7.6   CONCLUSIONS

The issue of PPCPs in the aquatic environment was introduced, as well as water and wastewater treatment technologies for the removal or degradation of PPCPs. Many findings on TiO_2 photocatalysts in AOP were summarized, from the reaction mechanism to the treatment efficiency of PPCPs, and the effects of experimental conditions and water quality. This chapter is an extremely important fundamental insight into photocatalytic membrane reactors for practical applications. In the future, many studies on the combination of photocatalyst and light source and reactor design will be reported. In addition, by steadily solving a wide variety of research problems, it is expected that it will be possible to determine what kind of water source (water quality), what kind of treatment system, and under what kind of treatment conditions it will be possible to obtain water for the intended use. If these things can be accomplished, it is believed that we can also reach a level where we can achieve one of the goals of World Health Organization (WHO), United Nations International Children's Emergency Fund (UNICEF), and others: an environment where "Everyone has the right to sufficient, continuous, safe, acceptable, physically accessible and affordable water for personal and domestic use". Therefore, the development of treatment technology using photocatalysts is expected to make a significant contribution to international society.

REFERENCES

[1] Zhang K, Parker KM. Halogen radical oxidants in natural and engineered aquatic systems. Environmental Science and Technology 2018; 52:9579–9594. https://doi.org/10.1021/acs.est.8b02219.

[2] Ozawa T. Generation of free radicals in living organisms by radiation and their functions. IAEA 50 years of INIS 1995; 30:5–13 (in Japanese).

[3] Nakamura S. Chemistry of Reactive Oxygen Species and Antioxidants. Journal of Nippon Medical School 2013; 9:164–169 (in Japanese).

[4] Wang J, Chu L. Irradiation treatment of pharmaceutical and personal care products (PPCPs) in water and wastewater: An overview. Radiation Physics and Chemistry 2016; 125:56–64. https://doi.org/10.1016/j.radphyschem.2016.03.012.

[5] Oturan MA, Aaron J-J. Advanced oxidation processes in water/wastewater treatment: Principles and applications. A review. Critical Reviews in Environmental Science and Technology 2014; 44:2577–2641. https://doi.org/10.1080/10643389.2013.829765.

[6] Malato S, Blanco J, Vidal A, Richter C. Photocatalysis with solar energy at a pilot-plant scale: An overview. Applied Catalysis B: Environmental 2002; 37:1–15. https://doi.org/10.1016/S0926-3373(01)00315-0.

[7] Iwasaki T. Trend of UV Technologies in Various Industries. Journal of Japan Society on Water Environment 2005; 28:234–247 (in Japanese).

[8] Shemer H, Kunukcu YK, Linden KG. Degradation of the pharmaceutical Metronidazole via UV, Fenton and photo-Fenton processes. Chemosphere 2006; 63:269–276. https://doi.org/10.1016/j.chemosphere.2005.07.029.

[9] Chowdhury P, Sarathy SR, Das S, Li J, Ray AK, Ray MB. Direct UV photolysis of pharmaceutical compounds: Determination of pH-dependent quantum yield and full-scale performance. Chemical Engineering Journal 2020; 380. https://doi.org/10.1016/j.cej.2019.122460.

[10] Al-Kandari H, Abdullah AM, Al-Kandari S, Mohamed AM. Synergistic effect of O_3 and H_2O_2 on the visible photocatalytic degradation of phenolic compounds using TiO_2/reduced graphene oxide nanocomposite. Science of Advanced Materials 2017; 9:739–746. https://doi.org/10.1166/sam.2017.3025.

[11] Zhong X, Cui C, Yu S. Identification of Oxidation Intermediates in Humic Acid Oxidation. Ozone: Science and Engineering 2018; 40:93–104. https://doi.org/10.1080/01919512.2017.1392845.

[12] Yoshino K., Iwasaki T., Ono K., Machida M., Kinoshita S., Matsubara J., Muroya M., Nishio N. YT. Sterilization of household papers such as tissues, toilet papers by using of UV radiation and hydrogen peroxide together sterilization of wastewaters after paper-making process by using of UV radiation and hydrogen peroxide together. Japan TAPPI Journal 1999; 53:1345–1352 (in Japanese).

[13] Tanaka T, Tsuzuki KTT. Reduction of organic matter in secondary treated municipal wastewater by chemical oxidation processes -comparison of methods involving ozone, ultraviolet radiation and TiO_2 catalyst-. Journal of Japan Society on Water Environment 1999; 22:926–931 (in Japanese).

[14] Serna-Galvis EA, Ferraro F, Silva-Agredo J, Torres-Palma RA. Degradation of highly consumed fluoroquinolones, penicillins and cephalosporins in distilled water and simulated hospital wastewater by UV_{254} and UV_{254}/persulfate processes. Water Research 2017; 122:128–138. https://doi.org/10.1016/j.watres.2017.05.065.

[15] Nebot Sanz E, Salcedo Dávila I, Andrade Balao JA, Quiroga Alonso JM. Modelling of reactivation after UV disinfection: Effect of UV-C dose on subsequent photoreactivation and dark repair. Water Research 2007; 41:3141–3151. https://doi.org/10.1016/j.watres.2007.04.008.

[16] Beck SE, Ryu H, Boczek LA, Cashdollar JL, Jeanis KM, Rosenblum JS, et al. Evaluating UV-C LED disinfection performance and investigating potential dual-wavelength synergy. Water Research 2017; 109:207–216. https://doi.org/10.1016/j.watres.2016.11.024.

[17] Kim I. Applicability of UV-based and O_3-based processes for the reduction of pharmaceuticals and personal care products (PPCPs). Graduate School of Engineering, Kyoto University, Doctoral Dissertation 2008.

[18] Kim I., Tanaka H., Yamashita N., Kobayashi Y., Okuda T., Iwasaki T. YK and TT. Batch test on the removal of pharmaceuticals by UV treatment. Proceedings of Environmental Engineering Research 2006; 43:47–56 (in Japanese).

[19] Wols BA, Hofman-Caris CHM, Harmsen DJH, Beerendonk EF. Degradation of 40 selected pharmaceuticals by UV/H_2O_2. Water Research 2013; 47:5876–5888. https://doi.org/10.1016/j.watres.2013.07.008.

[20] Esplugas S, Bila DM, Krause LGT, Dezotti M. Ozonation and advanced oxidation technologies to remove endocrine disrupting chemicals (EDCs) and pharmaceuticals and personal care products (PPCPs) in water effluents. Journal of Hazardous Materials 2007; 149:631–642. https://doi.org/10.1016/j.jhazmat.2007.07.073.

[21] Ganiyu SO, van Hullebusch ED, Cretin M, Esposito G, Oturan MA. Coupling of membrane filtration and advanced oxidation processes for removal of pharmaceutical residues: A critical review. Separation and Purification Technology 2015; 156:891–914. https://doi.org/10.1016/j.seppur.2015.09.059.

[22] Buxton GV, Greenstock CL, Helman WP, Ross AB. Critical Review of rate constants for reactions of hydrated electrons, hydrogen atoms and hydroxyl radicals ($\cdot OH/\cdot O^-$ in Aqueous Solution. Journal of Physical and Chemical Reference Data 1988; 17:513–886. https://doi.org/10.1063/1.555805.

[23] Chou M-S, Chang K-L. Decomposition of aqueous 2,2,3,3-tetra-fluoro-propanol by UV/O_3 process. Journal of Environmental Engineering 2007; 133:979–986. https://doi.org/10.1061/(ASCE)0733-9372(2007)133:10(979).

[24] Miklos DB, Remy C, Jekel M, Linden KG, Drewes JE, Hübner U. Evaluation of advanced oxidation processes for water and wastewater treatment – A critical review. Water Research 2018; 139:118–131. https://doi.org/10.1016/j.watres.2018.03.042.

[25] Lin AY-C, Lin C-F, Chiou J-M, Hong PKA. O_3 and O_3/H_2O_2 treatment of sulfonamide and macrolide antibiotics in wastewater. Journal of Hazardous Materials 2009; 171: 452–458. https://doi.org/10.1016/j.jhazmat.2009.06.031.

[26] Kim I, Kim S, Lee H, Tanaka H. Effects of adding UV and H_2O_2 on the degradation of pharmaceuticals and personal care products during O_3 treatment. Environmental Engineering Research 2011; 16:131–136. https://doi.org/10.4491/eer.2011.16.3.131.

[27] Lester Y, Avisar D, Gozlan I, Mamane H. Removal of pharmaceuticals using combination of $UV/H_2O_2/O_3$ advanced oxidation process. Water Science and Technology 2011; 64:2230–2238. https://doi.org/10.2166/wst.2011.079.

[28] Giri RR, Ozaki H, Ota S, Takanami R, Taniguchi S. Degradation of common pharmaceuticals and personal care products in mixed solutions by advanced oxidation techniques. International Journal of Environmental Science and Technology 2010; 7:251–260. https://doi.org/10.1007/BF03326135.

[29] Illés E, Szabó E, Takács E, Wojnárovits L, Dombi A, Gajda-Schrantz K. Ketoprofen removal by O_3 and O_3/UV processes: Kinetics, transformation products and ecotoxicity. Science of the Total Environment 2014; 472:178–184. https://doi.org/10.1016/j.scitotenv.2013.10.119.

[30] Wang Y, Li H, Yi P, Zhang H. Degradation of clofibric acid by UV, O_3 and UV/O_3 processes: Performance comparison and degradation pathways. Journal of Hazardous Materials 2019; 379. https://doi.org/10.1016/j.jhazmat.2019.120771.

[31] Somathilake P, Dominic JA, Achari G, Langford CH, Tay J-H. Degradation of Carbamazepine by Photo-assisted Ozonation: Influence of Wavelength and Intensity of Radiation. Ozone: Science and Engineering 2018; 40:113–121. https://doi.org/10.1080/0191 9512.2017.1398635.

[32] Kim I, Tanaka H, Iwasaki T, Takubo T, Morioka T, Kato Y. Classification of the degradability of 30 pharmaceuticals in water with ozone, UV and H_2O_2. Water Science and Technology 2008; 57:195–200. https://doi.org/10.2166/wst.2008.808.

[33] Gerrity D, Lee Y, Gamage S, Lee M, Pisarenko AN, Trenholm RA, et al. Emerging investigators series: Prediction of trace organic contaminant abatement with UV/H_2O_2: Development and validation of semi-empirical models for municipal wastewater effluents. Environmental Science: Water Research and Technology 2016; 2:460–473. https://doi.org/10.1039/c6ew00051g.

[34] Ohtani B. AR. Will common sense prevail?: Reduction Mechanism of Oxygen in Photocatalytic Reactions. Chemistry, 2008; 63:19–23 (in Japanese).

[35] Kernazhitsky L, Shymanovska V, Gavrilko T, Naumov V, Fedorenko L, Kshnyakin V, et al. Room temperature photoluminescence of anatase and rutile TiO_2 powders. Journal of Luminescence 2014; 146:199–204. https://doi.org/10.1016/j.jlumin.2013.09.068.

[36] Hirakawa T., Mera N., Sano T., Negishi N. TK. Decontamination of chemical warfare agent by photocatalysis. Yakugaku Zasshi 2006; 129:71–92 (in Japanese).

[37] Watanabe T., Sunada K. HK. Bactericidal Effect Utilizing Titanium Dioxide Photocatalysis. Inorganic Materials 1999; 6:532–540 (in Japanese).

[38] Irie HHK. Current Status and future development of photocatalytic technology. Material 2002; 41:242–246 (in Japanese).

[39] Tanizakii T., Hashimoto A., Matsuoka Y., Ishikawa S., Hanada Y. KK and SR. Photodegradation of organic compounds in tap water using high reactive titanium dioxide. Journal of Environmental Chemistry, 2005; 15:4847–853.

[40] Loeb SK, Alvarez PJJ, Brame JA, Cates EL, Choi W, Crittenden J, et al. The technology horizon for photocatalytic water treatment: Sunrise or sunset? Environmental Science and Technology 2019; 53:2937–2947. https://doi.org/10.1021/acs.est.8b05041.

[41] Kanakaraju D, Glass BD, Oelgemöller M. Titanium dioxide photocatalysis for pharmaceutical wastewater treatment. Environmental Chemistry Letters 2014; 12:27–47. https://doi.org/10.1007/s10311-013-0428-0.

[42] Awfa D, Ateia M, Fujii M, Johnson MS, Yoshimura C. Photodegradation of pharmaceuticals and personal care products in water treatment using carbonaceous-TiO_2 composites: A critical review of recent literature. Water Research 2018; 142:26–45. https://doi.org/10.1016/j.watres.2018.05.036.

[43] Mahmoud WMM, Rastogi T, Kümmerer K. Application of titanium dioxide nanoparticles as a photocatalyst for the removal of micropollutants such as pharmaceuticals from water. Current Opinion in Green and Sustainable Chemistry 2017; 6:1–10. https://doi.org/10.1016/j.cogsc.2017.04.001.

[44] Tong AYC, Braund R, Warren DS, Peake BM. TiO_2-assisted photodegradation of pharmaceuticals—A review. Central European Journal of Chemistry 2012; 10:989–1027. https://doi.org/10.2478/s11532-012-0049-7.

[45] van Doorslaer X, Demeestere K, Heynderickx PM, van Langenhove H, Dewulf J. UV-A and UV-C induced photolytic and photocatalytic degradation of aqueous ciprofloxacin and moxifloxacin: Reaction kinetics and role of adsorption. Applied Catalysis B: Environmental 2011; 101:540–547. https://doi.org/10.1016/j.apcatb.2010.10.027.

[46] Lin HHH, Lin AYC. Photocatalytic oxidation of 5-fluorouracil and cyclophosphamide via UV/TiO_2 in an aqueous environment. Water Research 2014; 48:559–568. https://doi.org/10.1016/j.watres.2013.10.011.

[47] Lutterbeck CA, Machado TL, Kümmerer K. Photodegradation of the antineoplastic cyclophosphamide: A comparative study of the efficiencies of UV/H_2O_2, $UV/Fe^{2+}/H_2O_2$ and UV/TiO_2 processes. Chemosphere 2015; 120:538–546. https://doi.org/10.1016/j.chemosphere.2014.08.076.

[48] Im J-K, Son H-S, Kang Y-M, Zoh K-D. Carbamazepine degradation by photolysis and titanium dioxide photocatalysis. Water Environment Research 2012; 84:554–561. https://doi.org/10.2175/106143012X13373550427273.

[49] Martínez C, Canle L M, Fernández MI, Santaballa JA, Faria J. Kinetics and mechanism of aqueous degradation of carbamazepine by heterogeneous photocatalysis using nanocrystalline TiO_2, ZnO and multi-walled carbon nanotubes-anatase composites. Applied Catalysis B: Environmental 2011; 102:563–571. https://doi.org/10.1016/j.apcatb.2010.12.039.

[50] Yang L, Yu LE, Ray MB. Degradation of paracetamol in aqueous solutions by TiO_2 photocatalysis. Water Research 2008; 42:3480–3488. https://doi.org/10.1016/j.watres.2008.04.023.

[51] Lalhriatpuia C, Tiwari D, Tiwari A, Lee SM. Immobilized Nanopillars-TiO_2 in the efficient removal of micro-pollutants from aqueous solutions: Physico-chemical studies. Chemical Engineering Journal 2015; 281:782–792. https://doi.org/10.1016/j.cej.2015.07.032.

[52] Safari GH, Hoseini M, Seyedsalehi M, Kamani H, Jaafari J, Mahvi AH. Photocatalytic degradation of tetracycline using nanosized titanium dioxide in aqueous solution. International Journal of Environmental Science and Technology 2015; 12:603–616. https://doi.org/10.1007/s13762-014-0706-9.

[53] Martínez C, Canle L. M, Fernández MI, Santaballa JA, Faria J. Aqueous degradation of diclofenac by heterogeneous photocatalysis using nanostructured materials. Applied Catalysis B: Environmental 2011; 107:110–118. https://doi.org/10.1016/j.apcatb.2011.07.003.

[54] Uyguner-Demirel CS, Birben NC, Bekbolet M. Elucidation of background organic matter matrix effect on photocatalytic treatment of contaminants using TiO_2: A review. Catalysis Today 2017; 284:202–214. https://doi.org/10.1016/j.cattod.2016.12.030.

[55] Xi W, Geissen S-U. Separation of titanium dioxide from photocatalytically treated water by cross-flow microfiltration. Water Research 2001; 35:1256–1262. https://doi.org/10.1016/S0043-1354(00)00378-X.

[56] Thiruvenkatachari R, Vigneswaran S, Moon IS. A review on UV/TiO_2 photocatalytic oxidation process. Korean Journal of Chemical Engineering 2008; 25:64–72. https://doi.org/10.1007/s11814-008-0011-8.

[57] Zheng X, Shen Z-P, Shi L, Cheng R, Yuan D-H. Photocatalytic membrane reactors (PMRs) in water treatment: Configurations and influencing factors. Catalysts 2017; 7. https://doi.org/10.3390/catal7080224.

[58] Mozia S. Photocatalytic membrane reactors (PMRs) in water and wastewater treatment. A review. Separation and Purification Technology 2010; 73:71–91. https://doi.org/10.1016/j.seppur.2010.03.021.

[59] Kim J, Lee CW, Choi W. Platinized WO$_3$ as an environmental photocatalyst that generates OH radicals under visible light. Environmental Science and Technology 2010; 44:6849–6854. https://doi.org/10.1021/es101981r.

[60] Choi W, Termin A, Hoffmann MR. The role of metal ion dopants in quantum-sized TiO$_2$: Correlation between photoreactivity and charge carrier recombination dynamics. Journal of Physical Chemistry 1994; 98:13669–13679. https://doi.org/10.1021/j100102a038.

[61] Banerjee S, Pillai SC, Falaras P, O'shea KE, Byrne JA, Dionysiou DD. New insights into the mechanism of visible light photocatalysis. Journal of Physical Chemistry Letters 2014; 5:2543–2554. https://doi.org/10.1021/jz501030x.

[62] Benotti MJ, Stanford BD, Wert EC, Snyder SA. Evaluation of a photocatalytic reactor membrane pilot system for the removal of pharmaceuticals and endocrine disrupting compounds from water. Water Research 2009; 43:1513–1522. https://doi.org/10.1016/j.watres.2008.12.049.

[63] Kim I, Tanaka H. Photodegradation characteristics of PPCPs in water with UV treatment. Environment International 2009; 35:793–802. https://doi.org/10.1016/j.envint.2009.01.003.

[64] Carlson JC, Stefan MI, Parnis JM, Metcalfe CD. Direct UV photolysis of selected pharmaceuticals, personal care products and endocrine disruptors in aqueous solution. Water Research 2015; 84:350–361. https://doi.org/10.1016/j.watres.2015.04.013.

[65] Voigt M, Jaeger M. On the photodegradation of azithromycin, erythromycin and tylosin and their transformation products – A kinetic study. Sustainable Chemistry and Pharmacy 2017; 5:131–140. https://doi.org/10.1016/j.scp.2016.12.001.

[66] Benitez FJ, Real FJ, Acero JL, Roldan G. Removal of selected pharmaceuticals in waters by photochemical processes. Journal of Chemical Technology and Biotechnology 2009; 84:1186–1195. https://doi.org/10.1002/jctb.2156.

[67] Pereira VJ, Weinberg HS, Linden KG, Singer PC. UV degradation kinetics and modeling of pharmaceutical compounds in laboratory grade and surface water via direct and indirect photolysis at 254 nm. Environmental Science and Technology 2007; 41:1682–1688. https://doi.org/10.1021/es061491b.

[68] Sun P, Lee W-N, Zhang R, Huang C-H. Degradation of DEET and caffeine under UV/Chlorine and simulated sunlight/Chlorine conditions. Environmental Science and Technology 2016; 50:13265–13273. https://doi.org/10.1021/acs.est.6b02287.

[69] Guo H, Ke T, Gao N, Liu Y, Cheng X. Enhanced degradation of aqueous norfloxacin and enrofloxacin by UV-activated persulfate: Kinetics, pathways and deactivation. Chemical Engineering Journal 2017; 316:471–480. https://doi.org/10.1016/j.cej.2017.01.123.

[70] López-Peñalver JJ, Sánchez-Polo M, Gómez-Pacheco CV, Rivera-Utrilla J. Photodegradation of tetracyclines in aqueous solution by using UV and UV/H$_2$O$_2$ oxidation processes. Journal of Chemical Technology and Biotechnology 2010; 85:1325–1333. https://doi.org/10.1002/jctb.2435.

[71] Iovino P, Chianese S, Prisciandaro M, Musmarra D. Triclosan photolysis: Operating condition study and photo-oxidation pathway. Chemical Engineering Journal 2019; 377. https://doi.org/10.1016/j.cej.2019.02.132.

[72] Alfiya Y, Friedler E, Westphal J, Olsson O, Dubowski Y. Photodegradation of micropollutants using V-UV/UV-C processes; Triclosan as a model compound. Science of the Total Environment 2017; 601–602:397–404. https://doi.org/10.1016/j.scitotenv.2017.05.172.

8 Membrane Bioreactor for Wastewater Treatment

Mengying Yang and Xinwei Mao

CONTENTS

8.1 Introduction...133
8.2 Membrane Technology Based on Removal Mechanisms............................135
    8.2.1 Pressure-Driven Membrane Technology 135
    8.2.2 Concentration-Driven Membrane Technology 137
    8.2.3 Thermal-Driven Membrane Technology 138
    8.2.4 Electrical-Driven Membrane Technology...138
    8.2.5 Integrated/ Hybrid Membrane Technology in MBR..........................139
8.3 Membrane Materials and Modules...140
8.4 MBR Treating Wastewater: Water Types and MBR Configurations............141
    8.4.1 Industrial Wastewater versus Domestic Wastewater........................141
    8.4.2 Conventional MBR Configurations for Nutrients Removal
        from Wastewater 142
        8.4.2.1 Two-Chamber MBR...142
        8.4.2.2 Single-Chamber MBR 143
        8.4.2.3 MBB-MBR..143
        8.4.2.4 MABR...144
    8.4.3 Novel MBRs for Nutrients Removal from Wastewater 145
        8.4.3.1 Microalgae-MBR (MMBR)...145
        8.4.3.2 Bio-Electrochemical Membrane Bioreactor (BEC-MBR) 145
        8.4.3.3 Anaerobic MBR (AnMBR)..146
    8.4.4 Small-Scale MBR Applications..148
8.5 Challenges and Future Directions...149
References...150

8.1 INTRODUCTION

The membrane bioreactor (MBR) is a promising technology for developing new approaches to remove a variety of contaminants from wastewater. For this application, MBRs typically consist of (1) a bioreactor zone containing microorganisms with functional genes that encode enzymes responsible for the biological degradation of target contaminants into harmless byproducts, and (2) a membrane filtration process either as a separate compartment or immersed in the bioreactor that separates the solids (e.g., particles and microorganisms) from treated effluent.

DOI: 10.1201/9781003260738-8

MBRs have several advantages compared with conventional wastewater treatment approaches. The coupling of membrane filtration with biological processes eliminates the need for sedimentation after the bioreactor, reduces overall system footprint by 50–70% and improves the effluent quality [1–2]. MBR technology has gained greater acceptance for secondary wastewater treatment (e.g., biological oxygen demand (BOD)) with more MBR facilities installeyworldwide [3–6]. In recent years, more MBRs have been developed for removal of nutrients and contaminants of emerging concerns [7].

Two major configurations of MBRs used in municipal wastewater treatment are side-stream membrane bioreactors (sMBR) and immersed membrane bioreactors (iMBR). In an sMBR, the membrane module is external to the bioreactor, and the transmembrane pressure (TMP) and flow configuration create high crossflow velocity along the membranes. The crossflow velocity also serves as the principal mechanism to prevent the deposition of foulants on the membrane and reduce cake layer formation [8]. In contrast, the membrane module in an iMBR is immersed in the bioreactor, and the treated water is withdrawn by vacuum. Rather than relying on crossflow velocity, aeration typically is used to scour the surface of the membrane in the iMBR to reduce cake layer formation. Over the decades, studies showed that iMBRs have been more widely accepted for domestic wastewater, whereas sMBRs have been used more for low-solid and/or high-strength wastewater flows, such as the treatment of pharmaceutical wastewater, landfill leachate, paper-mill effluent, and dairy wastewater [3, 9–12].

MBRs also could be categorized based on the membrane type or shape, including flat sheet (FS), hollow fiber (HF), and multi-tubular (MT). A FS-MBR, for example, comprises single or multiple membrane sheets mounted to plates or panels. Water being treated passes through two adjacent membrane assemblies, and the treated water (permeate) flows through the channel provided by the plates/panels. An HF-MBR consists of a bundle of hollow fibers installed in a pressure vessel. Water is applied to the outside of the fiber. Finally, in a MT-MBR, membranes are placed inside a support composed of porous tubes. Water flow passes from the inside to the outside of the tube [8]. Based on different biological treatment process, MBRs can also be divided into aerobic and anaerobic MBRs with various configurations, such as two-chamber MBR, single-chamber MBR, moving-bed biofilm MBR, membrane-aerated biofilm reactor, to remove and recover nutrients [13].

Despite promise, the MBR approach also faces several challenges before wider application can occur, such as the high capital and operational costs, process complexity, and membrane fouling (e.g., cake layer formation and pore clogging) [14, 15]. In this chapter, we first introduce different MBR technologies based on the separation mechanisms. Then various membrane materials and modules are discussed, followed by an in-depth discussion of MBR treatment feasibility, configuration, and performance with different wastewater types. Finally, we highlight the challenges and opportunities related to the broader application of MBR for wastewater treatment (Figure 8.1).

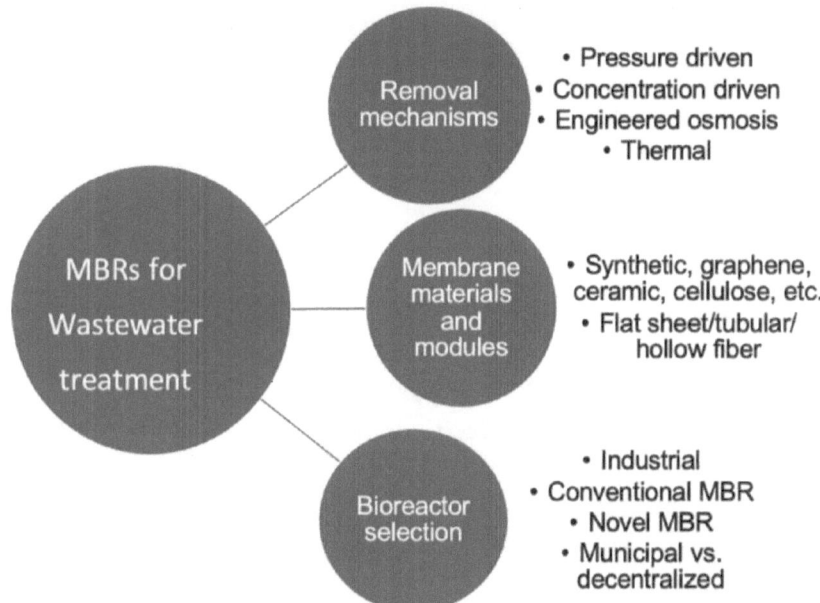

FIGURE 8.1   An overview of the workflow in this chapter.

Source: Original illustration.

8.2   MEMBRANE TECHNOLOGY BASED ON REMOVAL MECHANISMS

From the operation perspective, membrane techniques are a simple, economic, and easy operational method to separate unwanted pollutants from feed flow. To separate the organic and inorganic species through the membrane interphase, external driving force is generally required. To meet the diverse and complex MBR application, different membrane technology requires specific design of membrane and selection of suitable driving force and separation process. Based on the different types of driving force, membrane technology can be classified into pressure-driven, concentration-driven, thermal-driven, electrical driven, and hybrid technologies as shown in Figure 8.2 [8, 9] The mechanism, advantages and challenges of each type of membrane technology are described in following sections.

8.2.1   PRESSURE-DRIVEN MEMBRANE TECHNOLOGY

Pressure-driven membrane technology uses the differential pressure between two sides of membrane to promote the feed solution to pass through the membrane to the permeate phase. It is the most common technology in MBR applications with

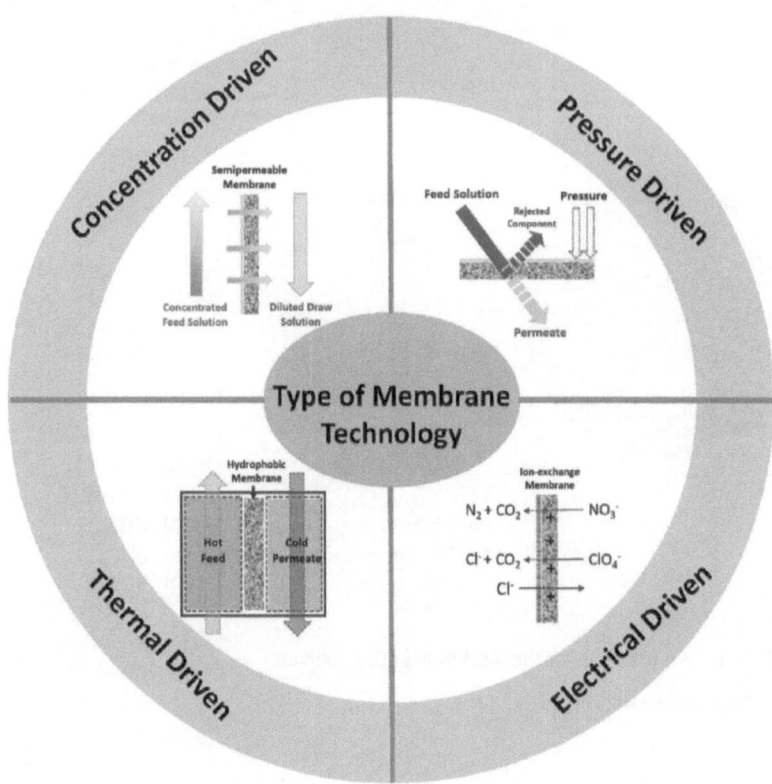

FIGURE 8.2 Types of membrane technology based on different driving forces: pressure driven membrane, concentration driven membrane, thermal driven membrane, and electrical driven membrane example were included inside the circle.

Source: Original illustration.

advantages including smaller footprint, less aeration-associated energy consumption, less sludge production, high operation stability, and maximized energy recovery [16]. The applied pressure usually determines the total operation cost of membrane separation process. Separation mechanism can include size exclusive, solution diffusion and electrostatic interactions [17]. Together with the membrane material and pore size, membrane separation performance and permeate quality can be also influenced by pre-treatment of feed water and fouling control practice [18]. According to the sieving effect of membranes, contaminant with different size can be selectively removed by tailoring the pore size of membrane.

Based on the applied pressure level and membrane pore size, membranes can be categorized into the microfiltration (MF), ultrafiltration (UF), nanofiltration (NF), and reverse osmosis (RO) membrane. MF and UF are extremely popular low-energy water filters. While the MF process can remove dissolved solids, turbidity, and microorganisms, the UF process can further separate the macromolecules, retain proteins, endotoxins, viruses, and silica. Wang et al. [19] used titanium dioxide (TiO_2) nanoparticles to modify the polyvinylidene fluoride (PVDF) UF membranes in anoxic–aerobic membrane bioreactor (A/O-MBR) process to treat landfill leachate. The modified membranes showed significantly improved antifouling performance in separation of mixed liquor suspended solids (MLSS). Due to the limitation of membrane pore size (>0.1 um), irons, dissolved substances, and smaller particles can still permeate through the MF and UF membranes. Therefore, to produce high-grade product water, NF and RO membranes are selected to reduce the biodegradable components, separate ions, and unwanted molecules. To achieve low salinity and produce high-grade permeate, Falizi et al [20]. evaluated the effluents quality desalinated by NF and RO process and compared it with the irrigation standards. It was found that RO-MBR mixture (2:1) was suitable for irrigation purposes because of the significant reduction in sodium and total dissolved solid (TDS) concentration by RO separation. For municipal wastewater reclamation, Recar et al [21]. evaluated the MBR-NF/RO process for reduction of contaminants of emerging concern from the Watch List in European Union Decision. While MBR efficiently removed methiocarb, tri-allate, clothianidin, and clarithromycin, only the permeate after NF/RO treatment satisfied the requirements for irrigation water reuse. The drawbacks of NF and RO include the requirement of high pressure, vulnerable membrane for fouling, and high operation cost since the flow needs to go against the concentration gradient by pressure.

8.2.2 Concentration-Driven Membrane Technology

Concentration-driven separation includes forward osmosis (FO), dialysis process, pervaporation, and gas separation. Driven by concentration gradient at isobaric and isothermal concentration, FO is more energy efficient than other conventional wastewater treatment methods. The FO-MBR combined process has advantages including higher contaminant rejection, lower operational hydraulic pressure, less frequent backwashing, and ease of fouling removal compared to pressure driven systems [22, 23].

FO processes can recover phosphorus from the feed water or remove trace organic contaminants from wastewater containing high total suspended solids. For example, Nawaz et al. [24], compared two kinds of FO membranes: (i) FS membrane made with cellulose tri-acetate active layer on polyester mesh and (ii) HF membrane constructed with polyamide skin layer. They found that FO-MBR flux can be influenced by osmotic backwashing, cross flow velocity of the feed solution (activated sludge), and particle size of activated sludge particles. While osmotic backwashing was not very effective for flux recovery, cross flow velocity needs to be optimized to maintain the permeate flux and particle size of sludge. Furthermore, FO-MBR has less irreversible membrane fouling than the RO-MBR, but the uncertain stability of membrane in high salinity environment can decrease the microbial kinetics and permeate flow [25].

Volatile components in liquid mixtures could be removed by gas separation process through the porous/nonporous membrane [26]. HF polymeric membrane is generally used. As the solution permeates the membrane due to concentration difference, vapor is formed and removed by either flowing inert medium or applying low pressure in the process. For example, gas-permeable membrane has been used in membrane-biofilm reactor (MBB-MBR) and or membrane-aerated biofilm reactor (MABR) for nitrogen removal [27]. Biofilms can naturally be formed on the gaseous substrate in a counter-diffusional manner. The membrane not only serves as a biofilm carrier but also as an oxygen supply material [28]. Compared to the traditional biofilm treatment method, MBB-MBR can treat a wide range of reduced, oxidized, and organic compounds [29]. Detailed discussion on the application of MBB-MBR and MABR will be provided in the following section.

8.2.3   THERMAL-DRIVEN MEMBRANE TECHNOLOGY

Thermal driven membrane technology has also been widely used in MBR applications. Membrane distillation (MD) is a thermally driven membrane separation process. It can be driven by the vapor pressure difference between hydrophobic membranes with micro pores and can be operated at low pressures (applied vacuum pressure is lower than the saturation vapor pressure) [30]. MD has advantages such as less capital cost, high permeate quality and less affected by feed salinity [31]. Moreover, MD separation has lower fouling potential compared to the high pressure-driven membrane process such as NF and RO when producing permeate of high quality. MD performance greatly depends on the intrinsic properties of membrane material. MD membranes are commonly fabricated by phase inversion method using PVDF, polypropylene (PP), polytetrafluoroethylene (PTFE) due to their intrinsic hydrophobic nature along with the good mechanical strength [32, 33]. MD has four major configurations: Direct contact membrane distillation (DCMD), air gap membrane distillation (AGMD), sweeping gas membrane distillation (SGMD), and vacuum membrane distillation (VMD). DCMD is the most widely used configuration in the desalination and concentration process due to the simple form [34]. Compared with other pressure-driven membrane bioreactors such as RO-MBR, MD-MBR has lower operation cost because of its stability and economical membrane selection (hydrophobic, microporous membrane). By utilizing the low-grade heat as a driven force, almost all macromolecules, colloids, volatile substances, salts (such as NaCl, $MgCl_2$, $CaCl_2$, CH_3COONa, and $Mg (CH_3COO)_2$) can be removed during the wastewater treatment. Shahzad et al. [34] investigated the application of DCMD in osmotic membrane bioreactor (Os-MBR) for concentrating the diluted synthetic wastewater. With optimized circulation velocity and operation temperature, DCMD with PTFE flat sheet membrane produced the high quality of permeate and achieved high draw solution recovery for aqueous salt solutions re-concentration in Os-MBR-MD system. The limitations of MD can be the relatively low permeate flux and water recovery ratio compared to a conventional desalination process like RO, high sensitivity to temperature polarization, high specific thermal energy consumption, proneness to membrane wetting, and significant heat loss in operation [35].

8.2.4   ELECTRICAL-DRIVEN MEMBRANE TECHNOLOGY

Electrodialysis is driven by electric potential through the ion exchange membrane (IEM). IEM can be classified based on different ionic charge groups in polymetric matrix. Cation exchange membranes contain negatively changed groups like sulfonate, phosphonate, or carboxyl groups while anion exchange membranes contain positively charged groups like quaternary ammonium bases or weakly basic secondary and tertiary amines [36]. IEMs can be fabricated using organic polymers, such as poly(styrene-co-divinylbenzene), poly(propylene) and poly(ethylene) or organic–inorganic composite materials with the membrane pore size ranges from a few nanometers to several micrometers [37, 38].

IEM can be used for desalination of seawater, pharmaceutical application, and organic acid removal in the food industry. In MBR application, nanoporous anion exchanger is used to remove nominal organic matter. Ion exchange membrane bioreactor (IEMBR) combines an ion-selective homogeneous membrane (nonporous) with bioreactor for nitrate removal [39]. IEMBR has great potential in removing not only macropollutants but also anionic micropollutants such as perchlorate and bromate. For example, Ricardo et al. [40] used a mixed anoxic microbial culture and ethanol as the carbon source and investigated the feasibility of IEMBR removing nitrate and perchlorate from drinking water. As a result of biofilm stratification, nitrate and perchlorate can be eliminated sequentially by heterotrophic denitrifying biofilm and perchlorate reducing biofilm separately. In addition, the electrical field can play a positive role in mitigating fouling in the MBR system. Wang et al. [41] demonstrated that irreversible membrane fouling can be reduced by locating membrane filtration in the electric field due to the decrease of soluble microbial products.

8.2.5 Integrated/ Hybrid Membrane Technology in MBR

To increase the separation efficiency and mitigate the membrane fouling, a novel hybrid membrane system has been designed and developed for MBR applications. It can be a combination of two or more membrane processes or a combination of the membrane process with other conventional processes. For example, the coagulation process is commonly combined with membrane filtration in MBR to increase the pollutant removal efficiency and diminish the membrane fouling. As an example, Park et al. [42] investigated the effect of coagulants (polyaluminium chloride and chitosan) addition on the membrane fouling and removal of pharmaceuticals and personal care products in MBR. They found the coagulation-MBR combination can reduce the membrane fouling, lengthen the membrane operation time, and improve the removal efficiency.

Prefiltration-membrane process is also popular in MBR design to extend the membrane lifespan by removing coarse materials using sand and packed bed materials [43]. For example, UF membrane can be integrated into the aeration tank to avoid the need for a clarifier and can improve the RO performance in the later step. Parlar et al. [44] demonstrated that the permeate quality increased regarding removing excess salinity from MBR effluent when NF pre-treatment was applied before the RO process due to the removal of organic/inorganic substances in the permeate in NF pre-treatment. Tay et al. [45] compared fouling mechanisms and foulant characteristics of NF-MBR+RO and UF-MBR+RO systems in treating domestic wastewater. It was found that NF-MBR achieved superior permeate quality compared to the UF-MBR because of the increased biodegradation and high rejection capacity of NF membrane. In the meantime, energy consumption between the NF-MBR+RO and

UF-MBR+RO system was comparable. Other integrated/hybrid MBR technologies will be further discussed in the following section.

8.3   MEMBRANE MATERIALS AND MODULES

FS and HF membranes are both broadly applied membrane modules in MBRs. Considering the high surface area and operational stability, HF module is most widely used configuration in MBR applications [46, 47] In terms of membrane materials, the most popular commercial membranes materials applied in MBRs are ceramic membranes, metallic membranes, polymeric membranes, and composite membranes (Figure 8.3). To meet the high-water flux requirement for MBR, membranes should be fabricated both to have a high surface porosity, or total surface pore cross-sectional area, and narrow pore size distribution to achieve ideal rejection efficiency as much as possible. On the other hand, the material of the membrane must have good resistance to thermal and chemical attack, temperature extremes and various pHs. In addition, to prolong the service life, membranes need strong mechanical strength to maintain structural integrity.

Ceramic and metallic membranes display good filtration performance with high chemical resistance and good durability to cleaning [48]. For example, Kimura et al. [49] investigated the influence of intensive mechanical cleaning on membrane antifouling performance in bench-scale MBR. While polymeric membranes cannot stand for the frequent chemically enhanced backwash, FS ceramic membranes were able to be stably operated for long periods under high flux conditions. However, the drawbacks like high cost and intrinsic property restriction (fragile nature for ceramic membranes) limit the large-scale application of ceramic and metallic membranes in wastewater treatment.

Commercialized membranes for wastewater treatment are predominantly made from synthetic organic polymers including PVDF, polyethersulfone (PES), polysulfone (PS), polyacrylonitrilic (PAN), polyethylene (PE), polyvinyl butyral (PVB), cellulose

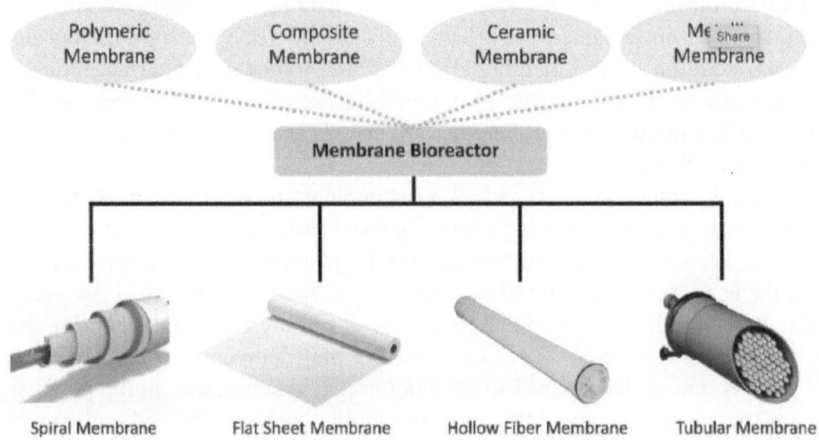

FIGURE 8.3   Typical membrane material and configuration used in MBR applications.

Source: Original illustration.

acetate (CA), polypropylene (PP), and PTFE due to their prominent mechanical properties, vast pore size variety, controllable preparation process, and consistent reproducibility [50–52]. However, polymer-based membranes commonly face problems such as the complicated and costly manufacturing process as well as their susceptibility to membrane fouling, which would require frequent maintenance [13, 53].

Composite membranes are typically composed of two or more materials: while one composite forms the support layer, and the others serve as the active surface layer [54, 55]. Considering hydrophobic polymer membranes are prone to adsorb foulants because of the lack of hydrogen bonding interactions between the membrane surface and water particles, some novel materials were used to design the performance-enhanced composite membranes for water treatment [56, 57]. Yang et al. [58] developed the nanostructured all-cellulose membrane for efficient ultrafiltration of wastewater. The hydrophilicity and negative surface charge of cellulose membrane mitigate the membrane fouling due to the weak attraction between the foulant and the hydrophilic groups on cellulose membrane surface. Wei et al. [59] prepared the multi-layered graphene oxide (GO) membrane with enhanced stability and long-term desalting performance. In the meantime, the authors achieved fast membrane preparation speed and extend the possibilities for GO membranes in practical applications.

Another important factor to consider when selecting the membrane material is to minimize the fouling potential [60]. For artificial polymers, the hydrophobic property makes them susceptible to fouling in the bioreactor liquors they are filtering, which makes the surface modification of the base material necessary and hence increase the total cost potentially. In addition to material properties, the pore size distribution of membranes that arises from phase inversion can also increase the fouling propensity. Large pores are likely to result in rapid fouling problems as more foulants can penetrate and remain within the membrane interior [61]. Membrane fouling challenge and mitigation strategy will be further discussed in the last section.

8.4   MBR TREATING WASTEWATER: WATER TYPES AND MBR CONFIGURATIONS

The selection of proper types of MBR for wastewater treatment depends on the wastewater quality and the treatment goal. Wastewater composition and biodegradability varies significantly between industrial wastewater and domestic wastewater. In this section, we will discuss how MBR was selected and applied in different types of wastewater treatment.

8.4.1   INDUSTRIAL WASTEWATER VERSUS DOMESTIC WASTEWATER

MBRs have been widely used in treating industrial wastewater, for example, MBR has been demonstrated in the treatment of wastewater containing pharmaceuticals and endocrine disruptors [62], separation of oil from aqueous media [63], degradation of phenol and phenolic compounds and removal of surfactants from wastewater [64, 65]. The major characteristics of industrial wastewater include high chemical oxygen demand (COD), color, toxicity, etc. Therefore, pre-treatment is needed before the

MBR treatment process to biodegrade the contaminants. Both aerobic and anaerobic MBRs have been used to treat industrial wastewater [66, 67]. On the other hand, MBRs have been applied in numerous domestic municipal wastewater treatment applications for organic matter (i.e., BOD) removal and pathogenic microorganisms (bacteria and virus) removal [68]. However, as more stringent water quality regulations are in place to control the nutrients loadings to bodies of water, the applications of MBR technology for large-scale tertiary domestic wastewater treatment (i.e., nutrients removal), as well as for decentralized and onsite systems, warrant more attention. In the following section, we will discuss the MRB applications in wastewater treatment processes for nutrients (N and P) removal and recovery based on various MBR configurations.

8.4.2 Conventional MBR Configurations for Nutrients Removal from Wastewater

8.4.2.1 Two-Chamber MBR

The sequencing anoxic/oxic MBR system is the most-commonly used MBR configuration for nitrogen removal from wastewater. The system consists of (i) a primary clarification tank prior to the MBR with or without the air supply and internal recycle (Figure 8.4); or (ii) a single chamber configuration, in which the primary clarification step is integrated into a single tank but separated from the aeration zone by baffles (Figure 8.4). Baffles can create alternative aerobic/anoxic conditions in separate zones that favor the nitrogen removal from domestic wastewater (Eq. 1–2). A continuous MBR system with a separate anoxic zone for denitrification and a zone for aeration is a common approach in upgraded nitrogen removal MBR systems in municipal

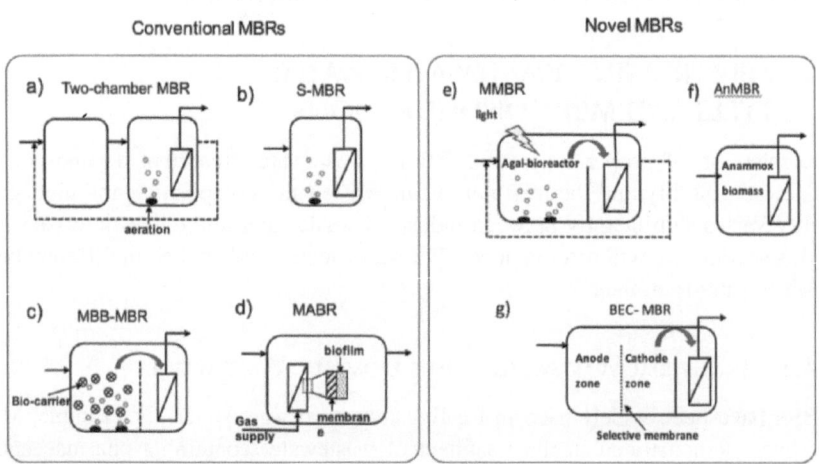

FIGURE 8.4   Conventional and novel MBR configurations for wastewater treatment (a) two-chamber MBR; (b) single chamber MBR for nitrogen removal; (c) moving-bed biofilm MBR; (d) membrane aerated biofilm reactor; e) microalgae MBR; (f) anaerobic MBR; and (g) bio-electrochemical MBR.

Source: Original illustration.

wastewater treatment plants [8, 69–71], where the process typically requires recycling of water/sludge from the aerobic zone to the anoxic zone. By changing the rate of sludge recycling, a high nitrogen removal efficiency (>90%) can be achieved [70–76].

$$NH_4^+ + 2O_2 \rightarrow NO_3^- + 2H_2O + 4H^+ \qquad \text{(Eq. 1)}$$

$$5C_6H_{12}O_6 + 24NO_3^- + 24H^+ \rightarrow 12N_2 + 42H_2O + 30CO_2 \qquad \text{(Eq. 2)}$$

A minimum COD to nitrogen ratio (g-COD/g-N, i.e., C:N ratio) of 3.5–4.5:1 is required to achieve adequate nitrogen removal [73–77]. A higher nitrogen removal rate (>90%) was observed when influent C: N ratio increased to above 10:1 [73, 74, 77]. In addition to nitrogen removal, two-chamber MBRs with anaerobic/oxic or anoxic/oxic zones have also been applied for phosphorus removal. Research suggests the internal liquid/sludge recycle ratio has a higher impact on phosphorus removal efficiency compared to nitrogen removal [69–71, 73, 78].

8.4.2.2 Single-Chamber MBR

Simultaneous biological nitrogen removal MBR (SBNR-MBR) consists of a single bioreactor with only one chamber, in which periodic aeration occurs and the effluent is withdrawn through a membrane module. An SBNR-MBR does not possess defined anaerobic and/or anoxic zones. Therefore, the reactor volume was greatly reduced (Figure 8.4). Hydraulic mixing and aeration provide cycling of the mixed liquor within the bioreactor, while diffusion resistance develops oxygen-sufficient and oxygen-deficient zones in the reactor to facilitate simultaneous nitrification and denitrification (SND) to occur [8, 79]. In addition to the small footprint, SBNR-MBR offers several other advantages, such as long solids retention time (SRT) to maintain the growth of the nitrification bacteria, flexibility of operating the system at various dissolved oxygen (DO) levels to facilitate SND, and simple system design and operation [8, 79–85]. Nitrogen removal efficiency by SBNR-MBR varies widely in the literature and ranges from 30% to 89%, depending on operating conditions [80–82, 84–85].

One critical factor affecting nitrogen removal using SBNR-MBR is the DO concentration, which is usually controlled via intermittent aeration [82, 83, 85]. At low DO levels, partial nitrification with denitrification might also occur within an SBNR-MBR process. Researchers have found that ammonium could be oxidized to nitrite and the resulting nitrite could be reduced to nitrogen gas by heterotrophic denitrification bacteria during the SBNR-MBR treatment [85, 86]. Incomplete nitrification in these studies was presumed to be caused by low DO levels, high ammonium concentration, or by inhibition from soluble microbial products (SMPs). SRT is another critical factor controlling the nitrogen removal performance in an SBNR-MBR. For example, SRT was found to have a direct effect on sludge age and granule/floc size and composition, which would influence the nitrification and denitrification kinetics [81]. SBNR-MBR could also be integrated with a moving bed biofilm (MBB) or a MABR.

8.4.2.3 MBB-MBR

The moving bed biofilm reactor (MBBR) is a technology that incorporates both suspended and attached growth activated sludge as biofilm carriers in a mixed motion within a wastewater treatment basin (Figure 8.4) [87]. These biofilm carriers could increase the biomass concentration by providing more surface area favorable for the selective development of slow-growing microorganisms [88]. The high population density of bacteria not only ensures high biodegradation rates within the system, but also offers higher treatment reliability and ease of operation by retaining high biomass concentrations, with less membrane fouling [89–91]. Other advantages of MBB-MBR for nutrient removal include no requirement for recycling the activated sludge stream, flexibility for increased loading, less time for establishment of enriched microbial populations and protection of the microbial community from disruption in case of high substrate loading rates [87, 92, 93]. The co-existence of outer zone and inner anoxic zone in the biofilm can also promote SND in the MBB-MBR single reactor [94].

Two-chamber MBB-MBR was the most reported configuration and has been used for treating high strength wastewater, such as anaerobic digestion reject-water and municipal wastewater [93–97]. Enhanced nitrogen removal (61–82%) has been reported for MBB-MBRs mainly because of SND [90, 96–99]. A relatively higher abundance of nitrifying bacteria in the microbial community was found in the carrier biofilms [96]. Adhesion characteristics, such as roughness of the carrier surface, and protein and polysaccharide concentration were found to be important in biofilm stability on the carrier surface [91]. Studies on membrane fouling during MBB-MBR have shown contradictory results. Slower membrane fouling in MBB-MBRs was observed in studies where low levels of SMPs were released compared to conventional MBRs, [88, 90] while other studies found more severe membrane fouling on the membrane surface due to the formation of a thick cake layer and the presence of more filamentous bacteria in the suspended solids [98, 100] Commercialized plastic carriers (e.g. polyethylene) have been applied in municipal and decentralized wastewater treatment processes as well as novel types of carriers, such as biodegradable polymer (PCL) [89].

8.4.2.4 MABR

The membrane aerated biofilm reactor (MABR) is a type of biofilm-based bioreactor, which contains both biofilm and membrane in a single bioreactor (Figure 8.4). Instead of using the membrane as a solid/liquid separation unit for the effluent, a hydrophobic permeable membrane is used to support the biofilm growth and deliver gaseous electron donor (e.g., hydrogen and methane) or electron acceptor (e.g., oxygen) directly to the biofilm in MABR. This configuration can greatly improve the substrate utilization efficiency [101]. Air or oxygen is supplied to the reactor through the pores of the membrane directly to the biofilm without the formation of bubbles, providing up to 100% gas transfer [102]. The MABR can be in different configurations, including shell, tube, HF, plate or frame configurations [102, 103]. The unique feature of MABR is that the membrane does not serve as the filtration unit as in conventional MBRs, and it does not function to retain the biomass in the reactor [8].

One advantage of using MABR for nitrogen removal is the significant increase in nitrification rate due to the thin biofilms and high bulk liquid DO concentration [102]. Nitrifying bacteria, which grow much slower than BOD degraders, can be more easily maintained in biofilm. Since oxygen and nutrients are provided from two opposite sides of the biofilm, higher nitrification and denitrification activity can be obtained compared to conventional biofilm reactors [104]. Another advantage of MABR is that efficient nitrogen removal can be achieved at low COD:TN (total nitrogen) ratios. A few studies reported 80~100% TN removal in domestic wastewater treatment using MABRs at low COD:TN ratios (1~4) through sequential nitrification/denitrification by optimized oxygen supply [103, 105]. Hydraulic retention time also plays an important role in nitrogen removal within an MABR. Hu, et al. [105], for example, found that nitrogen removal efficiency decreased significantly when HRT was reduced. This result was due to the high organic loading rate and excessive growth of biomass on the membrane.

8.4.3 Novel MBRs for Nutrients Removal from Wastewater

8.4.3.1 Microalgae-MBR (MMBR)

Algae can grow on available nutrients (nitrogen and phosphorus) in wastewater and the resulting biomass can be used to produce biodiesel, high-value chemicals, and/or agricultural products [106, 107]. In a photobioreactor, the main mechanism for nutrient removal is through the assimilation of nutrients by microalgae cells [108–110]. One of the challenges of using photobioreactors to treat wastewater is the dilute microalgae biomass maintained in the reactor, which may limit the treatment efficiency [108, 109]. The use of a microalgae membrane bioreactor (MMBR) could decouple the hydraulic retention time (HRT) and biomass retention time (i.e., SRT), which enables higher microalgae concentrations and makes downstream algae harvesting and treatment more efficient [108, 111–113]. A typical MMBR configuration contains: the main photobioreactor, a membrane module, a light provision system and the gas supplementation system (Figure 8.4). In most MMBR studies, the membrane module was immersed into the wastewater to separate the effluent and the biomass [108, 109] while other studies separated the photobioreactor and the MBR as independent treatment units [114]. HF and FS membranes are commonly used in different membrane modules for MMBRs [108, 112–114]. Internal cycling was also applied to the MMBR system in some studies to increase the mixing and improve the algal productivity and settleability [114].

The use of MMBR has been well investigated at lab-scale experiments and has only been examined recently in pilot-scale experiments for removal of nutrients from secondary sewage effluent [108–111]. Studies of using MMBR to treat primary effluent and synthetic domestic wastewater have also been reported [112, 114]. In addition, MMBR studies using both pure algae cultures (e.g., Chlorella vulgaris) and mixed algae cultures have been examined with respect to the nitrogen removal performance [108–110, 112] There are other studies that explored a mixed bacteria-microalgae inoculum and demonstrated enhanced nutrient removal. However, the complicated intra-species relationship among algae and bacteria seemed difficult to make the system run at a steady state [111, 114].

8.4.3.2 Bio-Electrochemical Membrane Bioreactor (BEC-MBR)

Microbial electrochemical technology is promising for its potential to produce electricity through the microbial metabolism of wastewater [115]. The bio-electrochemical membrane bioreactor (BEC-MBR) is an approach that incorporates the benefit of membrane processes, electrochemical processes, and biological processes to treat various types of wastewaters along with electricity production. In a BEC-MBR, the microbial fuel cell (MFC) unit is integrated with an MBR. The membrane unit can be installed externally to the MFC unit for liquid/solid separation or can be incorporated into the MFC cathode chamber (Figure 8.4) [116–120]. The anode and cathode compartments can be integrated in a single chamber or can be separated by a selective membrane, such as FO membranes, cation exchange membranes, proton exchange membranes or a stainless-steel separator [116, 118–121].

In a conventional BEC-MBR, organic carbon degradation takes place in the anode chamber (anoxic zone), while the cathode chamber (aerobic zone) is aerated to facilitate oxygen reduction on the cathode surface and generate electricity. However, higher energy may be consumed rather than recovered from wastewater due to the aeration process [121]. Nitrogen removal from BEC-MBRs has been studied in modified system configurations to promote autotrophic and heterotrophic denitrification in the cathode chamber [116, 117, 119, 122, 123]. Biofilms were developed on the surface of various types of cathodes. This process enabled autotrophic denitrifying bacteria to occupy the inner biofilm and utilize the electrode as the electron donor to reduce nitrate/nitrite, where nitrification bacteria were dominant in the outer layer of the biofilm and the bulk medium that could oxidize ammonium [117, 118, 123]. In lab-scale BEC-MBR studies, various materials have been investigated to serve both as a cathode for redox reactions and a filtration membrane, including electrically conductive UF or MF membranes [124, 125]. non-woven cloth separators [126], carbon MF membranes [127], and stainless-steel mesh with biofilm on its surface [123]. The high cost of membranes and the energy demand for aeration are two other limiting factors for large-scale implementation of BEC-MBR in wastewater treatment [115, 121, 125].

8.4.3.3   Anaerobic MBR (AnMBR)

An anaerobic MBR (AnMBR) operates without oxygen supply. This technology appears to be suitable for treatment of high carbon concentration wastewater, such as industrial wastewater from food processing and landfill leachate [128, 129]. Although AnMBR offers a few advantages over conventional aerobic processes, such as lower energy requirement, less biomass production, and generation of valuable biogas (e.g., methane), AnMBR treatment is susceptible to more serious membrane fouling compared to aerobic MBR [128–130]. Most AnMBR studies have been conducted in lab-scale systems, though a few case studies reported the application of AnMBR for domestic wastewater treatment at the pilot-scale [130, 131].

Most AnMBR studies for municipal wastewater treatment focused on COD removal and biogas production from high COD strength wastewater, while complete autotrophic nitrogen removal over nitrite (CANON) could be a promising solution for nutrient removal using an AnMBR [132]. Lab-scale studies have shown MBR to be a suitable experimental setup for the operation of the CANON process [132]. In this process, ammonium is first oxidized into nitrite by ammonium-oxidizing bacteria (AOB) (Eq. 3). Subsequently, nitrite and remaining ammonium are converted to nitrogen gas by anaerobic ammonium-oxidizing bacteria (AnAOB) (Eq. 4). This anaerobic ammonium oxidation (ANAMMOX) process is an innovative technological advancement in nitrogen removal from wastewater. The combined partial nitrification, ANAMMOX and denitrification process has also been applied to remove COD and nitrogen simultaneously from wastewater using intermittently aerated MBRs [133, 134]. Another application of

AnMBR for nitrogen removal is to grow microbial communities that are dominated by AnAOB. For example, van der Star et al. [135] reported that high purity of AnAOB (> 97%) in the biomass could be enriched in an AnMBR. The system start-up with MBR appeared to be more effective at shorter time periods compared to a sequencing batch reactor (SBR) for the enrichment of ANAMMOX bacteria due to the higher sludge retention time achieved in the MBR [136,137].

$$NH_4^+ + O_2 \rightarrow NO_2^- + H_2O + 2H^+ \qquad \text{(Eq. 3)}$$

$$NH_4^+ + NO_2^- \rightarrow N_2 + 2H_2O \qquad \text{(Eq. 4)}$$

A brief comparison of the MBR configurations is shown in Table 8.1. Two-chamber MBRs and single-chamber MBRs are the two common MBR configurations used

TABLE 8.1
A Summary of MBRs Used in Wastewater Treatment Processes

Type of MBR	Scale	Membrane module	Advantages	Major limitations	Reference
Two-chamber MBR	Lab/pilot/full	Flat sheet Hollow fiber	Easy to upgrade from conventional activated sludge system	May need external carbon and internal sludge recycle	[69–76]
S-MBR	Lab/pilot/full	Hollow fiber	Small footprint	May need external carbon and intermittent aeration for nutrients removal	[81–85]
MBB-MBR	Lab/pilot	Hollow fiber	Retain high biomass, higher treatment reliability and ease of operation with flexible loading, no requirement for recycling	Membrane fouling may be more severe than conventional MBRs	[87–98]
MABR	Lab/pilot	Hollow fiber	Improved oxygen utilization efficiency, Efficient nitrogen removal at low carbon to nitrogen ratios, potential to be integrated with AnMBRs	Biofilm management is critical to maintain high flux and nutrients removal performance	[101–105]
MMBR	Lab/pilot	Flat sheet Hollow fiber	Decouples HRT and SRT, which enables higher microalgae concentrations for harvesting	Difficult to maintain the system at steady state due to the complexity of the intraspecies relationship among algae and bacteria	[108–114]

(Continued)

TABLE 8.1
Continued

Type of MBR	Scale	Membrane module	Advantages	Major limitations	Reference
BEC-MBR	Lab	Electrically conductive UF/MF membranes	Potential electricity production, potential lower energy cost, potential lower membrane fouling	High cost of the membrane, energy demand for aeration.	[116–124]
AnMBR	Lab	Hollow fiber	Lower energy requirement, less biomass production, potential to decrease membrane fouling, ideal to grow AnAOB, potential to be integrated with MABR	Hard to achieve steady performance when treating real wastewater, most of studies are lab-scale	[131–137]

in full scale applications, while other novel MBRs have been proof-concepted in lab-scale and are being demonstrated in pilot-scale tests. Each type of MBR configuration has its unique advantages and major limitations. The selection of the MBR configuration is also affected by the influent water composition, membrane fouling potentials, treatment goal (i.e., effluent quality), and cost/benefit considerations.

8.4.4 SMALL-SCALE MBR APPLICATIONS

To date, only a limited number of studies have investigated the applications of MBR for small-scale, decentralized wastewater treatments. In contrast to larger systems, the design of small-scale, decentralized MBR systems must consider the greater fluctuations in wastewater flow and composition, environmental perturbations, and the decentralized production of a concentrate stream that requires disposal [138, 139]. Reduction in operation complexity, process reliability, and energy consumption are also critical challenges in the design for decentralized wastewater treatment [140, 141]. Membrane technologies have been used in source separation [142, 143], as well as grey and black water treatment for single houses [144–146]. MBR has been utilized to achieve nitrogen removal from septic tank effluent and it has been applied to small, decentralized communities [138–141]. The high quality of MBR treated effluent (elimination of pathogens) makes it feasible to consider various reuse options [139, 146, 147]

Two-chamber iMBR with intermittent aeration, which provides for anoxic/oxic zones for nitrogen removal, have been explored for applications in single homes or small clusters of homes [138–140, 145–147]. Like large-scale wastewater treatment systems, some small-scale treatment facilities applied internal recycling of sludge (30%- 300%) and liquid to improve nutrient removal [138, 139, 141]. MBB-MBRs have also been used to enhance biomass concentration in a system designed for grey water treatment [146]. Both FS and HF membranes have been used for filtration [138, 139, 145, 146]. Various types of membrane materials have been used for small-scale MBRs, including polyelectrolyte complex (PEC), PVDF, microporous ceramic filter, and nonwoven fabric bag [141, 148–150]. However, membrane fouling has not been well documented or characterized in small-scale MBR systems.

8.5   CHALLENGES AND FUTURE DIRECTIONS

Significant challenges remain for the wide implementation and adoption of membrane technology for wastewater treatment (Figure 8.5). First, the operational control of MBR systems can be challenging, especially when considering nutrients removal. For example, the application of SND is effective only within a narrow range of carbon-to-nitrogen ratio and certain range of dissolved oxygen concentrations. Other factors related to wastewater streams, such as variations in feed composition and flow rate, also make operational control of MBRs challenging. In addition, the use of MBR for small-scale or decentralized wastewater treatment can be particularly difficult due to the fluctuation of wastewater composition and various loadings at these scales (e.g., weekend versus weekday, vacations, etc.). Many new technologies and approaches offer promise for better operational control of MBR systems. Significant efforts have been made to translate existing bioprocess models (e.g., activated sludge model and associated models) for the prediction of nutrient removal using MBR. As these models develop, they offer the potential for increasing our understanding of how various process variables affect performance, information that can then be

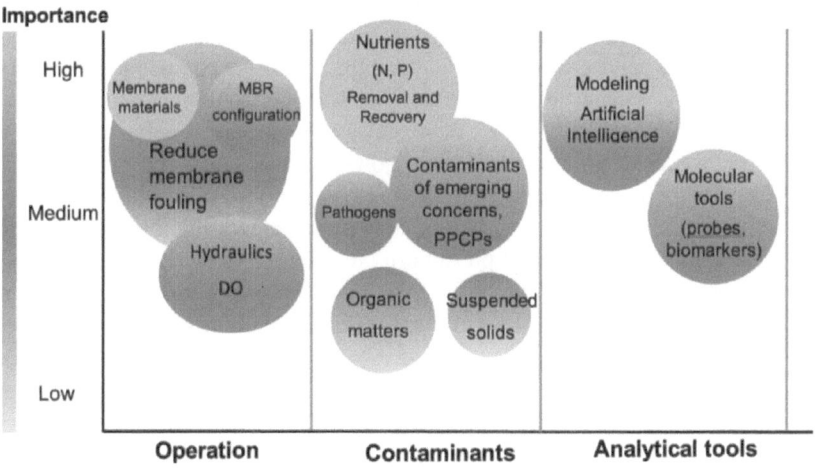

FIGURE 8.5   A bubble graph illustrating the significance of each contributing factor to the MBR technology in wastewater treatment.

Source: Original illustration.

integrated into control algorithms. New and more powerful tools (e.g., molecular tools, probes, microscopy, etc.) for understanding biofilms and relating the structure of biofilms to process performance are also emerging. These tools are providing unprecedented insight into the structure and function of biofilms, information that will be valuable to the operation and control of MBR systems. New developments in data analytics and artificial intelligence also hold promise for applications in the operation and control of MBR systems [152].

High cost is the other challenge MBR is facing for greater adoption of wastewater treatment. MBR technology remains more expensive compared to conventional activated sludge processes. The greater cost of MBR systems is due to the cost of installing, maintaining, and replacing membranes, the need for more extensive pre-treatment, higher degree of automation and higher energy requirements [151]. Understanding and preventing membrane fouling in MBR systems remains particularly difficult. The complexity of MBR systems for nutrient removal, in terms of microbial community structure, soluble product composition, and potential for biofilm formation, make prediction of membrane fouling particularly difficult. While significant progress has been made to understand fouling in such systems, additional research is needed to gain further insight and enable the translation of these insights into process improvements to minimize or eliminate fouling. Research progress continues to be made in developing new membrane materials (e.g., nanocellulose, graphene oxides) with novel functionalities (e.g., biomimetic, self-cleaning) and/or surface properties (super hydrophilic) that reduce fouling. New membrane cleaning approaches will also help spur improvements in MBR technology and the wider adoption of this approach for nutrient removal.

It is also important to highlight the changing nature of wastewater discharges, and in particular, the presence of pharmaceuticals and personal care products. The occurrence of these compounds in wastewater is rapidly changing as new chemicals, products, and therapies are developed. The impact of these compounds of microbial processes during wastewater treatment is just starting to be understood. Also, our understanding related to the ability of wastewater treatment systems and MBR systems to remove these compounds is also evolving. MBR technology not only offers a platform for the development of nutrient removal processes but may also provide approaches for nutrient recovery processes (e.g., by integrating ion exchange materials for ammonia recovery and phosphorus recovery). As materials and resources become more constrained in the future, MBR holds promise for the development of approaches such that wastewater can more effectively be utilized as a resource of water, nutrients and energy.

REFERENCES

[1] American Membrane Technology Association. *Membrane bioreactors (MBR)*. Florida: Stuart, 2013.
[2] Neoh, CH, Noor, ZZ, Mutamim, NSA, Lim, CK. Green technology in wastewater treatment technologies: Integration of membrane bioreactor with various wastewater treatment systems. Chem. Eng. J 2016; 283: 582–94.

[3] Andersson SP, Mberg EK, Grundestam J, Lindblom E. Extension of two large wastewater treatment plants in Stockholm using membrane technology. Water Pract. Technol 2016; 11 (4): 744–53.

[4] Judd S. The status of membrane bioreactor technology. Trends Biotechnol 2008; 26 (2): 109–16.

[5] Kraemer JT, Menniti AL, Erdal ZK, Constantine TA, Johnson BR, Daigger GT, Crawford GV. A practitioner's perspective on the application and research needs of membrane bioreactors for municipal wastewater treatment. Bioresour. Technol 2012; 122: 2–10.

[6] Krzeminski P, Leverette L, Malamis S, Katsou E. Membrane bioreactors—A review on recent developments in energy reduction, fouling control, novel configurations, LCA and market prospects. J. Membr. Sci 2017; 527: 207–27.

[7] Racar M, Dolar D, Karadakić K, Čavarović N, Glumac N, Ašperger D, Košutić K. Challenges of municipal wastewater reclamation for irrigation by MBR and NF/RO: Physico-chemical and microbiological parameters, and emerging contaminants. Sci Total Environ 2020; 722: 137959.

[8] Judd, S., Judd, C. *The MBR book principles and applications of membrane bioreactors for water and wastewater treatment introduction.* Oxford: Elsevier, 2011.

[9] Gander M, Jefferson B, Judd S. Aerobic MBRs for domestic wastewater treatment: A review with cost considerations. Sep. Purif. Technol 2000; 18 (2): 119–30.

[10] Yang W, Cicek N, Ilg J. State-of-the-art of membrane bioreactors: Worldwide research and commercial applications in North America. J. Membr. Sci 2006; 270 (1): 201–11.

[11] Buer T, Cumin J. MBR module design and operation. Desalination 2010; 250 (3): 1073–77.

[12] Falahti-Marvast H, Karimi-Jashni A. Performance of simultaneous organic and nutrient removal in a pilot scale anaerobic–anoxic–oxic membrane bioreactor system treating municipal wastewater with a high nutrient mass ratio. Int. Biodeterior. Biodegrad 2015; 104: 363–70.

[13] Mao X, Myavagh PH, Lotfikatouli S, Hsiao BS, Walker JHW. Membrane bioreactors for nitrogen removal from wastewater: A review. J Environ Eng 2020; 146 (5): 03120002

[14] Kraemer, J. T., A. L. Menniti, Z. K. Erdal, T. A. Constantine, B. R. Johnson, G. T. Daigger, G. V. Crawford. A practitioner's perspective on the application and research needs of membrane bioreactors for municipal wastewater treatment. Bioresour. Technol 2012; 122: 2–10.

[15] Qu, F., Liang, H., Zhou, J., Nan, J., Shao, S., Zhang, J., Li, G. Ultrafiltration membrane fouling caused by extracellular organic matter (EOM) from Microcystis aeruginosa: effects of membrane pore size and surface hydrophobicity. Journal of Membrane Science, 2014; 449: 58–66

[16] Li Y, Sim LN, Ho JS, Chong TH, Wu B, Liu Y. Integration of an anaerobic fluidized-bed membrane bioreactor (MBR) with zeolite adsorption and reverse osmosis (RO) for municipal wastewater reclamation: Comparison with an anoxic-aerobic MBR coupled with RO. Chemosphere 2020; 245: 125569.

[17] Echevarría C, Valderrama C, Cortina JL, Martín I, Arnaldos M, Bernat X, De la Cal A, Boleda MR, Vega A, Teuler A, Castellví E. Hybrid sorption and pressure-driven membrane technologies for organic micropollutants removal in advanced water reclamation: A techno-economic assessment. Journal of Cleaner Production 2020; 273: 123108.

[18] Hube S, Eskafi M, Hrafnkelsdottir KF, Bjarnadottir B, Bjarnadottir MA, Axelsdottir S, Wu B. Direct membrane filtration for wastewater treatment and resource recovery: A review. Sci Total Environ 2020; 710: 136375.

[19] Wang H, Ding K. Effect of Self-Made TiO_2 Nanoparticle Size on the Performance of the PVDF Composite Membrane in MBR for Landfill Leachate Treatment. Membranes (Basel) 2022; 12(2).

[20] Falizi NJ, Hacıfazlıoğlu MC, Parlar İ, Kabay N, Pek TÖ, Yüksel M. Evaluation of MBR treated industrial wastewater quality before and after desalination by NF and RO processes for agricultural reuse. Journal of Water Process Engineering 2018; 22: 103–8.

[21] Racar M, Dolar D, Karadakic K, Cavarovic N, Glumac N, Asperger D, Kosutic K. Challenges of municipal wastewater reclamation for irrigation by MBR and NF/RO: Physico-chemical and microbiological parameters, and emerging contaminants. Sci Total Environ 2020; 722: 137959.

[22] Aftab B, Khan SJ, Maqbool T, Hankins NP. High strength domestic wastewater treatment with submerged forward osmosis membrane bioreactor. Water Sci Technol 2015; 72(1): 141–9.

[23] Nawaz MS, Parveen F, Gadelha G, Khan SJ, Wang R, Hankins NP. Reverse solute transport, microbial toxicity, membrane cleaning and flux of regenerated draw in the FO-MBR using a micellar draw solution. Desalination 2016; 391: 105–11.

[24] Nawaz MS, Parveen F, Khan SJ, Hankins NP. Impact of osmotic backwashing, particle size distribution and feed-side cross-flow velocity on flux in the forward osmosis membrane bioreactor (FO-MBR). Journal of Water Process Engineering 2019; 31: 100861.

[25] Mulligan CN, Yong RN, Gibbs BF. Surfactant-enhanced remediation of contaminated soil: A review. Engineering Geology 2001; 60(1): 371–80.

[26] Abedini R, Nezhadmoghadam A. Application of membranes in gas separation processes: Its suitability and mechanisms. Pet Coal 2010; 52: 69–80.

[27] Nerenberg R. The membrane-biofilm reactor (MBfR) as a counter-diffusional biofilm process. Current Opinion in Biotechnology 2016; 38: 131–6.

[28] Wu Y, Wu Z, Chu H, Li J, Ngo HH, Guo W, Zhang N, Zhang H. Comparison study on the performance of two different gas-permeable membranes used in a membrane-aerated biofilm reactor. Sci Total Environ 2019; 658: 1219–27.

[29] Valladares Linares R, Fortunato L, Farhat NM, Bucs SS, Staal M, Fridjonsson EO, Johns ML, Vrouwenvelder JS, Leiknes T. Mini-review: Novel non-destructive in situ biofilm characterization techniques in membrane systems. Desalination and Water Treatment 2016; 57(48–49): 22894–901.

[30] Alsaadi AS, Francis L, Amy GL, Ghaffour N. Experimental and theoretical analyses of temperature polarization effect in vacuum membrane distillation. Journal of Membrane Science 2014; 471: 138–48.

[31] Luo H, Wang Q, Zhang TC, Tao T, Zhou A, Chen L, Bie X. A review on the recovery methods of draw solutes in forward osmosis. Journal of Water Process Engineering 2014; 4: 212–23.

[32] Choudhury MR, Anwar N, Jassby D, Rahaman MS. Fouling and wetting in the membrane distillation driven wastewater reclamation process—A review. Adv Colloid Interface Sci 2019; 269: 370–99.

[33] Tibi F, Charfi A, Cho J, Kim J. Fabrication of polymeric membranes for membrane distillation process and application for wastewater treatment: Critical review. Process Safety and Environmental Protection 2020; 141: 190–201.

[34] Shahzad MA, Khan SJ, Siddique MS. Draw solution recovery using direct contact membrane distillation (DCMD) from osmotic membrane bioreactor (Os-MBR). Journal of Water Process Engineering 2019; 30: 100484.

[35] Ghaffour N, Soukane S, Lee JG, Kim Y, Alpatova A. Membrane distillation hybrids for water production and energy efficiency enhancement: A critical review. Applied Energy 2019; 254: 113698.

[36] Pismenskaya N, Tsygurina K, Nikonenko V. Recovery of nutrients from residual streams using ion-exchange membranes: Current state, bottlenecks, fundamentals and innovations. Membranes 2022; 12(5).

[37] Jiang S, Sun H, Wang H, Ladewig BP, Yao Z. A comprehensive review on the synthesis and applications of ion exchange membranes. Chemosphere 2021; 282: 130817.

[38] Ran J, Wu L, He Y, Yang Z, Wang Y, Jiang C, Ge L, Bakangura E, Xu T. Ion exchange membranes: New developments and applications. Journal of Membrane Science 2017; 522: 267–91.

[39] Al-Rashed WS, Lakhouit A. Nitrate removal from drinking water using different reactor/membrane types: A comprehensive review. Int. Res. J. Pub. Environ. Health 2022; 9(1): 1–9.

[40] Ricardo AR, Carvalho G, Velizarov S, Crespo JG, Reis MAM. Kinetics of nitrate and perchlorate removal and biofilm stratification in an ion exchange membrane bioreactor. Water Research 2012; 46(14): 4556–68.

[41] Wang Y, Jia H, Wang J, Cheng B, Yang G, Gao F. Impacts of energy distribution and electric field on membrane fouling control in microbial fuel cell-membrane bioreactor (MFC-MBR) coupling system. Bioresource Technology 2018; 269: 339–45.

[42] Park J, Yamashita N, Tanaka H. Membrane fouling control and enhanced removal of pharmaceuticals and personal care products by coagulation-MBR. Chemosphere 2018; 197: 467–76.

[43] Bera SP, Godhaniya M, Kothari C. Emerging and advanced membrane technology for wastewater treatment: A review. J Basic Microbiol 2022; 62(3–4): 245–59.

[44] Parlar I, Hacıfazlıoğlu M, Kabay N, Pek TÖ, Yüksel M. Performance comparison of reverse osmosis (RO) with integrated nanofiltration (NF) and reverse osmosis process for desalination of MBR effluent. Journal of Water Process Engineering 2019; 29: 100640.

[45] Tay MF, Liu C, Cornelissen ER, Wu B, Chong TH. The feasibility of nanofiltration membrane bioreactor (NF-MBR)+reverse osmosis (RO) process for water reclamation: Comparison with ultrafiltration membrane bioreactor (UF-MBR)+RO process. Water Research 2018; 129: 180–9.

[46] Liu M, Yang M, Chen M, Yu D, Zheng J, Chang J, Wang X, Ji C, Wei Y. Numerical optimization of membrane module design and operation for a full-scale submerged MBR by computational fluid dynamics. Bioresource Technology 2018; 269: 300–8.

[47] Ghernaout, D. New Configurations and Techniques for Controlling Membrane Bioreactor (MBR) Fouling. Open Access Library Journal, 2020; 7: 1–18.

[48] Singh H, Saxena P, Puri YM. The manufacturing and applications of the porous metal membranes: A critical review. CIRP Journal of Manufacturing Science and Technology 2021; 33: 339–68.

[49] Kimura K, Uchida H. Intensive membrane cleaning for MBRs equipped with flat-sheet ceramic membranes: Controlling negative effects of chemical reagents used for membrane cleaning. Water Research 2019; 150: 21–8.

[50] Ismail NH, Salleh WNW, Ismail AF, Hasbullah H, Yusof N, Aziz F, Jaafar J. Hydrophilic polymer-based membrane for oily wastewater treatment: A review. Separation and Purification Technology 2020; 233: 116007.

[51] Koo CH, Mohammad AW, Suja' F, Meor Talib MZ. Review of the effect of selected physicochemical factors on membrane fouling propensity based on fouling indices 2012; 287.

[52] Abdullah N, Yusof N, Lau WJ, Jaafar J, Ismail AF. Recent trends of heavy metal removal from water/wastewater by membrane technologies. Journal of Industrial and Engineering Chemistry 2019; 76: 17–38.

[53] Krzeminski P, Leverette L, Malamis S, Katsou E. Membrane bioreactors – A review on recent developments in energy reduction, fouling control, novel configurations, LCA and market prospects. Journal of Membrane Science 2017; 527: 207–27.

[54] *Membrane bioreactors: WEF manual of practice No. 36*. 1st edition. New York: McGraw-Hill Education; 2012.

[55] Lari S, Parsa SAM, Akbari S, Emadzadeh D, Lau WJ. Fabrication and evaluation of nanofiltration membrane coated with amino-functionalized graphene oxide for highly efficient heavy metal removal. International Journal of Environmental Science and Technology 2022;19(6): 4615–26.

[56] Yang M, Hadi P, Yin X, Yu J, Huang X, Ma H, Walker H, Hsiao BS. Antifouling nanocellulose membranes: How subtle adjustment of surface charge lead to self-cleaning property. Journal of Membrane Science 2021; 618: 118739.

[57] Lotfikatouli S, Hadi P, Yang M, Walker HW, Hsiao BS, Gobler C, Reichel M, Mao X. Enhanced anti-fouling performance in Membrane Bioreactors using a novel cellulose nanofiber-coated membrane. Separation and Purification Technology 2021; 275: 119145.

[58] Yang M, Lotfikatouli S, Chen Y, Li T, Ma H, Mao X, Hsiao BS. Nanostructured all-cellulose membranes for efficient ultrafiltration of wastewater. Journal of Membrane Science 2022; 650: 120422.

[59] Wei Y, Gao X, Wang J, Chen J, Mi B, Tian X, Gao C, Zhang Y. Facile and extensible preparation of multi-layered graphene oxide membranes with enhanced long-term desalting performance. Journal of Membrane Science 2021; 638: 119695.

[60] Zhao C, Xu X, Chen J, Wang G, Yang F. Highly effective antifouling performance of PVDF/graphene oxide composite membrane in membrane bioreactor (MBR) system. Desalination 2014; 340: 59–66.

[61] Alkhatib A, Ayari MA, Hawari AH. Fouling mitigation strategies for different foulants in membrane distillation. Chemical Engineering and Processing—Process Intensification 2021; 167: 108517.

[62] Westerhoff P, Yoon Y, Snyder S, Wert E. Fate of endocrine-disruptor, pharmaceutical, and personal care product chemicals during simulated drinking water treatment processes. Environ. Sci. Technol 2005; 39(17): 6649–63.

[63] Padaki M, Murali RS, Abdullah MS, Misdan N, Moslehyani A, Kassim MA, Hilal N, Ismail, AF. Membrane technology enhancement in oil-water separation. A review. Desalination 2015; 357: 197–207.

[64] Barrios-Martinez A, Barbot E, Marrot B, Moulin P, Roche N. Degradation of synthetic phenol-containing wastewaters by MBR. J. Membr. Sci 2006; 281(1–2): 288–96.

[65] Gonzalez S, Petrovic M, Barcelo D. Removal of a broad range of surfactants from municipal wastewater—Comparison between membrane bioreactor and conventional activated sludge treatment. Chemosphere 2007; 67(2): 335–43.

[66] Song W, Xie B, Huang S, Zhao F, Shi X. Chapter 6 — Aerobic membrane bioreactors for industrial wastewater treatment. In Current Developments in Biotechnology and Bioengineering. London: Elsevier; 2020, pp. 129–45.

[67] Nguyen TB, Bui XT, Vo TDH, Cao NDT, Dang BT, Tra VT, Tran HT, Tran LL, Ngo HH. Chapter 7 — Anaerobic membrane bioreactors for industrial wastewater treatment. In Current Developments in Biotechnology and Bioengineering. London: Elsevier; 2020, pp. 167–96

[68] Marti E, Monclus H, Jofre J, Rodriguez-Roda I, Comas J, Balcazar, JL. Removal of microbial indicators from municipal wastewater by a membrane bioreactor (MBR). Bioresour. Technol 2011; 102(8): 5004–9.

[69] Chae SR, Shin HS. Characteristics of simultaneous organic and nutrient removal in a pilot-scale vertical submerged membrane bioreactor (VSMBR) treating municipal wastewater at various temperatures. Process Biochem 2007; 42(2): 193–8.

[70] Kim HG, Jang HN, Kim HM, Lee DS, Eusebio RC, Kim HS, Chung TII. Enhancing nutrient removal efficiency by changing the internal recycling ratio and position in a pilot-scale MBR process. Desalination 2010; 262(1–3): 50–6.

[71] Song KG, Cho J, Cho KW, Kim SD, Ahn KH. Characteristics of simultaneous nitrogen and phosphorus removal in a pilot-scale sequencing anoxic/anaerobic membrane bioreactor at various conditions. Desalination 2010; 250(2): 801–4.

[72] Perera MK, Englehardt JD, Tchobanoglous G, Shamskhorzani R. Control of nitrification/denitrification in an onsite two-chamber intermittently aerated membrane bioreactor with alkalinity and carbon addition: Model and experiment. Water Res 2017; 115: 94–110.

[73] Abegglen C, Ospelt M, Siegrist H. Biological nutrient removal in a small-scale MBR treating household wastewater. Water Res 2008; 42(1–2): 338–46.

[74] Bracklow U, Drews A, Gnirss R, Klamm S, Lesjean B, Stuber J, Barjenbruch M, Kraume M. Influence of sludge loadings and types of substrates on nutrients removal in MBRs. Desalination 2010; 250(2): 734–9.

[75] Tan TW, Ng HY. Influence of mixed liquor recycle ratio and dissolved oxygen on performance of pre-denitrification submerged membrane bioreactors. Water Res 2008; 42(4–5): 1122–32.

[76] Falahti-Marvast H, Karimi-Jashni A. Performance of simultaneous organic and nutrient removal in a pilot scale anaerobic-anoxic-oxic membrane bioreactor system treating municipal wastewater with a high nutrient mass ratio. Int. Biodeterior. Biodegradation 2015; 104: 363–70.

[77] Chen W, Sun FY, Wang XM, Li XY. A membrane bioreactor for an innovative biological nitrogen removal process. Water Sci. Technol 2010; 61(3): 671–6.

[78] Yuan LM, Zhang CY, Zhang YQ, Ding Y, Xi DL. Biological nutrient removal using an alternating of anoxic and anaerobic membrane bioreactor (AAAM) process. Desalination 2008; 221(1–3): 566–75.

[79] Daigger GT, Littleton HX. Simultaneous Biological Nutrient Removal: A State-of-the-Art Review. Water Environ. Res 2014; 86(3): 245–57.

[80] Ahmed Z, Lim BR, Cho J, Song KG, Kim KP, Ahn KH Biological nitrogen and phosphorus removal and changes in microbial community structure in a membrane bioreactor: Effect of different carbon sources. Water Res 2008; 42(1–2): 198–210.

[81] Hocaoglu SM, Insel G, Cokgor EU, Orhon D. Effect of low dissolved oxygen on simultaneous nitrification and denitrification in a membrane bioreactor treating black water. Bioresour. Technol 2011a; 102(6): 4333–40.

[82] Hocaoglu SM, Insel G, Cokgor EU, Orhon D. Effect of sludge age on simultaneous nitrification and denitrification in membrane bioreactor. Bioresour. Technol 2011b; 102(12): 6665–72.

[83] Hocaoglu SM, Atasoy E, Baban A, Insel G, Orhon D. Nitrogen removal performance of intermittently aerated membrane bioreactor treating black water. Environ. Technol 2013; 34(19): 2717–25.

[84] Insel G, Erol S, Ovez S. Effect of simultaneous nitrification and denitrification on nitrogen removal performance and filamentous microorganism diversity of a full-scale MBR plant. Bioprocess Biosyst Eng 2014; 37(11): 2163–73.

[85] Sarioglu M, Insel G, Artan N, Orhon D. Model evaluation of simultaneous nitrification and denitrification in a membrane bioreactor operated without an anoxic reactor. J. Membr. Sci 2009; 337(1–2): 17–27.

[86] Giraldo EJP, Liu Y, Muthukrishnan S. Presence and Significance of ANAMMOX spcs and Ammonia Oxidizing Archea, AOA, in Full Scale Membrane Bioreactors for Total Nitrogen Removal. Proceedings of the Water Environment Federation/ International Water Association Conference on Nutrient Recovery and Management; 2011, pp. 607–16.

[87] Ivanovic I, Leiknes TO. The biofilm membrane bioreactor (BF-MBR)-a review. Desalin. Water Treat 2012; 37(1–3): 288–95.

[88] Lee WN, Kang IJ, Lee CH. Factors affecting filtration characteristics in membrane-coupled moving bed biofilm reactor. Water Res 2006; 40(9): 1827–35.

[89] Chu LB, Wang JL. Nitrogen removal using biodegradable polymers as carbon source and biofilm carriers in a moving bed biofilm reactor. Chem. Eng. J 2011; 170(1): 220–5.

[90] Luo Y, Jiang Q, Ngo HH, Nghiem LD, Hai FI, Price WE, Wang J, Guo W. Evaluation of micropollutant removal and fouling reduction in a hybrid moving bed biofilm reactor–membrane bioreactor system. Bioresour. Technol 2015; 191, 355–9.

[91] Tang B, Yu C, Bin L, Zhao Y, Feng X, Huang S, Fu F, Ding J, Chen C, Li P, Chen Q. Essential factors of an integrated moving bed biofilm reactor–membrane bioreactor: Adhesion characteristics and microbial community of the biofilm. Bioresour. Technol 2016; 211, 574–83.

[92] Artiga P, Oyanedel V, Garrido JM, Mendez R. An innovative biofilm-suspended biomass hybrid membrane bioreactor for wastewater treatment. Desalination 2005; 179(1–3): 171–9.

[93] Zekker I, Rikmann E, Tenno T, Lemmiksoo V, Menert A, Loorits L, Vabamäe P, Tomingas M, Tenno T. Anammox enrichment from reject water on blank biofilm carriers and carriers containing nitrifying biomass: Operation of two moving bed biofilm reactors (MBBR). Biodegradation 2012; 23(4), 547–60.

[94] Yang S, Yang F, Fu, Lei R. Comparison between a moving bed membrane bioreactor and a conventional membrane bioreactor on organic carbon and nitrogen removal. Bioresour. Technol 2009; 100(8): 2369–74.

[95] Duan L, Li S, Han L, Song Y, Zhou B, Zhang J. Comparison between moving bed-membrane bioreactor and conventional membrane bioreactor systems. Part I: Membrane fouling. Environ. Earth Sci 2015; 73(9): 4881–90.

[96] Leyva-Diaz JC, Gonzalez-Martinez A, Munio MM, Poyatos JM. Two-step nitrification in a pure moving bed biofilm reactor-membrane bioreactor for wastewater treatment: nitrifying and denitrifying microbial populations and kinetic modeling. Appl. Microbiol. Biotechnol 2015; 99(23): 10333–43.

[97] Yang S, Yang F, Fu Z, Wang T, Lei R. Simultaneous nitrogen and phosphorus removal by a novel sequencing batch moving bed membrane bioreactor for wastewater treatment. J. Hazard. Mater 2010; 175(1–3): 551–7.

[98] Yang QY, Yang T, Wang HJ, Liu KQ. Filtration characteristics of activated sludge in hybrid membrane bioreactor with porous suspended carriers (HMBR). Desalination 2009; 249(2): 507–14.

[99] Leyva-Diaz JC, Munio MM, Gonzalez-Lopez J, Poyatos JM. Anaerobic/anoxic/oxic configuration in hybrid moving bed biofilm reactor-membrane bioreactor for nutrient removal from municipal wastewater. Ecol. Eng 2016; 91: 449–58.

[100] Lee J, Ahn WY, Lee CH. Comparison of the filtration characteristics between attached and suspended growth microorganisms in submerged membrane bioreactor. Water Res 2001; 35(10): 2435–45.

[101] Nerenberg R. The membrane-biofilm reactor (MBfR) as a counter-diffusional biofilm process. Curr Opin Biotechnol 2016; 38: 131–6.

[102] Casey E, Glennon B, Hamer G. Review of membrane aerated biofilm reactors. Resour Conserv Recycl 1999; 27(1–2): 203–15.

[103] Downing LS, Nerenberg R. Total nitrogen removal in a hybrid, membrane-aerated activated sludge process. Water Res 2008; 42(14): 3697–708.

[104] Sun S.-P, Nàcher CP, Merkey B, Zhou Q, Xia SQ, Yang DH, Sun JH, Smets BF. Effective biological nitrogen removal treatment processes for domestic wastewaters with low C/N ratios: A review. Environ. Eng. Sci 2010; 27(2): 111–26.

[105] Hu SW, Yang FL, Sun C, Zhang JY, Wang TH. Simultaneous removal of COD and nitrogen using a novel carbon-membrane aerated biofilm reactor. Res. J. Environ. Sci 2008; 20(2): 142–48.

[106] Johnson KR, Admassu W. Mixed algae cultures for low cost environmental compensation in cultures grown for lipid production and wastewater remediation. J. Chem. Technol. Biotechnol 2013; 88(6): 992–8.

[107] Markou G, Georgakakis D. Cultivation of filamentous cyanobacteria (blue-green algae) in agro-industrial wastes and wastewaters: A review. Applied Energy 2011; 88(10): 3389–01.

[108] Gao F, Yang ZH, Li C, Wang Y, Jin W, Deng Y. Concentrated microalgae cultivation in treated sewage by membrane photobioreactor operated in batch flow mode. Bioresour. Technol 2014; 167: 441–6.

[109] Gao F, Li C, Yang ZH, Zeng GM, Mu J, Liu M, Cui W. Removal of nutrients, organic matter, and metal from domestic secondary effluent through microalgae cultivation in a membrane photobioreactor. J. Chem. Technol. Biotechnol 2016; 91(10): 2713–9.

[110] Praveen P, Heng JYP, Loh KC. Tertiary wastewater treatment in membrane photobioreactor using microalgae: Comparison of forward osmosis & microfiltration. Bioresour. Technol 2016; 222, 448–57.

[111] Han T, Lu HF, Ma SS, Zhang YH, Liu ZD, Duan N. Progress in microalgae cultivation photobioreactors and applications in wastewater treatment: A review. Int. J. Agric. Biol 2017; 10(1): 1–29.

[112] Marbelia L, Bilad M R, Passaris I, Discart V, Vandamme D, Beuckels A, Muylaert K, Vankelecom IFJ. Membrane photobioreactors for integrated microalgae cultivation and nutrient remediation of membrane bioreactors effluent. Bioresour. Technol 2014; 163: 228–35.

[113] Tang TY, and Hu ZQ. A comparison of algal productivity and nutrient removal capacity between algal CSTR and algal MBR at the same light level under practical and optimal conditions. Ecol. Eng 2016; 93, 66–72.

[114] Choi H. Intensified Production of Microalgae and Removal of Nutrient Using a Microalgae Membrane Bioreactor (MMBR). Appl. Biochem. Biotechnol 2015; 175(4): 2195–205.

[115] Logan BE, Wallack MJ, Kim KY, He W, Feng Y, Saikaly PE. Assessment of Microbial Fuel Cell Configurations and Power Densities. Environ. Sci. Technol 2015; 2(8): 206–14.

[116] Tian Y, Ji C, Wang K, Le-Clech P. Assessment of an anaerobic membrane bio-electrochemical reactor (AnMBER) for wastewater treatment and energy recovery. J. Membr. Sci 2014; 450, 242–8.

[117] Zhou GW, Zhou YH, Zhou GQ, Lu L, Wan XK, Shi, HX. Assessment of a novel overflow-type electrochemical membrane bioreactor (EMBR) for wastewater treatment, energy recovery and membrane fouling mitigation. Bioresour. Technol 2015; 196: 648–55.

[118] Ma J, Wang Z, He D, Li Y, Wu Z. Long-term investigation of a novel electrochemical membrane bioreactor for low-strength municipal wastewater treatment. Water Res 2015; 78: 98–110.

[119] Hou DX, Lu L, Sun DY, Ge Z, Huang X, Cath TY, Ren ZJ. Microbial electrochemical nutrient recovery in anaerobic osmotic membrane bioreactors. Water Res 2017; 114: 181–8.

[120] Li H, Zuo W, Tian Y, Zhang J, Di SJ, Li LP, Su XY. Simultaneous nitrification and denitrification in a novel membrane bioelectrochemical reactor with low membrane fouling tendency. Environ. Sci. Pollut. Res 2017; 24(6): 5106–17.

[121] Nakhate PH, Joshi NT, Marathe KV. A critical review of bioelectrochemical membrane reactor (BECMR) as cutting-edge sustainable wastewater treatment. Rev. Chem. Eng 2017; 33(2): 143–61.

[122] Tian Y, Li H, Li LP, Su XY, Lu YB, Zuo W, Zhang J. In-situ integration of microbial fuel cell with hollow-fiber membrane bioreactor for wastewater treatment and membrane fouling mitigation. Biosens. Bioelectron 2015; 64: 189–95.

[123] Wang, Y. K., Sheng, G. P., Li, W. W., Huang, Y. X., Yu, Y. Y., Zeng, R. J., and Yu HQ. Development of a Novel Bioelectrochemical Membrane Reactor for Wastewater Treatment. Environ. Sci. Technol 2011; 45(21): 9256–61.

[124] Huang L, Li X, Ren Y, Wang X. Preparation of conductive microfiltration membrane and its performance in a coupled configuration of membrane bioreactor with microbial fuel cell. Rsc Advances 2017; 7(34): 20824–32.

[125] Malaeb L, Katuri KP, Logan BE, Maab H, Nunes SP, Saikaly PE. A hybrid microbial fuel cell membrane bioreactor with a conductive ultrafiltration membrane biocathode for wastewater treatment. Environ. Sci. Technol 2013; 47(20): 11821–28.

[126] Wang YK, Sheng GP, Shi BJ, Li WW, Yu HQ. A novel electrochemical membrane bioreactor as a potential net energy producer for sustainable wastewater treatment. Sci. Rep 2013; 3: 1864

[127] Zuo K, Liang S, Liang P, Zhou X, Sun D, Zhang X, Huang X. Carbon filtration cathode in microbial fuel cell to enhance wastewater treatment. Bioresour. Technol 2015; 185: 426–30.

[128] Lin H, Peng W, Zhang M, Chen J, Hong H, Zhang Y. A review on anaerobic membrane bioreactors: Applications, membrane fouling and future perspectives. Desalination 2013; 314: 169–88.

[129] Dvorak L, Gomez M, Dolina J, Cernin A. Anaerobic membrane bioreactors a mini review with emphasis on industrial wastewater treatment: Applications, limitations and perspectives. Desalin. Water Treat 2016; 57(41): 19062–76.

[130] Skouteris, G, Hermosilla D, Lopez P, Negro C, Blanco A. Anaerobic membrane bioreactors for wastewater treatment: A review. Chem. Eng. J 2012; 198: 138–48.

[131] Saddoud A, Ellouze M, Dhouib A, Sayadi S. A comparative study on the anaerobic membrane bioreactor performance during the treatment of domestic wastewaters of various origins. Environ. Technol 2006; 27(9): 991–9.

[132] Zhang XJ, Li D, Liang YH, Zhang YL, Fan D, Zhang J. Application of membrane bioreactor for completely autotrophic nitrogen removal over nitrite (CANON) process. Chemosphere 2013; 93(11): 2832–38.

[133] Abbassi R, Yadav AK, Huang S, Jaffe PR. Laboratory study of nitrification, denitrification and anammox processes in membrane bioreactors considering periodic aeration. J. Environ. Manage 2014; 142: 53–9.

[134] Wang G, Xu XC, Gong Z, Gao F, Yang FL, Zhang HM. Study of simultaneous partial nitrification, ANAMMOX and denitrification (SNAD) process in an intermittent aeration membrane bioreactor. Process Biochem 2016; 51(5): 632–41.

[135] van der Star WRL, Miclea AI, van Dongen U, Muyzer G, Picioreanu C, van Loosdrecht MCM. The membrane bioreactor: A novel tool to grow anammox bacteria as free cells. Biotechnol. Bioeng 2008; 101(2): 286–94.

[136] Suneethi S, Joseph K. ANAMMOX process start up and stabilization with an anaerobic seed in Anaerobic Membrane Bioreactor (AnMBR). Bioresour. Technol 2011; 102(19): 8860–67.

[137] Tao Y, Gao DW, Fu Y, Wu WM, Ren NQ. Impact of reactor configuration on anammox process start-up: MBR versus SBR. Bioresour. Technol 2012; 104: 73–80.

[138] Abegglen C, Ospelt M, Siegrist H. Biological nutrient removal in a small-scale MBR treating household wastewater. Water Res 2008; 42(1–2): 338–46.

[139] Chong MN, Ho ANM, Gardner T, Sharma AK, Hood B. Assessing decentralised wastewater treatment technologies: correlating technology selection to system robustness, energy consumption and GHG emission. J. Water Clim. Chang 2013; 4(4): 338–47.

[140] Tai CS, Snider-Nevin J, Dragasevich J, Kempson J. Five years operation of a decentralized membrane bioreactor package plant treating domestic wastewater. Water Pract. Technol 2014; 9(2): 206–14.

[141] Verrecht B, James C, Germain E, Birks R, Barugh A, Pearce P, Judd S. Economical evaluation and operating experiences of a small-scale MBR for nonpotable reuse. J Environ Eng 2011; 138(5): 594–600.

[142] Pronk W, Palmquist H, Biebow M, Boller M. Nanofiltration for the separation of pharmaceuticals from nutrients in source-separated urine. Water Res 2006; 40(7): 1405–12.

[143] Udert KM, Wachter M. Complete nutrient recovery from source-separated urine by nitrification and distillation. Water Res 2012; 46(2): 453–64.

[144] Abdel-Kader AM. Assessment of an MBR system for segregated household wastewater by using simulation mathematical model. Proceedings of the 13th International Conference on Environmental Science and Technology, T. D. Lekkas, ed., 2013

[145] Fountoulakis MS, Markakis N, Petousi I, Manios T. Single house on-site grey water treatment using a submerged membrane bioreactor for toilet flushing. Sci. Total Environ 2016; 551: 706–11.

[146] Jabornig S, Favero E. Single household greywater treatment with a moving bed biofilm membrane reactor (MBBMR). J. Membr. Sci 2013; 446, 277–85.

[147] Wu TT, Englehardt JD. Mineralizing urban net-zero water treatment: Field experience for energy-positive water management. Water Res 2016; 106: 352–63.

[148] Atasoy E, Murat S, Baban A, Tiris M. Membrane Bioreactor (MBR) treatment of segregated household wastewater for reuse. Clean-Soil Air Water 2007; 35(5): 465–72.

[149] Matulova Z, Hlavinek P, Drtil M. One-year operation of single household membrane bioreactor plant. Water Sci. Technol 2010; 61(1): 217–26.

[150] Ren X, Shon HK, Jang N, Lee YG, Bae M, Lee J, Cho K, Kim IS. Novel membrane bioreactor (MBR) coupled with a nonwoven fabric filter for household wastewater treatment. Water Res 2010; 44(3): 751–60.

[151] Krzeminski P, Leverette L, Malamis S, Katsou E. Membrane bioreactors—A review on recent developments in energy reduction, fouling control, novel configurations, LCA and market prospects. J. Membr. Sci 2017; 527: 207–27.

[152] Lowe M, Qin R, Mao X. A review on machine learning, artificial intelligence, and smart technology in water treatment and monitoring. Water 2022; 14 (9): 1384.

9 Integration of Advanced Oxidation Processes as Pre-Treatment for Anaerobically Digested Palm Oil Mill Effluent

E. L. Yong, Z. Y. Yong, M. H. D. Othman, and H. H. See

CONTENTS

9.1   Introduction...161
9.2   POME Characteristics and Their Effects to the Environment.....................162
9.3   Anaerobically Treated POME and the Respective Biogas Production.........165
9.4   Advanced Oxidation Processes in the Treatment of POME........................168
     9.4.1   Ozonation...168
     9.4.2   Fenton 170
9.5   Conclusion 173
References..173

9.1 ' INTRODUCTION

Malaysia is another well-known top palm oil producer after Indonesia. In 2021, Malaysia exported a total of RM102,428 million palm oil and its related products and crude palm oil constituted approximately 20% [1]. The production of crude palm oil involves five major sequential processes, namely sterilization, threshing, digestion, pressing and clarification generating wastes such as empty fruit bunches, mesocarp fibre, palm kernel shells and palm oil mill effluent [2]. Empty fruit bunches, mesocarp fibre and palm kernel shell are normally incinerated but palm oil mill effluent (POME) is an obnoxious waste product that is difficult to handle. The volume of POME generated in every tonne of crude palm oil produced is 2.5 to 3 tonnes composed of water (94–96%), suspended solids (2–4%) and oil (2%) [3].

The content of POME is considered non-toxic as no chemicals are added during the extraction process. Despite that, POME is still highly polluted owing to its high organic loading and oily properties [4–8]. Its presence in the environment would

DOI: 10.1201/9781003260738-9

deplete not only the dissolved oxygen level in any receiving water bodies but also accelerate the eutrophication process. A study revealed that POME affected the early life stages of highly tolerant fish species (Nile tilapia) by slowing down hatching rate, decreasing survival rate, increasing heart rate and causing deformities in fish larvae [9]. Thus, treatment of POME is required to meet the final discharge limit as enacted in the Environmental Quality (Prescribed Premises) (Crude Palm Oil) Regulations 1977, amended in 1982 [10]. Nonetheless, current final discharge still poses problems where plankton diversity and fish reproduction were affected [11]. Hence, the limit of a few parameters including biochemical oxygen demand (BOD), total suspended solids and oil & grease is expected to be further lowered in the future [12]. Apart from water pollution, POME also releases methane along its degradation process. Methane is one of greenhouse gaseous that is 23 times more potent than carbon dioxide upsetting the earth climate [3]. In every tonne of POME, an estimation of 1011 kg carbon dioxide equivalent of biogas is generated [3]. If the biogas emitted could be captured, the biogas emission can be reduced to 192.5 kg carbon dioxide equivalent [3].

Anaerobic digestion has been rigorously researched to capture and ensure the consistency of biogas production. It is a preferred technology in ameliorating high organic loading wastewater. It is also a cheaper option compared to aerobic, another biological treatment process, that demands high energy consumption in providing oxygen to sustain the microorganisms. Topping up with the immense amount of sludge to be disposed of downstream of the aerobic treatment train, anaerobic treatment technology yielded a relatively small amount of sludge. A simple computation to analogize aerobic and anaerobic treatment processes can be made through the transformation of organic matter in terms of chemical oxygen demand (COD) [13]. Assuming that 100% of COD enters the aerobic treatment system, only 40–50% is transformed into carbon dioxide while 50–60% is assimilated by the microorganisms to produce new cells, which later become sludge. It is reported that only 5–10% of the COD will remain in the effluent. In an efficient anaerobic treatment, 7090% of the COD is turned into biogas. Sludge produced is merely 5–15% with 10–30% of COD left in the effluent. Although attractive, issues pertaining to long treatment time and inconsistency in the biogas production still exist. In tackling such issues, advanced oxidation processes have emerged but the mechanisms behind the degradation of POME are seldom reviewed. This chapter, therefore, delved into the POME properties as well as fundamental theories of both anaerobic digestion and advanced oxidation processes in order to better understand the application of pretreatment for POME treatment.

9.2   POME CHARACTERISTICS AND THEIR EFFECTS TO THE ENVIRONMENT

The earliest revelation of POME physicochemical characteristics can be found in the 1980s [4, 5, 14]. The characteristics of POME have remained quite consistent for the past decades except the concentration range becomes wider probably due to the intensive studies conducted since the year 2000 (Table 9.1). Judging from

TABLE 9.1

The Characteristics of POME in the 1980s and 2000s

Parameters	1980s [4–6, 14]	2000s [2, 7, 16–18]
pH	3.5–5.2	3.4–5.2
Biochemical oxygen demand, BOD (mg/L)	20,000–35,000	10,250–64,440
Chemical oxygen demand, COD (mg/L)	30,000–64,950	15,000–97,000
Oil and grease, O&G (mg/L)	4,800–12,000	2,000–14,110
Total solid, TS (mg/L)	4,240–50,350	4,980–79,000
Total suspended solids, TSS (mg/L)	19,000–27,450	3,115–59,350
Total dissolved solids, TDS (mg/L)	20,120	65–36,800
Volatile solid, VS (mg/L)	36,500–40,100	4,260–72,000
Volatile suspended solids, VSS (mg/L)	21,740–22,790	4,500–34,500
Volatile fatty acids, VFA (mg/L)	1,440–1,660	1,273–2,980
Total nitrogen, TN (mg/L)	546–944	180–1400
Total Kjeldahl nitrogen, TKN (mg/L)	500–1,050	749–3,200
Ammoniacal nitrogen (mg/L)	26–84	4–80
Total phosphorus (mg/L)	68–250	35–180
Lignin (mg/L)	4,700	130–28,140
Pectin (mg/L)	3,410	328
Holocellulose (mg/L)	7,310	12,060

the characteristics presented in Table 9.1, the sluggish brown waste liquid is highly viscous, emits unpleasant odor and can cause devastating water pollution issues if discharged to surface water without prior treatment. For instance, high BOD and COD concentrations would immediately diminish the dissolved oxygen content via biodegradation thereby leading to the death of aquatic species. The presence of POME in stagnant or slow flowing water bodies enables the release of a massive amount of greenhouse gases, especially methane, into the atmosphere. It was estimated that approximately 32 to 48 kg of methane per year was contributed by POME in one hectare of oil palm plantation [15]. This amount is equivalent to 0.8–1.2 Mg CO_2 equivalent of greenhouse gas emitted per hectare per year [15].

On top of releasing greenhouse gases, POME which contains high nutrient levels, i.e., nitrogen and phosphorus could also lead to severe algal bloom, a process known as eutrophication, in waterways. Although the ammoniacal nitrogen seems to be the main culprit for the algal to thrive, the dominant contributing factor is organic nitrogen content in POME. The TKN value shown in Table 9.1 disclosed that ammoniacal nitrogen only constituted 2.5% to 8% whereas the remaining nitrogenous percentage is organic nitrogen. Once entered the aquatic environment, organic nitrogen is transformed into ammoniacal nitrogen through an ammonification process, which later turns into nitrate via the nitrification process.

Another parameter of concern in POME is total solids. Total solids contain both total suspended solids and total dissolved solids. The suspended solids made up 62–75% of the total solids. Their presence in the environment will not only affect the clarity of the water and decrease the amount of light passing through the water but also absorb heat from sunlight. Heat absorption by the suspended solids will be transferred to the surrounding water and subsequently decrease the dissolved oxygen level in the water [19]. In addition, warmer water can cause stratification where a water body is divided into two unmixed layers, causing low dissolved oxygen level in the bottom layer [20] and affecting aquatic organisms. A study also showed that high total suspended solids led to rapid development of fish embryos [21]. While it may be beneficial, the embryos suffered from lower survival rates resulting in less viable fish embryos and larvae.

In view of this, Environmental Quality (Prescribed Premises) (Crude Palm Oil) Regulations 1977 stipulated under Environmental Quality Act 1974 has limited the effluent discharge quality of POME into Malaysian waterways. The current and future standard discharge limit for each parameter set by the Malaysian Department of Environment is listed in Table 9.2.

To meet the discharge standard listed in Table 9.2, all oil palm millers are required to treat POME prior to discharge. Over the past decades, 85% of palm oil millers in Malaysia have been adopting an open ponding type of treatment system [22]. Open ponding treatment is still very much preferred owing to the simplicity and low cost to construct and maintain. However, the duration of treatment is long and requires many ponds to bring down the concentration of each parameter.

Previous study has divulged that the COD and BOD concentrations were 1,725 and 610 mg/L, respectively, even though POME has passed through eight ponds equating to a total hydraulic retention time of 60 days [23]. The large land area utilized to treat POME makes it an unsustainable and inefficient method. Moreover, these ponds mostly are facultative in nature thereby contributing to greenhouse gaseous emissions. The remaining 15% would employ conventional anaerobic treatment systems [6, 14]. This method seems promising as a renewable energy source, i.e., methane, could be harnessed. Nonetheless, the difficulty in collecting the gas, expensive start-up and inconsistency of methane gas production makes the progress to convert from a ponding treatment system to anaerobic treatment technology very challenging. Therefore, anaerobic treatment of POME has progressed significantly in order to improve the treatment of POME as a renewable energy source.

TABLE 9.2
The Present and Future Standard Discharge Limit of POME

Parameters	Present [10]	Future [12]
pH	5.0–9.0	5.0–9.0
Temperature (°C)	45	45
Biochemical oxygen demand, BOD (mg/L)	100	20
Chemical oxygen demand, COD (mg/L)	–	–
Oil and grease, O&G (mg/L)	50	5
Total suspended solids, TSS (mg/L)	400	200
Total nitrogen, TN (mg/L)*	200	200
Ammoniacal nitrogen (mg/L)*	150	150

*Value of filtered sample

9.3   ANAEROBICALLY TREATED POME AND THE RESPECTIVE BIOGAS PRODUCTION

The anaerobic digestion in POME generally proceeds in four treatment stages, namely, hydrolysis, acidogenesis, acetogenesis and methanogenesis [13]. Hydrolysis is an important step to hydrolyze complex organic matter consisting of pectin (fibres), lignin, lipids, protein, cellulose and carbohydrates into long chain fatty acids, sugars and amino acids [13, 24]. The hydrolysis stage by-products are further degraded in the subsequent acidogenesis stage into short chain carboxylic acids frequently termed as volatile fatty acids, alcohols, carbon dioxide, ammonia and hydrogen sulfide. Acetate and hydrogen are produced simultaneously but at a much slower rate. The acceleration in the generation of acetate and hydrogen occurs in the acetogenesis stage. These two by-products are the precursors of methane formation in the subsequent methanogenesis step. One critical by-product in the generation of acetate, hydrogen and methane under strictly anaerobic conditions often overlooked would be carbon monoxide [25]. Coupling carbon monoxide with water, hydrogen, carbon dioxide and methane would form simultaneously. However, this will only happen in a thermal environment [26]. The anaerobic conversion pathways are summarized in Figure 9.1.

Anaerobic digestion system consists of three types of configurations, i.e., low-rate, single-stage high-rate and two-stage high-rate digesters [13]. As POME is categorized as high strength wastewater, the discussion on low-rate digesters will be omitted from this chapter. Two types of anaerobic digesters frequently researched in the treatment of POME were suspended and attached growth type. Up-flow anaerobic sludge blanket (UASB), expanded granular sludge bed (EGSB), anaerobic baffled reactor (ABR), anaerobic sequencing batch reactor (ASBR), continuous stirred tank reactor (CSTR) and anaerobic fluidized bed reactor (AFBR) are the suspended growth type, whereas anaerobic filtration and up-flow anaerobic sludge fixed-film (UASFF) belong to the attached growth type.

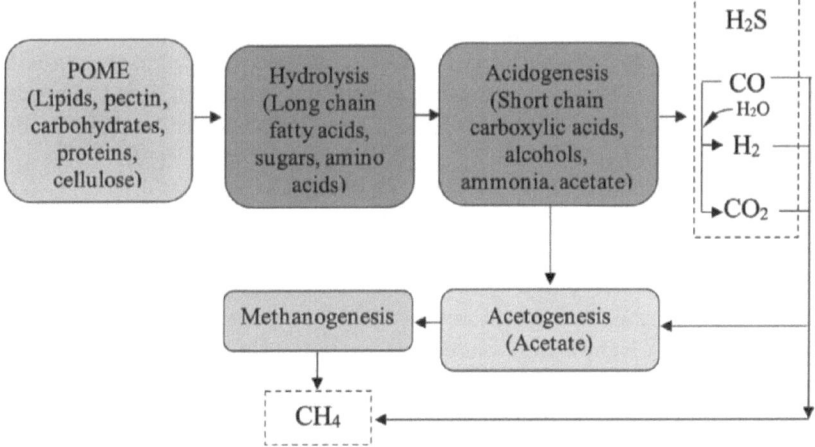

FIGURE 9.1   Four conversion pathways in an anaerobic digestion.

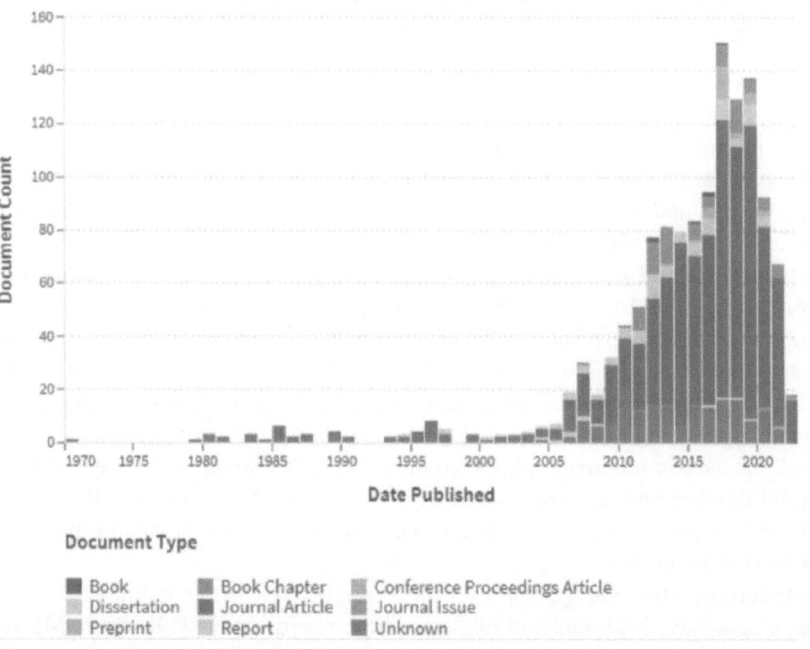

FIGURE 9.2 The published work related to anaerobic treatment technology for palm oil mill effluent since 1970. (https://www.lens.org/lens/search/scholar/analysis?q=anaerobic%20palm%20oil%20mill%20effluent&preview=true)

A quick search of Lens.org revealed that only 1299 documents related to the anaerobic treatment of POME were found worldwide as of June 2022 (Figure 9.2). Surprisingly, it took about 25 years for the research to progress intensely since the first article published in the 1970s.

In the early phase of the research in attaining biogas or more specifically methane from POME, the focus was very much concentrated on a single-stage treatment system. UASB, CSTR and ASBR were most often investigated and temperature was found to play a pivotal role. At 28°C in an UASB digester, methane produced was 0.012 to 0.063 L CH_4/g $COD_{removed}$ [27]. When the temperature of the UASB digester was elevated to 35°C to 38°C, methane was increased to 0.26–0.33 L CH_4/g COD_{POME} [8, 28, 29] and 0.14–0.35 L H_2/g $COD_{removed}$ [30] was obtained concurrently. However, methane yield became lower when the temperature rose to 55°C where it was reduced to 0.034 L CH_4/g COD_{POME} [31]. Although POME was deoiled with the intention to improve methane production in UASB, the gas generated was not significantly increased, i.e., 0.042 L CH_4/g COD_{POME} [31]. This coincided with the decreasing survival of the methane producing microorganisms above 38°C. Meanwhile, production of hydrogen was not affected by the temperature raised but the hydraulic retention time (HRT) would impact the biogas yield. It was demonstrated that too high HRT or too low HRT would decrease the hydrogen generation [32].

Similar trends for methane and hydrogen were also observed in CSTR [33–37] and ASBR [27, 37–40] type digester. The gases generated, however, were not consistent. Methane, for instance, could achieve as high as 0.454 L-CH_4/g COD_{POME} [28] but sometimes it could go as low as 0.013 L-CH_4/g-COD_{POME} [27]. No part of the system is superior to another. The inconsistency of the gas production may be due to anaerobic digestion is a two-phase process where the first phase, i.e., hydrogen producing phase, requires high temperature (55°C–60°C) while the methane producing phase requires lower temperature (35°C–40°C) in the second phase. It was not until 2014 that the treatment of POME progressed to two-stage treatment but only a handful of published works can be found over the span of 8 years (Figure 9.3).

Here, two reactors in series are normally constructed. The first reactor will undergo hydrolysis and acidogenesis processes, focusing on the production of hydrogen. Meanwhile, the second reactor will concentrate on the acetogenesis and methanogenesis processes for the generation of methane. Previous study has shown that the respective amount of hydrogen and methane produced in two-stage systems were 90% and 34% higher than that of single-stage anaerobic treatment systems [41]. Nonetheless, the amount of gases produced still fluctuated between 0.033 and 0.215 L H_2/g COD_{POME} for hydrogen and 0.156 and 0.315 L CH_4/g COD_{POME} for methane [41–44]. Comparing the two phases, the hydrogen producing phase possesses a larger variation of gas produced than the methane-producing phase. This is

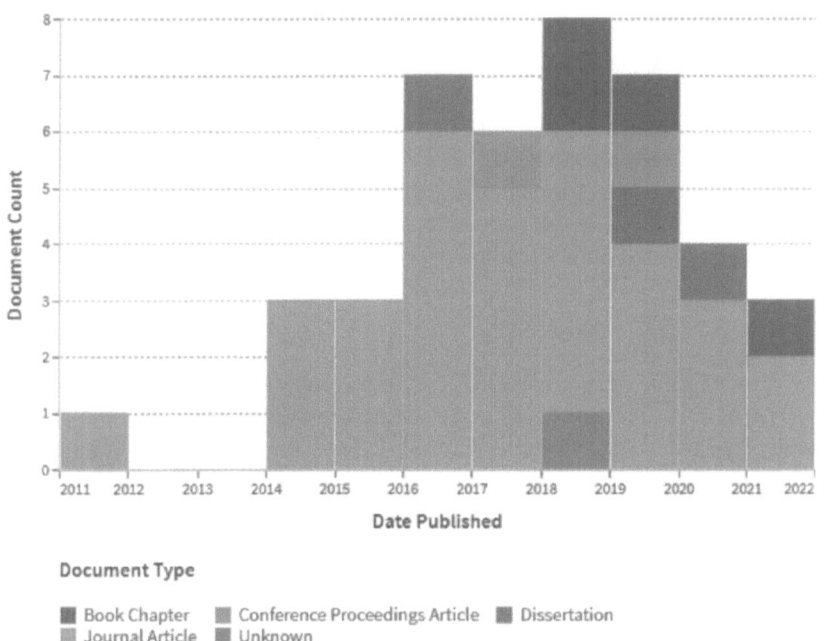

FIGURE 9.3   The amount of published work on two-stage anaerobic treatment for POME. (https://www.lens.org/lens/search/scholar/analysis?q=(%22two-stage%20 anaerobic%22%20%20%22palm%20oil%20mill%20effluent%22)&p=0&n=10& s=_score&d=%2B&f=false&e=false&l=en&authorField=author& dateFilterField=publishedYear&orderBy=%2B_score&presentation= false&preview=true&stemmed=true&useAuthorId=false)

mainly because the hydrolysis process is limited by the presence of high lignocel-lulosic compounds including lignin, pectin and holocellulose in POME (Table 9.1). These compounds consist of hydrophobic heteropolymers making them resistant towards biodegradation thereby limiting efficient conversion of the lignocellulosic compounds to biogas [45, 46]. Additionally, lignocellulose also comprises compounds such as phenolics, furans, and polyphenols that interfere with the microbial biodegradation [47]. For this, pretreatment of POME would serve as an essential step to enhance biogas production, and the methods involved are physical, chemical, coagulation-flocculation, biological and adsorption [7]. Among them, only chemical methods focusing on advanced oxidation processes of pretreatment methods will be discussed in this chapter.

9.4   ADVANCED OXIDATION PROCESSES IN THE TREATMENT OF POME

As discussed in the previous section, lignocellulosic content is the limiting factor in converting POME into biogas. The presence of lignocellulosic compounds, specifically hemicellulose, would decrease the enzymatic hydrolysis rate by protecting the cellulose from cellulolytic enzymes [48, 49]. Therefore, in order to increase the biodegradation of POME, delignification via the removal of hemicellulose needs to be achieved first. It is well-known that advanced oxidation processes are the most effective in breaking the hemicellulose structure by the non-selective oxidant, hydroxyl radicals [50]. For instance, past studies have revealed that the application of Fenton process can increase the transformation of cellulose into fermentable sugar [51] and subsequently enhanced the biogas production by a few folds in lignocellulosic biomass [47, 52]. This phenomenon occurs mainly due to the removal of inhibitory compounds encapsulating the cellulose [50]. To date, POME is mainly based on ozonation and Fenton [2, 7, 53, 54]. Therefore, only ozonation and the Fenton oxidation process will be discussed in the following session.

9.4.1 OZONATION

Ozone is an unstable oxidant generating hydroxyl radicals, which are later involved in a series of chain reactions catalyzing the decomposition of ozone in an aqueous medium [55]. The decomposition of ozone follows the pseudo first-order kinetics that proceed in two stages: the fast initial stage and a slower second stage. Four reactions occur simultaneously with complex organic compounds. They are direct reaction, initiation, promotion and inhibition as depicted in Figure 9.4. Direct react occurs when organic compounds are directly oxidized by ozone into stable by-products. Meanwhile, initiation reaction is the most important route for the formation of hydroxyl radical where ozone will react selectively with hydroxide ion or some chemical groups containing high electron density such as aromatic, olefinic and aliphatic amines groups [56]. As hydroxide ion relates closely to pH, high pH (>8.0) value will hasten ozone degradation. However, in the presence of organic compounds, the effect of hydroxide ion on the overall initiation reaction is negligible.

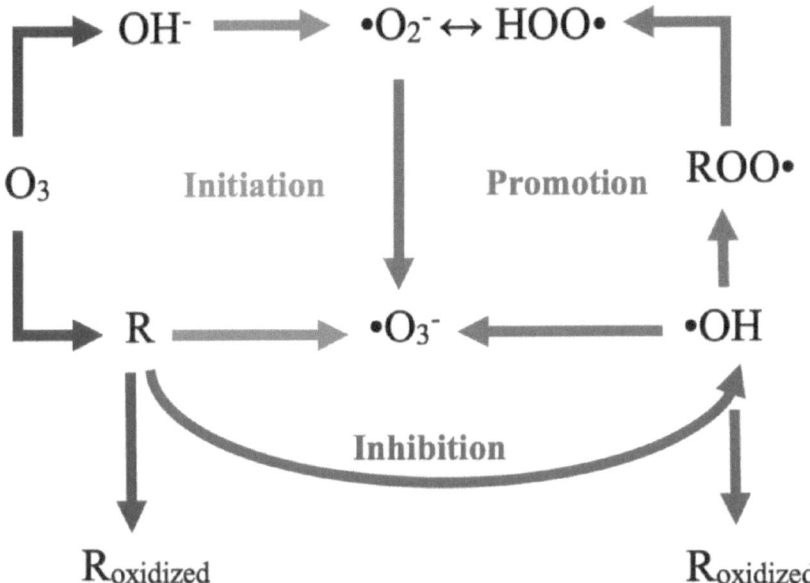

FIGURE 9.4   Simultaneous reactions in an ozonation process in the presence of organic.

Contrasting with ozone, hydroxyl radical is a non-selective oxidant. It possesses extremely high reaction rates with both organic and inorganic compounds, except with ammonia, manganese(II), iron(II) and carbonate [57]. Therefore, the oxidation of these inorganic compounds frequently proceeds with ozone. Owing to its non-selectivity, hydroxyl radical will also react with ozone, forming a series of chain reactions which promote the degradation of ozone. Some C-H functional groups such as formate ion will also participate in this promotion cycle [56]. To cease the promotion reaction, inhibitors like tertiary butanol are normally added. Inhibitors react with hydroxyl radical producing by-products that do not participate in the chain reactions of the decomposition process.

Lignocellulose compounds contain various functional groups playing the role as initiator, promoter and inhibitor. Evidence has disclosed that ozone reacted with acid insoluble lignin forming acid soluble lignin [58]. The by-products produced from this oxidation process may continue to react with the remaining ozone inducing the formation of carboxylic acids. Under low pH, the undissociated carboxylic acids can percolate through the cell membrane and decrease the intracellular pH, thereby affecting the functionality of the cell [59]. Hence, pH adjustment is essential before the subsequent anaerobic digestion process.

In the case of POME, the only initiation pathway of ozone reaction is by the lignocellulosic compounds because of its low pH. After being pre-treated with ozone, enhancement of biodegradability was observed via the elevation of concentration in BOD_5 and soluble COD [60]. This coincided with the decline in the total COD upon ozonation. Reduction of alkalinity was also observed due to the consumption of inhibitors, i.e., carbonates and bicarbonates [56, 57]. With the improvement on the biodegradability of POME, a significant increase in biogas production was discerned. Chaiprapat and Laklam [60] discovered that ozonated POME maintained a low volatile fatty acids production, whereas in non-ozonated POME there was an accumulation of volatile fatty acids. The accumulation of volatile fatty acids decreased the overall pH of the system making it unsuitable for methane producing microorganisms to grow.

A few other studies also observed the elevation of biogas production after POME underwent ozonation treatment. Ahmad [61] showed that methane gas generated only reached 245 mL/g $COD_{removed}$ in non-ozonated condition but the ozonated POME achieved 899 to 955 mL/g $COD_{removed}$. Methane gas production increases with the increasing ozonation time and dosage. Additionally, biomass build-up was also noticed in ozonated POME. Nonetheless, prolonged ozonation time (>70 h) and greater dosage (>5 mg/L/h) would accumulate polyhydroxyalkanoic acid that was toxic to the microorganisms.

The improvement of biogas production was not limited to methane but also hydrogen, another potential renewable energy created under anaerobic treatment process. Hydrogen yield as high as 182 mL/g $COD_{removed}$ was achieved for ozonated POME, whereas raw POME only produced 120 mL of H_2/g $COD_{removed}$ under mesophilic condition (37°C) [62]. But as the temperature increased to thermophilic (55°C) and extreme thermophilic (70°C), there was a significant drop in the hydrogen gas production. It is also important to note that the amount of hydrogen production would be affected by the consumption of hydrogen by bacteria accumulating propionic acid and ethanol as the by-products [63]. This denotes those different operational parameters including temperature, ozonation dosage and time as well as the amount of volatile fatty acids could affect both methane and hydrogen production.

9.4.2 FENTON

Fenton is another oxidation process employed in POME treatment targeting organic degradation [64–66] as well as for color removal [67, 68]. Briefly, the Fenton reaction generates hydroxyl radicals from the reaction between dissolved iron(II) (Fe^{2+}) and hydrogen peroxide (H_2O_2). Along the process, it also produces other radicals such as hydroperoxyl (•OOH) and superoxide (•O_2^-) propagating a series of chain reactions [69]. Simultaneously, Fe^{2+} is oxidized in iron(III) (Fe^{3+}) forming precipitate. Reaction (R1) is an initiation reaction for the formation of hydroxyl radicals. The rest of the reactions (R2 to R8) are propagating reactions. Therefore, the ratio as well as amount of Fe^{2+} concentration and H_2O_2 is crucial for the effectiveness in enhancing degradation of organics and removing color.

$$Fe^{2+} + H_2O_2 \rightarrow Fe^{3+} + OH^- + OH \tag{R1}$$

$$OH + H_2O_2 \rightarrow OOH / O_2^- + H_2O \tag{R2}$$

$$Fe^{3+} + OOH / O_2^- \rightarrow Fe^{2+} + O_2 + H^+ \tag{R3}$$

$$Fe^{2+} + OH \rightarrow Fe^{3+} + OH^- \tag{R4}$$

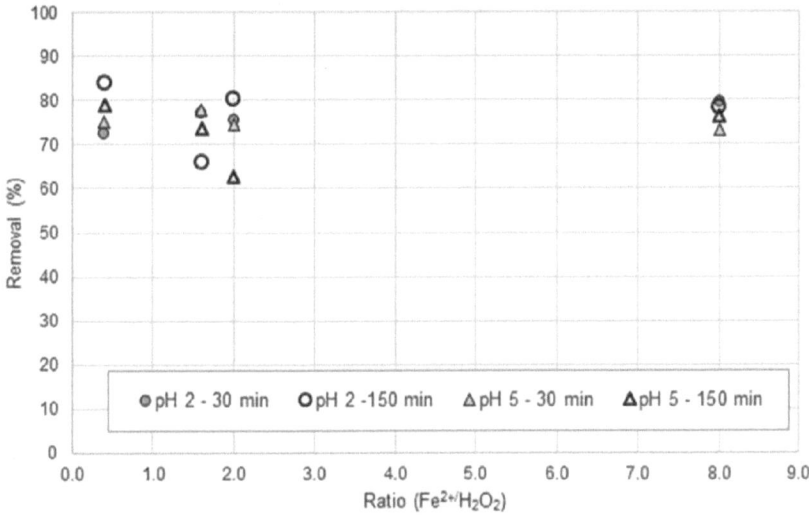

FIGURE 9.5 The organic removal percentage in terms of COD using different Fe^{2+}/H_2O_2 ratio at different pH level and time (Data was extracted from Saeed et al. (64) and reproduced).

$$Fe^{2+} + OOH / O_2^- \rightarrow Fe^{3+} + H_2O_2 \tag{R5}$$

$$OOH / O_2^- + OOH / O_2^- \rightarrow H_2O_2 \tag{R6}$$

$$OOH / O_2^- + OH \rightarrow H_2O + O_2 \tag{R7}$$

$$OH + OH \rightarrow H_2O_2 \tag{R8}$$

Saeed and co-workers [64] found that the removal of organics in POME was 12.6% without Fenton treatment owing to the natural photo oxidation. Although the removal of organics in POME using the Fenton method could increase to as high as 85.0%, it largely dependent on Fe^{2+}/H_2O_2 ratio, pH and duration of treatment. Figure 9.5 depicts the removal of organic matter measured in terms of COD at different Fe^{2+}/H_2O_2 ratio and pH. Based on the data provided, the removal ranged between 60% and 83.9%. At a high Fe^{2+}/H_2O_2 ratio of 8.0, duration of treatment and pH did not seem to affect the removal, whereas pH did not pose any significant impact on the treatment performance at 30 min. However, a more pronounced variation was observed on the removal percentage when the treatment time was extended to 150 min. This might be due to the degradation profile of the Fenton reaction that follows a pseudo-first order kinetic.

Interestingly, when the condition of treatment was set at pH 3.5 with 90-min reaction (Figure 9.6), the overall removal percentage demonstrated that increasing Fe^{2+}/H_2O_2 ratio increased the organic removal percentage. For all the treatment conditions except those of pH 2.0, a reduction of pH was noticed. This contradicted the

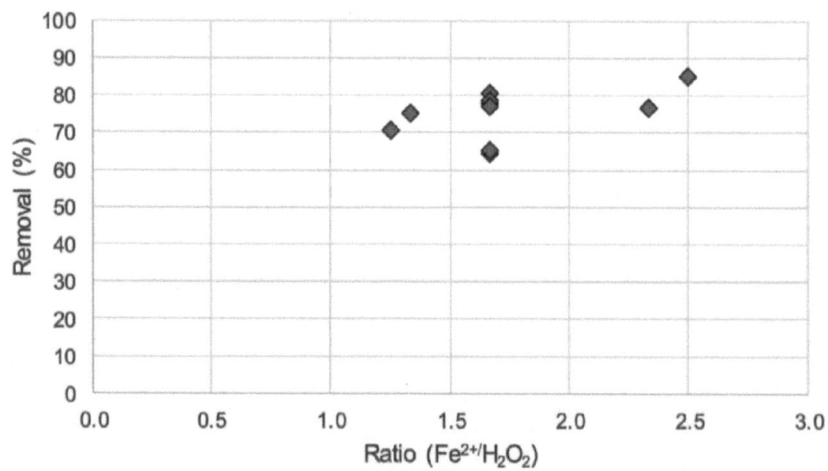

FIGURE 9.6   The percentage of organic removal in terms of COD using different Fe^{2+}/H_2O_2 ratio at pH 3.5 and 90 min (Data was extracted from Saeed et al. (64) and reproduced).

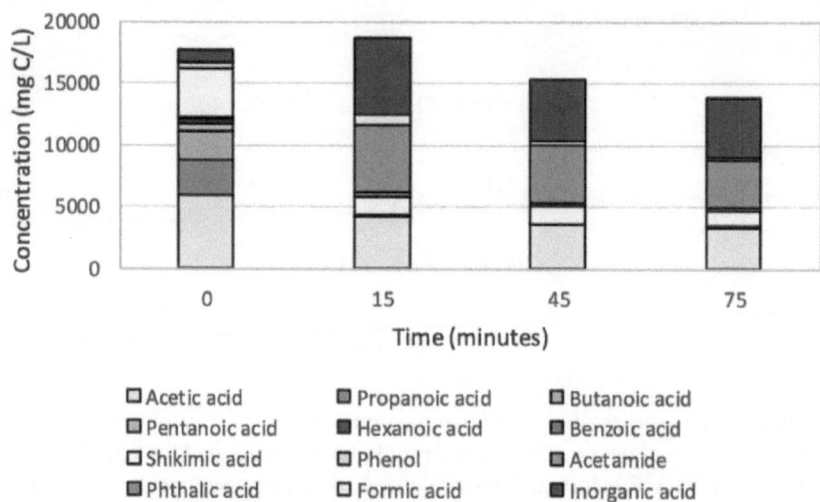

FIGURE 9.7   The different carboxylic and inorganic acid concentrations over time during Fenton treatment process (Data was extracted from Gamaralalage (65) and reproduced).

fundamental theory that pH should increase during the Fenton treatment owing to the release of hydroxide ions. Here, only pH 2.0 showed elevation of pH between 0.2 and 0.6 units [64].

Reduction in pH was explained via the formation of carboxylic acids and the increment of inorganic carbon concentration after the Fenton oxidation process [65]. As shown in Figure 9.7, carboxylic acids such as phthalic and formic acids were the intermediate products formed as a result of the oxidation of aromatic compounds in POME. The concentration of these acids, however, decrease after its formation denoting their concurrent degradation by hydroxyl radical generated from the Fenton reaction. Meanwhile, other organic acids such as pentanoic acid, butanoic acid, hexanoic acid and propanoic acid achieved complete removal. Another accumulation observed during the Fenton reaction was inorganic carbon. Because the pH of POME is between 3.5 and 4.5, the inorganic carbon formed was in the form of carbonic acid. This may also contribute to the reduction of pH after the Fenton treatment.

9.5 CONCLUSION

The diversion of POME anaerobic treatment technologies from single- to two-stage treatment did not manage to solve the inconsistency in biogas (methane and hydrogen) production owing to the presence of lignocellulosic compounds. To address this issue, pretreatment employing advanced oxidation processes has emerged as a promising technology in breaking down the insoluble lignin to soluble lignin that subsequently promotes biodegradation. Two frequently researched oxidation technologies are ozonation and Fenton. While the former enhanced methane production, the latter was targeted on organic removal. Ozonated POME could yield as high as 955 mL/g $COD_{removed}$, which was approximately 4 times higher than non-ozonated POME. On top of that, hydrogen gas production was also elevated to 182 mL/g $COD_{removed}$ compared to 120 mL of H_2/g $COD_{removed}$ in non-ozonated condition. In the Fenton reaction, organic degradation in terms of COD ranged between 60% to 83.9%. The organic removal efficiency is largely dependent on the Fe^{2+}/H_2O_2 ratio employed. Contrasting with the theoretical principles where the Fenton would increase the pH after oxidation, vice versa condition was observed for POME owing to the production of carboxylic acids and inorganic acids. Both methods have exhibited great potential to be adopted on an industrial scale. Ozonation method could be easily applied in full-scale but the energy consumption is high and unsustainable. Meanwhile, the Fenton method is relatively inexpensive. Yet, the treated effluent may contain high amounts of dissolved iron(III) because of the lowered pH after the Fenton treatment. Hence, deeper understanding on the reaction mechanisms and the cost-effectiveness is required if the industry has decided to embrace either ozonation or Fenton treatment.

REFERENCES

[1] (MPOB) MPOB. Monthly export of oil palm products—2021. Available from: https://bepi.mpob.gov.my/index.php/en/export/export-2021/monthly-export-of-oil-palm-products-2021.
[2] Cheng YW, Chong CC, Lam MK, Ayoub M, Cheng CK, Lim JW, Yusup S, Tang YY, Bai JM. Holistic process evaluation of non-conventional palm oil mill effluent (POME) treatment technologies: A conceptual and comparative review. J Hazard Mater 2021; 409: 124964.
[3] Hosseini SE, Abdul Wahid M. Pollutant in palm oil production process. J Air & Waste Manage Assoc 2015; 65(7):773–781.
[4] Ng WJ, Wong KK, Chin KK. Two-phase anaerobic treatment kinetics of palm oil wastes. Water Res 1985; 19(5):667–670.

[5] Ho CC, Tan YK, Wang CW. The distribution of chemical constituents between the soluble and the particulate fractions of palm oil mill effluent and its significance on its utilization/treatment. Agric Wastes 1984; 11:61–72.

[6] Yeoh BG. A kinetic-based design for thermophilic anaerobic treatment of a high-strength agroindustrial wastewater. Environ Technol Let 1986; 7:509–520.

[7] Ratnasari A, Syafiuddin A, Boopathy R, Malik S, Mehmood MA, Amalia R, Pratsyo DD, Zaidi NS. Advances in pretreatment technology for handling the palm oil mill effluent: Challenges and prospects. Bioresour Technol 2022; 344:126239.

[8] Borja R, Banks CJ. Anaerobic-digestion of palm oil mill effluent using an up-flow anaerobic sludge blanket reactor. Biomass Bioenerg 1994; 6(5):381–389.

[9] Muliari M, Zulfahmi I, Akmal Y, Karja NWK, Nisa C, Sumon KA, Rahman MM. Toxicity of palm oil mill effluent on the early life stages of Nile tilapia (Oreochromis niloticus, Linnaeus 1758). Environ Sci and Pollut Res 2020; 27:30592–30599.

[10] Department of Environment (DOE). Environmental Quality (Prescribed Premises) (Crude Palm Oil) Regulations 1977 (Amendment) 1982.

[11] Hashiguchi Y, Zakaria MR, Maeda T, Mohd Yusoff MZ, Hassan MA, Shirai Y. Toxicity identification and evaluation of palm oil mill effluent and its effects on the planktonic crustacean Daphnia magna. Sci Tot Environ 2020; 710:136277.

[12] Zainal NH, Jalani NF, Mamat R, Astimar AA. A review on the development of palm oil mill effluent final discharge polishing treatments. J Oil Palm Res 2017; 29(4):528–540.

[13] Chernicharo CadL. Biological Wastewater Treatment: Anaerobic Reactors (Volume 4): London: IWA Publishing; 2007.

[14] Chin KK. Anaerobic treatment kinetics of palm oil sludge. Water Res 1981; 15(2):199–202.

[15] Reijnders L, Huijbregts MAJ. Palm oil and the emission of carbon-based greenhouse gases. J Clean Prod 2008; 16(4):477–482.

[16] Choong YY, Chou KW, Norli I. Strategies for improving biogas production of palm oil mill effluent (POME) anaerobic digestion: A critical review. Renew Sust Energ Rev 2018; 82:2993–3006.

[17] Lee ZS, Chin SY, Lim JW, Witoon T, Cheng CK. Treatment technologies of palm oil mill effluent (POME) and olive mill wastewater (OMW): A brief review. Environ Technol Inno 2019; 15:100377.

[18] O-Thong S, Boe K, Angelidaki I. Thermophilic anaerobic co-digestion of oil palm empty fruit bunches with palm oil mill effluent for efficient biogas production. Appl Energ 2012; 93:648–654.

[19] Wetzel RG. Limnology: Lake and River Ecosystems. 3rd ed. San Diego: Academic Press; 2001.

[20] Langland M, Cronin T. A summary report of sediment processes in Chesapeake Bay and watershed. New Cumberland, PA, US Geological Survey; 2003. Report No.: 03–4123

[21] Mueller JS, Grabowski TB, Brewer SK, Worthington TA. Effects of temperature, total dissolved solids, and total suspended solids on survival and development rate of larval Arkansas River shiner. J Fish Wildl Manag 2017; 8(1):79–88.

[22] Ma AN, Ong ASH. Pollution control in palm oil mills in Malaysia. J Am Oil Chem Soc 1985; 62:261–266.

[23] Loh SK, Lai ME, Ngatiman M, Lim WS, Choo Y, Zhang Z, et al. Zero discharge treatment technology of palm oil mill effluent. J Oil Palm Res 2013; 25:273–281.

[24] Menzel T, Neubauer P, Junne S. Role of microbial hydrolysis in anaerobic digestion. Energies 2020; 13.

[25] Conrad R, Thauer RK. Carbon monoxide production by Methanobacterium thermoautotrophicum. FEMS Microbiol Let 1983; 20:229–233.

[26] Andreides D, Olsa Fliegerova K, Pokorna D, Zabranska J. Biological conversion of carbon monoxide and hydrogen by anaerobic culture: Prospect of anaerobic digestion and thermochemical processes combination. Biotechnol Adv 2022; 58:107886.

[27] Chaiprapat S, Prasertsan P, Chaisri R. Effect of organic loading rate on methane and volatile fatty acids productions from anaerobic treatment of palm oil mill effluent in UASB and UFAF reactors. Songklanakarin J Sci Technol 2007; 2:311–323.

[28] Ibrahim MM, Jemaat Z, Hamid Nour A. Performance and Kinetic Evaluation of Palm Oil Mill Effluent (POME) Digestion in a Continuous High Rate Up-Flow Anaerobic Sludge Blanket (UASB) bioreactor. Mater Sci Forum 2021; 1025:141–149.

[29] Ahmad A, Ghufran R, Abd Wahid Z. Effect of cod loading rate on an upflow anaerobic sludge blanket reactor during anaerobic digestion of palm oil mill effluent with butyrate. J Environ Eng Landsc 2012; 20(4):256–264.

[30] Singh L, Siddiqui MF, Ahmad A, Rahim MHA, Sakinah M, Wahid ZA. Application of polyethylene glycol immobilized Clostridium sp LS2 for continuous hydrogen production from palm oil mill effluent in upflow anaerobic sludge blanket reactor. Biochem Eng J 2013; 70:158–165.

[31] Fang C, O-Thong S, Boe K, Angelidaki I. Comparison of UASB and EGSB reactors performance, for treatment of raw and deoiled palm oil mill effluent (POME). J Hazard Mater 2011; 189(1–2):229–234.

[32] Mahmod SS, Azahar AM, Tan JP, Jahim JM, Abdul PM, Mastar MS, et al. Operation performance of up-flow anaerobic sludge blanket (UASB) bioreactor for biohydrogen production by self-granulated sludge using pre-treated palm oil mill effluent (POME) as carbon source. Renew Energ 2019; 134:1262–1272.

[33] Choorit W, Wisarnwan P. Effect of temperature on the anaerobic digestion of palm oil mill effluent. Electron J Biotechn 2007; 10(3):376–385.

[34] Irvan I, Trisakti B, Wongistani V, Tomiuchi Y. Methane emission from digestion of palm oil mill effluent (POME) in a thermophilic anaerobic reactor. International J Sci Eng 2012; 3(1):32–35.

[35] Wong YS, Teng TT, Ong SA, Norhashirnah M, Rafatullah M, Leong JY. Methane gas production from palm oil wastewater-An anaerobic methanogenic degradation process in continuous stirrer suspended closed anaerobic reactor. J Taiwan Inst Chem E 2014; 45(3):896–900.

[36] Mohd Yusoff MZ, Abdul Rahman N, Abd. Aziz S, Chong ML, Hassan MA, Shirai Y. The effect of hydraulic retention time and volatile fatty acids on biohydrogen production from palm oil mill effluent under non-sterile condition. Aust J Basic Appl Sci 2010; 4(4):577–587.

[37] Seengenyoung J, O-Thong S, Prasertsan P. Comparison of ASBR and CSTR reactor for hydrogen production from palm oil mill effluent under thermophilic condition. Biosci Biotechnol 2014; 5(3):177–183.

[38] Prasertsan P, O-Thong S, Birkeland NK. Optimization and microbial community analysis for production of biohydrogen from palm oil mill effluent by thermophilic fermentative process. Int J Hydrogen Energ 2009; 34(17):7448–7459.

[39] Badiei M, Jahim JM, Anuar N, Abdullah SRS. Effect of hydraulic retention time on biohydrogen production from palm oil mill effluent in anaerobic sequencing batch reactor. Int J Hydrogen Energ 2011; 36(10):5912–5919.

[40] O-Thong S, Prasertsan P, Intrasungkha N, Dhamwichukorn S, Birkeland N. Improvement of biohydrogen production and treatment efficiency on palm oil mill effluent with nutrient supplementation at thermophilic condition using an anaerobic sequencing batch reactor. Enzyme Microb Technol 2007; 41(5):583–590.

[41] Mamimin C, Singkhala A, Kongjan P, Suraraksa B, Prasertsan P, Imai T, O-Thong S. Two-stage thermophilic fermentation and mesophilic methanogen process for biohythane production from palm oil mill effluent. Int J Hydrogen Energ 2015; 40(19):6319–6328.

[42] Krishnan S, Singh L, Sakinah M, Thakur S, Wahid ZA, Alkasrawi M. Process enhancement of hydrogen and methane production from palm oil mill effluent using two-stage thermophilic and mesophilic fermentation. Int J Hydrogen Energ 2016; 41(30):12888–12898.

[43] Krishnan S, Singh L, Sakinah M, Thakur S, Nasrul M, Otieno A, Wahid ZA. An investigation of two-stage thermophilic and mesophilic fermentation process for the production of hydrogen and methane from palm oil mill effluent. Environ. Prog. Sustain. Energy 2017; 36(3):895–902.

[44] Krishnan S, Singh L, Sakinah M, Thakur S, Wahid ZA, Ghrayeb OA. Role of organic loading rate in bioenergy generation from palm oil mill effluent in a two-stage up-flow anaerobic sludge blanket continuous-stirred tank reactor. J Clean Prod 2017; 142:3044–3049.

[45] Kucharska K, Rybarczyk P, Hołowacz I, Łukajtis R, Glinka M, Kaminski M. Pretreatment of lignocellulosic materials as substrates for fermentation processes. Molecules 2018; 23:2937.

[46] Carrere H, Antonopoulou G, Affes R, Passos F, Battimelli A, Lyberatos G, Ferrer I. Review of feedstock pretreatment strategies for improved anaerobic digestion: From lab-scale research to fullscale application. Bioresour Technol 2016; 199:386–397.

[47] Behera S, Arora R, Nandhagopal N, Sachin K. Importance of chemical pretreatment for bioconversion of lignocellulosic biomass. Renew Sust Energ Rev 2014; 36:91–106.

[48] Hendriks A, Zeeman G. Pretreatments to enhance the digestibility of lignocellulosic biomass. Bioresour Technol 2009; 100:10–18.

[49] Jeoh T, Ishizawa CI, Davis MF, Himmel ME, Adney WS, Johnson DK. Cellulase digestibility of pretreated biomass is limited by cellulose accessibility. Biotechnol Bioeng 2007; 98(1):112–122.

[50] Arimi MMM, Mecha CA, Kiprop AK, Ramkat R. Recent trends in application of advanced oxidation processes (AOPs) in bioenergy production: Review. Renew Sust Energ Rev 2020; 121:109669.

[51] Ninomiya K, Takamatsu H, Ayaka O, Kenji T, Nobuaki S. Sonocatalytic-Fenton reaction for enhanced OH radical generation and its application to lignin degradation. Ultrason Sonochem 2013; 20:1092–1097.

[52] Michalska K, Miazek K, Kryzstek L, Ledakowicz S. Influence of pretreatment with Fenton's reagent on biogas production and methane yield from lignocellulosic biomass. Bioresour Technol 2012; 119:72–78.

[53] Aziz MMA, Anuar Kassim K, ElSergany M, Syed Anuar S, Jorat ME, Yaacob H, et al. Recent advances on palm oil mill effluent (POME) pretreatment and anaerobic reactor for sustainable biogas production. Renew Sust Energ Rev 2020; 119:109603.

[54] Khadaroo SNBA, Poh PE, Gouwanda D, Paul G. Applicability of various pretreatment techniques to enhance the anaerobic digestion of Palm oil Mill effluent (POME): A review. J Environ Chem Eng 2019; 7:103310.

[55] Hoigne J, Bader H. The role of hydroxyl radical reactions in ozonation processes in aqueous solutions. Water Res 1976; 10:377–386.

[56] von Sonntag C, von Gunten U. *Chemistry of Ozone in Water and Wastewater Treatment: From Basic Principles to Application.* London: IWA Publishing; 2012.

[57] von Gunten U. Ozonation of drinking water: Part I. Oxidation kinetics and product formation. Water Res 2003; 37:1443–1467.

[58] Travaini R, Otero MDM, Coca M, Da-Silva R, Bolado S. Sugarcane bagasse ozonolysis pretreatment: Effect on enzymatic digestibility and inhibitory compound formation. Bioresour Technol 2013; 133:332–339.

[59] Travaini R, Martin-Juarez J, Lorenzo-Hernando A, Bolado-Rodriguez S. Ozonolysis: An advantageous pretreatment for lignocellulosic biomass revisited. Bioresour Technol 2016; 199:2–12.

[60] Chaiprapat S, Laklam T. Enhancing digestion efficiency of POME in anaerobic sequencing batch reactor with ozonation pretreatment and cycle time reduction. Bioresour Technol 2011; 2011:4061–4068.

[61] Ahmad A. Effect of ozonation on biodegradation and methanogenesis of palm oil mill effluent treatment for the production of biogas. Ozone: Sci Eng 2019; 41(5):427–436.

[62] Tanikkul P, Juntarakod P, Pisutpaisal N. Optimization of biohydrogen production of palm oil mill effluent by ozone pretreatment. Int J Hydrogen Energ 2019; 44:5203–5211.

[63] Tanikkul P, Boonyawanich S, He M, Pisupaisal N. Thermophilic hydrogen recovery from palm oil mill effluent. Int J Hydrogen Energ 2019; 44:5176–5181.

[64] Saeed MO, Azizli KAM, Isa MH, Ezechi EH. Treatment of POME using Fenton oxidation process: Removal efficiency, optimization, and acidity condition. Desalin Water Treat 2016; 57(50):23750–23759.

[65] Gamaralalage D, Sawai O, Nunoura T. Degradation behavior of palm oil mill effluent in Fenton oxidation. J Hazard Mater 2019; 364:791–799.

[66] Gamaralalage D, Sawai O, Nunoura T. Effect of reagents addition method in Fenton oxidation on the destruction of organics in palm oil mill effluent. J Environ Chem Eng 2020; 8(4).

[67] Saeed MO, Azizli K, Isa MH, Bashir MJK. Application of CCD in RSM to obtain optimize treatment of POME using Fenton oxidation process. J Water Process Eng 2015; 8:E7–E16.

[68] Razali NF, Chin HC, Zakaria S, Sajab MS, Tobe T, Tsuda M. Palm Oil Mill Effluent (POME) Decolorize through continuous fenton oxidation process using limonite as catalyst. Sains Malaysiana 2020; 49(1):69–79.

[69] Gupta P, Lakes A, Dziubla T. A free radical primer. In: Dziubla T, editor. *Oxidative Stress and Biomaterials*. London: Academic Press; 2016. pp. 1–33.

10 Electrocoagulation and Its Application in Food Wastewater Treatment

Mohammed J.K. Bashir and Koo Li Sin

CONTENTS

10.1 Introduction...179
10.2 Electrocoagulation Process..183
10.3 Operating Parameters...186
10.4 Application 193
10.5 Hybrid Technologies..194
    10.5.1 Electrochemical-Peroxidation or Peroxi-Electrocoagulation 195
    10.5.2 Photo-Electrocoagulation 195
    10.5.3 Sono-Electrocoagulation...195
10.6 Trends, Opportunities, Challenges...196
    10.6.1 Economic Aspects...196
    10.6.2 Ecological Aspects..197
    10.6.3 Technical Aspects 198
10.7 Conclusion 199
References...199

10.1 INTRODUCTION

Food wastewater are unavoidably generated in any form of food handling, processing, packing and storing. Wastewater from food manufacturing and processing can be a cause of very serious environmental and economic problems with respect to its treatability and disposal. The environmental consequences in inadequate removal of the pollutants from the food wastewater discharged into rivers can have a eutrophication issue developed within the aquatic environment due to the disposal of biodegradable oxygen-consuming materials. If this condition were sustained for an extended period of time, the ecological balance of the receiving stream, river or lake would be disturbed. Continual depletion of the oxygen in these waters would also lead to the development of obnoxious odours and unsightly scenes. The cost for treating the wastewater is contributed by the specific characteristics of wastewater like daily volume of discharge and the relative strength of the wastewater. The burden of the treatment cost is one of the big concerns among food industry players in terms of

DOI: 10.1201/9781003260738-10

179

TABLE 10.1

Wastewater Treatment Options Available to Remove Various Categories of Pollutants in Food and Agricultural Wastewater

Pollutants in wastewater	Management options
Dissolved organic species	Biological treatment; Adsorption; Land Application; Recovery and utilization
Dissolved inorganic species	Ion exchange; Reverse osmosis; Evaporation / distillation; Adsorption
Suspended organic materials	Physicochemical treatment; Biological treatment; Land application; Recovery and utilization
Suspended inorganic materials	Pre-treatment (screen); Physicochemical treatment (sedimentation, flotation, filtration, coagulation).

wastewater management and also compliance with current environmental policy and regulations. Table 10.1 spells out the treatment options available for food wastewater based on their characteristics [1].

The components present in food wastewater are almost all organic in nature. Organic matter are substances containing compounds comprising mainly the elements carbon I, hydrogen (H) and oxygen (O). The carbon atoms in the organic matter, also called carbonaceous compounds may be oxidized both chemically and biologically to yield carbon dioxide (CO_2) and energy. Thus, the characteristics of food wastewater can be viewed as a set of physicochemical and biological parameters that are vital in designing and managing food wastewater treatment facilities. Table 10.2 describes the common characteristics of food wastewaters using chemical oxygen demand (COD), biological oxygen demand (BOD_5), color, pH, total suspended solids (TSS), Oil and Grease. For instance, being one of the top global palm oil production, Malaysia generated 18.12 million tons of crude palm oil (CPO) in 2021 alone [2]. With every ton of CPO produced, a huge amount of water is used for extraction processes on fresh fruit bunches and around 50% of the water is disposed of as effluents [3]. The effluents, viscous brownish colloidal mixtures with unpleasant and distinct offensive odour, are high in organic matter and nutrients that are nontoxic but they carry potential to promote algal growth and eutrophication.

Electrocoagulation (EC) is a cost-efficient electrolysis-based technology in wastewater treatment, as it combines the benefits of coagulation, flotation, and electrochemistry. In EC systems, metal material, typically iron or aluminium, is dissolved from the anode in situ, promoting coagulation, whereas electroflotation (EF) by hydrogen gas produced at the cathode occurs simultaneously. A variety of minor side reactions like neutralization of solution pH may take place, which is mainly due to cathodic OH− ion generation [4]. These processes in EC promote degradation of pollutants in wastewater. From Figure 10.1, the conceptual framework explains how each foundation area (i.e., coagulation, flotation, electrochemistry) brings a certain perspective to EC, as represented by each lobe of this Venn diagram [5]. Central to understanding EC as a whole are the contact pattern (i.e., mixing) and process kinetics. The first of these describes how the various species (coagulant, bubbles, pollutant particles)

TABLE 10.2

Characteristics of Different Food Industry Wastewaters

Wastewaters		Palm Oil	Olive Oil	Fishery Processing	Sugarcane/Distillery	Dairy	Meat & Poultry	Coffee Processing
Reference		[26]	[27]	[28]	[29]	[30]	[15]	[14]
pH	–	3.4–5.2	4.7–5.7	3.8–7	3.55–4.55	4–11	6–8	7.5
Color	–	Dark Brown	Dark Brown	–	Dark Brown	White	–	Brown
COD	mg/L	15,000–100,000	16,500–190,000	300–90,000	20,000–150,000	500–102,100	2,000–8,000	1,400–1,990
BOD$_5$	mg/L	10,250–43,750	41,300–46,000	40–78,000	7,000–50,000	240–60,000	750–4,000	850–920
TSS	mg/L	5,000–54,000	32,000–300,000	2,000–3,000	300–16,400	60–22,150	800–3,300	3,400–3,550
Oil & Grease	mg/L	130–18,000	200–10,000	20–4,000	–	20–3,110	250–1,300	–

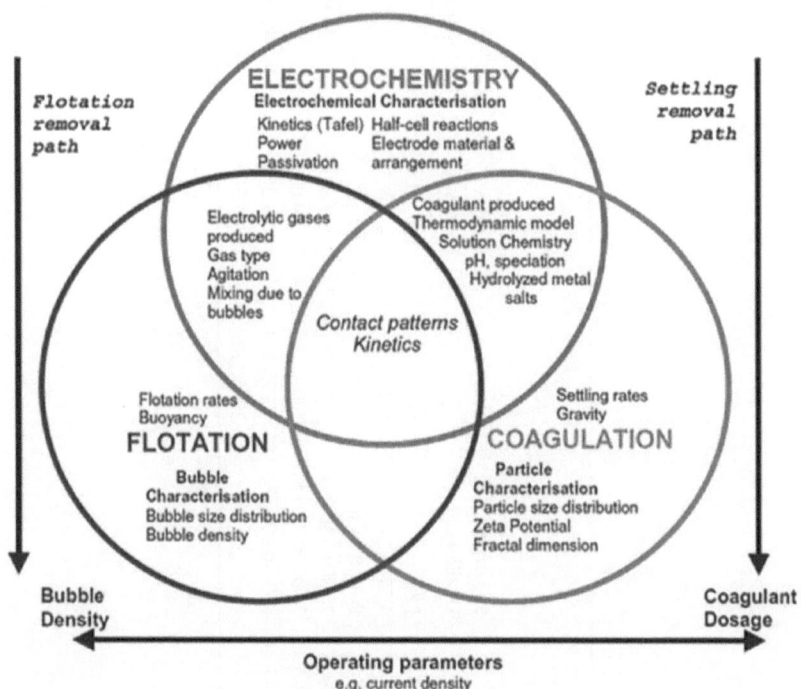

FIGURE 10.1   Conceptual framework for electrocoagulation.

(*Source:* P. K. Holt, G. W. Barton, and C. A. Mitchell, Chemosphere, 2005, 59, 355–367 [5]).

move and are brought into proximity with each other, while the latter describes the rate at which interactions between the various species occurs. It is a combination of physico-chemical processes occurring within an EC reactor that shifts the dominant pollutant separation mechanism between (gravity driven) settling and (buoyancy driven) flotation. The vertical arrows represent these two removal paths of flotation and settling. Current density shifts the relative importance between the flotation and coagulation lobes, as it determines both the coagulant dosage and bubble production rates, as well as influencing the extent of mixing within a reactor. The decisions about EC reactor design cannot be made in isolation, as there is an inseparable link between design and operational parameters brought by the complex interactions between the three foundation technologies. During the past two decades, an increase in global interest in EC applications in wastewater treatment has emerged. EC has many advantages, as it has great potential in clearing different organic pollutants in wastewater, not limiting to tannery, textile and colored wastewater; pulp and paper industry wastewater; oily wastewater; wastewater containing heavy metals, nutrients, cyanide, and other elements. Table 10.3 summarizes the advantages and disadvantages of EC in wastewater treatment [6]. In addition, EC does not require any ancillary chemical, temperatures and high pressures for the reaction to commence. In this

TABLE 10.3

The Advantages and Disadvantages of Using EC to Treat Wastewaters

Advantages	Disadvantages
Low energy requirement	Regular replacement of anodes and cathodes
Small equipment footprint	Requires a source of electricity
Suitable for decentralised/rural operation	Passivation of electrode reduce process efficiency
No requirement for hazardous chemicals	
Low risk of secondary contamination	
Low capital cost	
Low operating expense	
Reduction in sludge volume	
Can remove a variety of contaminants	

chapter, the focus on EC applications on food industry wastewater will be discussed together with their opportunities and challenges in the future.

10.2 ELECTROCOAGULATION PROCESS

Pollutant particles contained in wastewater are often in colloidal form. Colloids are stable, charged aggregates of atoms and molecules. They are very small in size (1–1,000 nm), and thus their surface area to mass ratio is enormous. The stability of colloids is significantly affected by particle surface charges, which may be formed in many ways (ionization of functional surface groups, ion adsorption, dissolution of ionic solids, or isomorphous substitution due to lattice imperfections) [4]. Due to their stability, it is hard to remove colloids from wastewater unless their size is increased. Therefore, to remove colloids from suspension, they need to be first chemically destabilized. EC is the process of destabilizing emulsified, suspended, or dissolved pollutants in an aqueous medium by introducing an electric current into the medium [7]. It can be used for wastewater treatment where Direct Current (DC) is applied to the electrodes, usually made of iron (Fe) or aluminum (Al), and the electrolyte is wastewater. In its simplest form, the EC process is studied in the laboratory experiments as shown in Figure 10.2. When the EC cell is connected to an external power source, the anode material is electrochemically corroded due to oxidation. The conductive metal plates are commonly known as sacrificial electrodes. A stirrer is set to keep the liquid and slurries uniform in the reactor. During electrolysis, the positive side undergoes anodic reactions, while on the negative side, cathodic reactions are encountered. Consumable metal plates, such as Al or Fe, are usually used as sacrificial electrodes to continuously produce ions in the water. The released ions neutralize the charges of the particles, thereby initiating coagulation. The released ions remove unwanted contaminants either by chemical reaction and precipitation or by causing the colloidal materials to coalesce, which can then be removed by flotation. In addition, as water containing colloidal particulates, oils, or other contaminants moves through the applied electric field, there may be ionization, electrolysis, hydrolysis, and free-radical formation which can alter the physical and

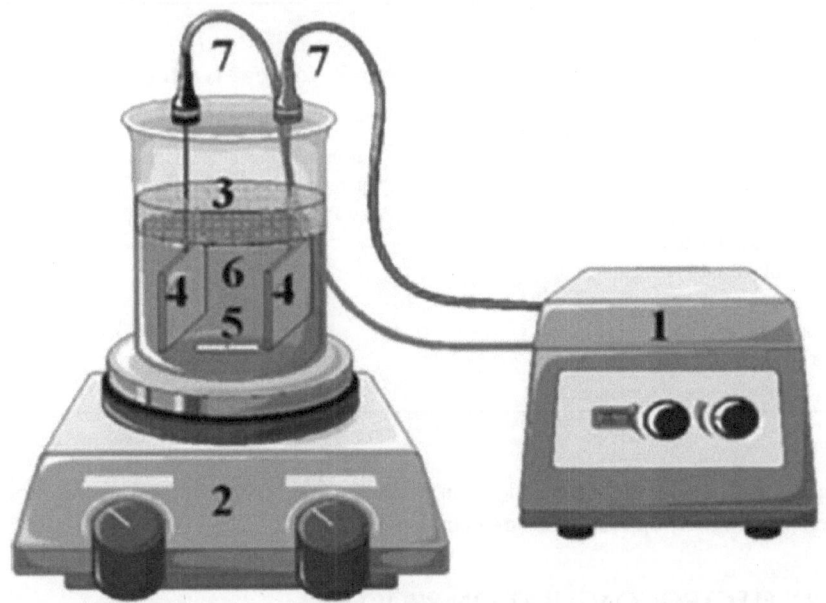

FIGURE 10.2 Experimental setup of electrocoagulation reactor (1) DC power supply, (2) magnetic stirrer, (3) cover, (4) Al electrodes, (5) magnetic bar-stirrer, (6) wastewater and (7) electric wire.

chemical properties of water and contaminants. As a result, the reactive and excited state causes contaminants to be released from the water and destroyed or made less soluble [8]. In other terms, the EC process is based on coagulation (destabilization, agglomeration, and further separation of pollutant particles) brought about by iron or aluminum derived *in situ* from electrochemical dissolution of the anode metal. At the same time, hydrogen gas bubbles and OH^- ions are formed at the cathode surface, promoting EF and solution neutralization respectively [4]. The gas bubbles may float some portion of the coagulated pollutants to the surface. Figure 10.3 demonstrates the conceptual diagram of an electrocoagulation reactor which includes a DC power supply, an anode, a cathode, and electrolyte [5].

In an EC cell, various electrochemical reactions occur simultaneously at the anodes and the cathodes [4]. These reactions include main and side reactions.

At the anode:

$$Al(s) \rightarrow Al^{3+} (aq) + 3e^- \qquad E_0 = + 1.66V \qquad (1)$$

$$Fe(s) \rightarrow Fe^{2+} (aq) + 2e^- \qquad E_0 = + 0.44V \qquad (2)$$

$$Fe(s) \rightarrow Fe^{3+}(aq) + 3e^- \qquad E_0 = + 0.04V \qquad (3)$$

$$2Fe^{2+} (aq) + 0.5O_2(g) + H_2O(l) \rightarrow 2Fe^{3+} (aq) + 2OH^- \qquad (4)$$

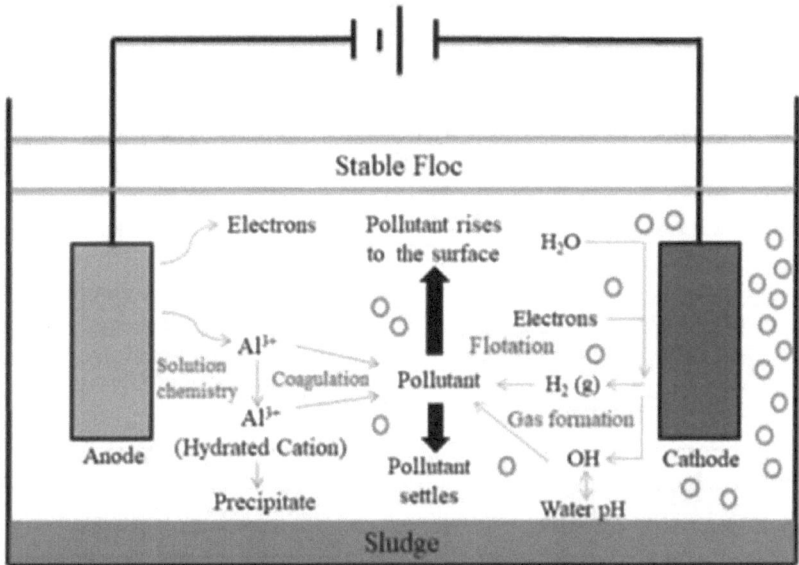

FIGURE 10.3   Base reactions and interactions occurring in an EC system.

(*Source:* Rakhmania et al. 2022 [11]).

At the cathode:

$$2H_2O(l) + 2e^- \rightarrow H_2(g) + 2OH^- \qquad E_0 = -0.83V \qquad (5)$$

In an EC system, a variety of minor side reactions may take place. These include direct reduction of metal ions on the cathodes and anodic oxygen gas generation. The OH⁻ ion production, as presented in Equation (5), is mostly due to which the EC process may neutralize the water treated. This means that after EC treatment, the wastewater's pH will be increased for acidic influent and will be decreased for alkaline influent. This change in pH typically occurs at a significantly greater rate in the beginning of an EC treatment than toward its end, where a steady state is reached. The increase in wastewater's pH has been typically found to be greater when iron anodes are used, compared with aluminum anodes, with the explanation that aluminum complexes have more hydroxide ions than iron, thus leading to a lower final pH value. However, different results have also been reported occasionally. The decrease in solution pH has been proposed to be due to the generation of aluminate ($Al(OH)^{4-}$), which will then consume alkalinity from the wastewater [4].

After the anode metal is dissolved in the wastewater, it will immediately undergo rapid and spontaneous hydrolyzation, producing amorphous hydroxides and/or polyhydroxides. For iron, these species include (but are not limited to), for example, $FeOH^{2+}$, $Fe(OH)_2^+$, $Fe_2(OH)_2^{4+}$, $Fe(OH)_4^-$, $Fe(H_2O)_5OH^{2+}$, $Fe(H_2O)_4(OH)_2^+$, $Fe(H_2O)_8(OH)_2^{4+}$, and $Fe_2(H_2O)_6(OH)_4^{2+}$. Correspondingly for aluminum, species such as $Al(OH)^{2+}$, $Al(OH)_2^+$, and $Al(OH)_4^-$ and $Al_6(OH)_{15}^{3+}$, $Al_7(OH)_{17}^{4+}$, $Al_8(OH)_{20}^{4+}$, $Al_{13}O_4(OH)_{24}^{4+}$, and $Al_{13}(OH)_{34}^{5+}$ are produced. These species will then finally transform into solid $Fe(OH)_3$ and $Al(OH)_3$ precipitate according to complex precipitation kinetics. Those precipitates form flocs that combine wastewater contaminants as well as metal hydroxides formed by hydrolysis and a range of coagulant species. These flocs are easily removed from treated wastewater by sedimentation and by H_2 flotation. Thus, the hydroxides flocs normally act as adsorbents or traps for pollutants. Wastewater's pH strongly affects the formation of the metal complexes. The well-known polymeric form of aluminum is Keggin cation $(Al_{13}O_4(OH)_{24})^{7+}$. The Keggin cation is regarded as the main unit of larger precipitated and soluble complexes, and is commonly known as the Al_{13} ion. The Al_{13} ion is typically denoted a the most effective and stable polymeric aluminum species for water and wastewater treatment, since it has been found to achieve the highest charge neutralization compared with other aluminum species [4].

The major factors that affect the EC process include initial pH, electrode material, distance between electrodes, applied current density, conductivity of the electrolyte, type of power supply, electrode configuration, mixing speed and treatment time.

10.3 OPERATING PARAMETERS

The treatment efficiency of the EC process is typically determined by the following parameters:

- Initial pH

The initial pH affects the contaminant distribution, coagulants produced in the EC process, electrode dissolution and the conductivity of the wastewater. Generally, pH of the wastewater targeted for EC is adjusted with addition of alkaline such as sodium hydroxide solution and acids like dilute sulfuric or hydrochloric acids. Varying the initial pH helps to optimize the EC process to determine the optimum initial pH for the specific EC process [9]. Different types of food wastewater vary in terms of optimum initial pH, such as the EC works listed in Table 10.4. Besides, the pH of the wastewater changes during the operating process which is dependent on the type of electrode material. Aluminium and iron electrodes are commonly used in EC treatment. Aluminium electrodes work well in acidic and neutral pH as the major ion species $Al(OH)_3$ is formed between pH 4 and pH 9.5. $Al(OH)_3$ is capable of trapping colloids and contaminants effectively from its precipitation. Iron electrodes operate well in acidic, neutral and alkaline pH. As shown in Figure 10.4, the amorphous hydroxides for Fe(III) and Al(III) will form under different pH solutions. In highly alkaline wastewater, the least removal efficiency occurs when $Al(OH)_4$ and $Fe(OH)_4$ ins form, which are poor coagulants. Due to the generation of hydroxide and hydrogen ions at the cathodes, the pH of the EC-treated wastewater may also increase slightly. Based on [10]'s study on sugar industry wastewater, the maximum of 60% for COD reduction and a 70% of color reduction were found at a pH of 6 at a time of 120 min and there was a slow decrease in COD removal from pH 7, 8, and 10.

TABLE 10.4

Application of EC Process in Food Wastewater

Type of wastewater	Electrodes/ configuration ***	Inter-electrode distance	Electrolyte/ coagulant/ oxidant *	Current density/ voltage	Initial pH	EC time (min)	Removal efficiency **	Reference
Poultry Slaughterhouse	Fe rods	–	Na_2SO_4	50 mA/cm^2	7	38	88.64% COD	[16]
Dairy	Al-Al Bipolar	–	KCl	60V	7.2–7.7	60	98.84% COD 97.95% BOD 97.75% TSS	[19]
Synthetic dairy	Al-Al Bipolar	–	NaCl	0.65 A/m^2	6	20	80% COD 100% TSS	[31]
Palm Oil Mill Effluents (POME)	Al-Al	30 mm	–	40.21 mA/cm^2	4.4	45.67	71.3 % COD 96.8 % Color 100 % TSS	[22]
POME	Steel wool Monopolar Parallel	8.6 mm	–	542 mA/cm^2	4.4	44.97	95.03% COD 94.52% BOD 96.12% TSS	[17]
Canola Oil Mill Effluent	Al-Al	10 mm	NaCl	0.91–13.66 mA cm^{-2}	–	10–70	90% COD	[32]
Sunflower Oil Mill Effluent	Al-Al	10 mm	NaCl	9.16 mA/cm^2	6	19.06	90% COD	[33]
Olive Oil Mill Effluent	Fe rods	–	PAC/H_2O_2	75 mA/cm^2	6.5	60	78% COD	[18]
Street Food – Hainan Chicken Rice	Al-Al Monopolar Parallel	40 mm	NaCl	20 mA/cm^2	5–7	<10	95% COD 84% FOG	[20]
Sugar Industry Wastewater	Fe-Fe Monopolar Parallel	20mm	–	156 A/m^2	6	120	82% COD 84% Color	[10]

(Continued)

TABLE 10.4
Continued

Type of wastewater	Electrodes/configuration ***	Inter-electrode distance	Electrolyte/coagulant/oxidant *	Current density/voltage	Initial pH	EC time (min)	Removal efficiency **	Reference
Brewery effluent	Al-Al	20 mm	NaCl	75 A/m²	7	60	96.7% Color 46.5% COD	[13]
Distillery	Fe-Fe Bipolar	30 mm	–	0.13 A/dm²	5	240	72% COD	[23]
Coffee Processing	SS/Fe Bipolar	10 mm	–	120 A/m²	7.5	90	95% Color 87% COD	[14]
Cashew Nut effluent	Fe-Fe	10 mm	–	100 mA/cm²	7.4	15	50% COD	[34]
Almond Wastewater	Al-Fe	10 mm	NaCl/Dalfloc	5 mA/cm²	5.7	15	81% COD 80% BOD	[21]
Fruit-juice production wastewater	Fe-Fe	7 mm	–	200 A/m²	4.1	25	61.3% COD	[35]

Note: *sodium chloride (NaCl); potassium chloride (KCl); Polyaluminium chloride (PAC); hydrogen peroxide (H_2O_2); sodium sulphate (Na_2SO_4), **chemical oxygen demand (COD); biological oxygen demand (BOD); total suspended solid (TSS); fat and grease (FOG), ***iron (Fe); aluminium (Al); Stainless Steel (SS).

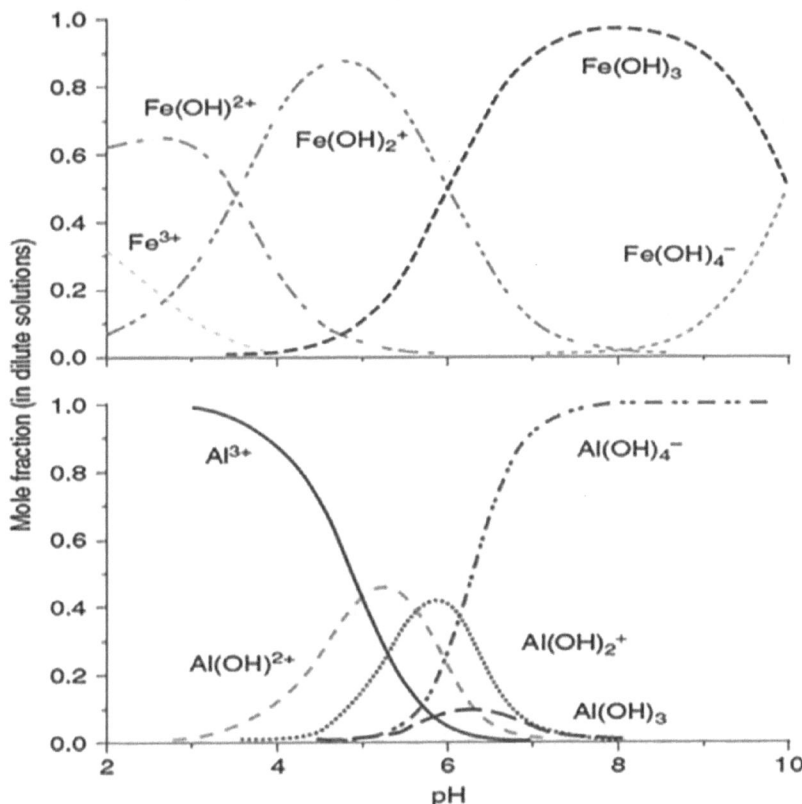

FIGURE 10.4   Mole fractions of dissolved hydrolysis products in equilibrium with amorphous hydroxides for Fe(III) and Al(III) in dilute solutions.

(*Source:* Reprinted from J. Duan and J. Gregory, Adv. Colloid Interfac. Sci., 2003, 100–102, 475–502 [25]).

- Electrolysis time

Faraday's law states that the amount of metal ions released from an anode is proportional to the electrolysis time and current intensity [11]. Electrolysis time affects the treatment efficiency of the EC process which may increase or decrease with current density or pH of the sample. With longer times, higher removal rate of pollutants is achieved due to more metal coagulants and flocs are generated at a constant current density. However, the EC efficiency forms a plateau at specific time frames as the pollutant removal rates become constant due to occupied active sites forming coagulant pollutant flocs [9]. Furthermore, longer reaction times lead to greater electricity consumption and thus, it is important to search for optimal electrolysis times for the EC process.

• Conductivity and supporting electrolyte

The conductivity of the EC process can be enhanced by the supporting electrolyte. A larger current can pass through the electrolytic solution at a smaller voltage, reducing the power consumption of the overall EC process, making it economical by saving time and resources to achieve the same efficiency. The common electrolyte used is sodium chloride (NaCl) or sodium sulphate (Na_2SO_4). Water is a polar solvent, thus ionic compounds, such as NaCl, dissolve and dissociate to form Na^+ and Cl^- ions, which are electrolytes. When voltage is applied across the electrodes, the positively charged ions (Na^+) move to the cathodes and the negatively charged ions (Cl^-) move to the anode and increase the current. High conductivity reduces the electrical resistance of the wastewater, decreases cell voltage and reduces energy consumption. However, addition of electrolytes such as NaCl and Na_2SO_4 at high concentrations is a disadvantage as highly saline wastewater can be the result and may not meet environmental discharge requirements and therefore need additional treatment.

• Mixing speed

The stirring speed is generally kept constant for most of the EC process, at the range of 100–300 rpm. This is to ensure homogeneity in the reactor mixture and enhance pollutant removal rate by imparting velocity through agitation. When the stirring speed was low at 100 rpm, coagulation was limited by collisions and attachments between flocs. It was noted that when mixing speed is increased from 100–150 rpm, it enhances the overall COD removal efficiency. However, when high-speed stirring was increased beyond the optimum range, flocs disintegrated in the reactor and formed small flocs that are hard to remove from water, leading to a decrease in COD removal efficiency. Hence, it is vital to optimize stirring speed as it affects removal efficiency and floc properties.

• Current density

Current density determines both coagulant dosage and bubble generation rates, and strongly influences both mass transfer at the electrodes and solution mixing. The amount of metal dissolved or deposited is dependent on the quantity of electricity that passes through electrolytic solution. The optimal current density always involves a trade-off between operational costs and efficient use of temperature, solution pH and flow rate. At high current densities, the extent of anodic dissolution increases, and in turn, the amount of hydroxo-cationic complexes increases too, which results in an increase in the removal efficiency of the COD and color. This is due to the high amounts of ions produced on the electrodes promoting destabilization of contaminant molecules. For instance, in [9]'s study of EC treatment of dairy wastewater, it can be seen that 60.65%, 77.25%, 81.29%, 88.85%, 89.37%, and 98.84% of COD were removed for applied voltages of 10, 20, 30, 40, 50, and 60 V, respectively, after 60 minutes electrolysis. It was found that as the applied voltage was increased, the required time for the EC process decreased. Also, the amount of oxidized metal increased, resulting in a greater amount of precipitate for the removal of pollutants. Although high current densities allow high removal, it does not benefit in running the EC reactor from an economic perspective. The high electrical power can reduce the lifespan of the electrodes as well as increase the cost of electricity. Furthermore, even though the applied current highly corresponds to the metal ion dissociation and release of ions in the solution, an excess of current may negatively affect the removal efficiency of EC by enabling secondary reactions and over dosage of coagulants can cause charge reversal of the colloids [9]. Therefore, it is crucial to determine optimum current density.

- Type of power supply

Generally, Direct current (DC) power is used generally in EC process, which may cause the formation of an impermeable oxide film on the cathode, which causes the passivation of the anode, leading to an increase of the electrolytic cell resistance, a decrease the ionic transfer and also a rise in energy consumption. The impact of passivation can be reduced by several methods: mechanical cleaning; addition of a sufficient amount of chloride ions, which can assist in breaking down the passive layer on electrodes; and using alternating current (AC). AC can slow down electrode consumption, thus, have higher efficiency and energy reductions.

- Electrode configuration

Electrodes can be organized in different arrangements, length and spacing in order to maximise the removal efficiencies. The electrodes can be arranged in a monopolar or bipolar mode. The most distinctive electrode connections are classified into monopolar-serial (MP-S), monopolar-parallel (MP-P), and bipolar series (BP-S) as shown in Figure 10.5. In the MP-S connection, the individual couple anode-cathode is linked internally where it is not linked with the outer electrodes, whereas in MP-P connection, a particular sacrificial anode is linked with another anode directly; similar arrangements for cathodes also exist. In the BP-S connection,

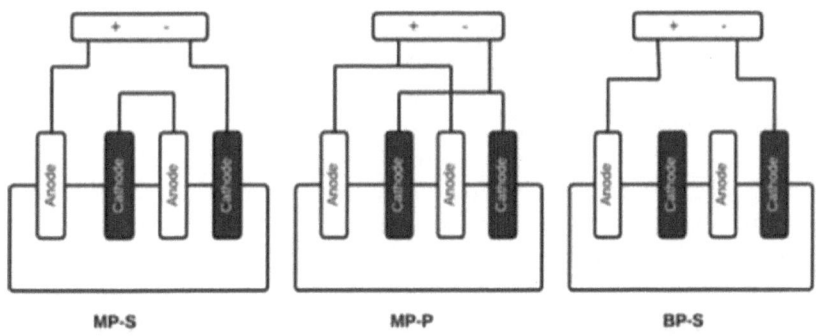

FIGURE 10.5   Different configurations of electrodes, adapted from Moussa et al. (2017).

a particular electrode present changed polarity at particular electrode edges which are subjected to the electrode charge [12]. According to [13], different anode/cathode arrangement of iron (Fe) and aluminium (Al) was studied in brewery effluent and found out that Al/Al serves the highest color and COD removal efficiency between Fe/Fe, Fe/Al, Al/Fe and Al/Al. Besides, stainless steel (SS) and iron electrode combinations (Fe-Fe-SS-SS, Fe-SS-Fe-SS, SS-SS-Fe-Fe, and SS-Fe-SS-Fe) are investigated in the research of [14] on coffee processing industrial wastewater and discovered that SS-SS-Fe-Fe electrode combinations showed higher color and COD removal.

- Material of electrode

The performance of EC process depends on the choice of electrode material as well as the consideration of cost-effectiveness and material availability. The efficiency is mainly judged by the electrode dissociation rate, pollutant removal percentage and the coagulant needed [9]. Various electrode materials have been used in the EC process, such as iron, aluminium, steel, zinc, copper, magnesium, silver and many others. Besides, others such as tin (IV) oxide (SnO_2), lead (IV) oxide (PbO_2), graphite, nickel, and boron-doped diamond (BDD) electrodes have the advantage of greater chemical resistance and higher efficiency in the treatment of cyanide bearing wastewaters [8]. Among them, Al and Fe electrodes have been reported to be commonly used for wastewater treatment due to availability, low cost, and better dissolution. Even though both metals, Fe and Al, are on a similar level in the metal reactivity series, differences in their ion dissolution rates result in varying outcomes in the EC process. The performance of iron electrodes on COD removal is weaker in comparison with aluminium electrodes because of weak settleability of the Fe^{2+} ion. Fe^{2+} is the common ion produced on site of electrolysis of iron electrodes [15]. It has high solubility at acidic or neutral pH and could be oxidized readily into Fe^{3+} by dissolved oxygen in water and is hard to settle. Despite aluminium electrodes showing a higher COD removal efficiency in many studies in comparison with Fe electrodes, Al produces a significant amount of sludge. Both Al and Fe have their own distinct properties with respective mechanisms in their chemical properties and dissociation. More factors such as optimum working pH and reaction time must be considered as well in parameter optimization as they vary for different electrodes, along with the characteristics of the targeted wastewater [9].

- Inter electrode distance

The space between the electrodes affects the reactions in the electrolysis reactor as it determines the electrostatic field between the anode and the cathode. The distance between the electrodes varies from 1 cm to 11.5 cm [11]. During electrolysis, the solution close to the cathode becomes more concentrated because of the different mobilities of the ions present, and this effect can also be reduced by agitation of the bulk solution. Narrower gaps enhance mass transfer and decrease ohmic loss. A narrow spacing of less than 10 mm is accompanied with low-energy consumption. However, the inter-electrode gap gets partially filled with gases and solid particles during electrolysis when the distance is too close, which increases its electrical resistance. An increase in interelectrode spacing increases cell voltage, bringing in an increase of the power consumption. Hence, it is important to run EC at an optimum inter-electrode distance.

10.4 APPLICATION

A list of wastewaters and its EC treatment conditions are shown in Table 10.4. Poultry slaughterhouse wastewater originating from slaughterhouse and meat processing contains high concentrations of organic substances. In the study of [16] for poultry slaughtering wastewater, the peroxy-electrocoagulation was investigated by addition of H_2O_2 with different concentrations, COD was decreased from 8,800 mg/L to colloids [425 mg/L which corresponds to 95.48% of removal efficiency, with an operation cost of $9 per m^3 of wastewater treated. Among POME treatment methods, [17] has shown that potential removal using EC was achieved at 542 mA/cm^2 of current density, 44.97 minutes of treatment time, 8.60 mm of inter-electrode distance and 4.37 of pH value, resulted in removal efficiency of 97.21% COD, 99.26% BOD and 99.00% TSS, respectively. EC method is also applicable to other vegetable oil mill wastewater such as olive oil, canola oil, sunflower seed oil, and *etc*. According to [18], the wastewater from olive oil production is violet-dark brown up to black in color with strong specific olive oil smell, having high degree of organic pollution (COD values up to 220 g/L), pH between 3 and 5.9 (slightly acidic), high content of polyphenols (up to 80 g/L) and high content of solid matter (total solids up to 20 g/L). The removal efficiency of COD for olive oil mill wastewater was in the range of 62–86%, whereas oil-grease and turbidity removal was 100% at the current density range of 20–75 mA/cm^2 depending on the concentrations of H_2O_2 and coagulant aid. Besides, [19]'s study has shown that the effectiveness of EC to treat dairy wastewater is up to 98.84% COD removal at the applied voltage of 60 V and 60-min treatment time. Sugarcane, from which sugar is produced, generally produces wastewater containing high percentages of total solids, COD, BOD, chlorides, sulfates, and oil and grease, but lower percentages of proteins, carbohydrates, and heavy metal [10]. It was observed that 82% reduction in COD and an 84% reduction in color can be achieved for sugar industry wastewater using iron plate at the optimum condition of pH 6, an electrode distance of 20 mm, and a current density of 156 A/m^2. Moreover, wastewater from street food in Thailand, Hainan chicken rice was studied by [20], and the highest COD and Fat and Grease (FOG) removal is found to be 95% and 84% respectively, at condition of current density 20 mA/cm^2, short reaction time less than 10 mins with adding of the sodium chloride. [13] demonstrated that EC process allowed removal of 96.7% color and 46.5% COD on brewery effluents together with the enhanced process of sono-electrocoagulation which ensured 99.2% color and 60.5% COD removal efficiency at pH 7.0, current density of 100 A/m^2, electrode combination of Al/Al, and reaction time of 60 min. Coffee processing wastewaters are generally dark brown in color which contains a large number of persistent pigments called melanoidins which are highly toxic, recalcitrant and non-biodegradable [14] has shown that 95% color and 85.58% COD removal in coffee processing wastewater by EC process at 120 min of reaction time and current density at 120 A/m^2. Furthermore, the nut processing industry includes many various types, including pistachio, almond, and cashew nut, etc produces nut rich effluent which is high in organic nutrients. In the case of the almond processing industry, several processes are required including cracking and blanching. In the blanching step, the resulting product contains high levels of organic compounds, suspended solids, turbidity, COD and color [11]. Based on [21]'s research, 81% COD removal was attained at the condition of initial pH5.7, a current density of 5 mA/cm^2 and 15 mins reaction time.

10.5   HYBRID TECHNOLOGIES

Several researchers had studied the impact of many hybrid treatments with EC processes to enhance its performance which is reviewed and summarised in Table 10.5.

TABLE 10.5
Research Studies on Hybrid Wastewater Treatment With EC Process

Type of wastewater	Hybrid EC Technology	Research Outcome	Reference
Distillery industrial effluent	photo-EC, ECP, photo-ECP	COD removals for photo-EC, ECP and photo-ECP are 78%, 85% and 82% at conditions of current density 0.13 A/dm^2, initial pH 7, H_2O_2 concentration 234 mg/L, stirring speed 100 rpm and reaction time 240 min.	[23]
POME	ECP	71.3 % COD removal and 96.77% color removal were achieved using Al electrodes with optimum conditions of pH 4.4, current density 40.21 mA/cm^2 and 45.67 min reaction time.	[22]
Canola oil refinery effluent	EC-EO	The optimal conditions were achieved for 98.6% sCOD removal at 13.27 mA/cm^2 current density and 403 min operation time. The anode/cathode used for EC is Al/Al, and for EO is BDD/SS.	[32]
Sunflower oil refinery effluent	EC-EO	90–95 % of organic wastes were removed at pH 5.8, current density 8.58 mA/cm^2 and operation time 243 min. The anode/cathode used for EC is Al/Al, and for EO is BDD/SS.	[33]
Cashew Nut effluent	EC-EO	80% COD removal is obtained at 100 mA/cm^2, pH 7.4, 180 reaction time. The anode/cathode used for EC is Fe/Fe, and for EO is BDD/SS.	
Brewery effluent	Sono-EC	98.2% COD removal and 53.7% color removal at a current density (75 A/m^2), pH (7.0), electrode combination (Al/Al), inter-electrode distance (2 cm), reaction time (60 min), and ultrasonic frequency (37 kHz).	[13]
Slaughterhouse Wastewater	ECP	95.48% COD removal efficiency was attained at a current density of 50 mA/cm^2, an initial pH of 3, a flow rate of 0.027 L/min, 0.2 M H_2O_2 and 0.5 g/L PAC.	[16]

10.5.1 ELECTROCHEMICAL-PEROXIDATION OR PEROXI-ELECTROCOAGULATION

It is the hybrid process combining both EC and advanced oxidation is also known as electrochemical peroxidation (ECP). ECP is an advanced oxidation process which involves addition of oxidant, namely hydrogen peroxide (H_2O_2) while operating EC process. For example, under consideration of cost efficiency and removal efficiency in the research of [22], optimized condition of 40.2 mA/cm^2, 45.67 min, 4.43, 0.5 g/L for current density, contact time, initial pH, and H_2O_2 dosage were selected for ECP treatment of POME. Removals obtained were 71.3%, 96.8%, and 100% for COD, colour, and TSS, correspondingly. Combination treatment of ECP as well as coagulant aid are proven hybrid systems for wastewater treatment, at the same time, saving time and energy costs. However, addition of coagulant aid suggests the need for further polishing steps for chemical by-product removal with additional unit operation. Also, the requirement of a powerful oxidant, H_2O_2 in ECP, makes the process less environmentally friendly.

10.5.2 PHOTO-ELECTROCOAGULATION

The performance of EC can be improved by UV radiation, through higher production rate of ·OH from the photoreduction of $Fe(OH)^{2+}$ and photodecomposition of complexes from Fe^{3+} reactions. The study of [23] has demonstrated that photo-EC processes can effectively treat the distillery wastewater up to 78% COD removal at the conditions of current density 0.13 A/dm^2, initial pH 5, stirring speed 100 rpm and reaction time 240 min. A 16 W low-pressure mercury lamp was placed above the EC cell.

10.5.3 SONO-ELECTROCOAGULATION

The process consists of ultrasound and EC techniques. Ultrasound irradiation can cause the formation, growth, and disintegration of small bubbles that concentrate the acoustic energy into tiny reaction zones, leading to extreme conditions in a short time, allowing the formation of radicals [24]. Moreover, ultrasound irritation affects particle size, solubilization, formation of refractory compounds, and organic material structure [13]. The use of ultrasound is a proven technique to prevent passive film formation on the electrode surface, thereby reducing the mass transfer resistance. Cavitation is the phenomenon during ultrasonic due to friction between the liquid and the surface of the electrodes. Cavitation with creating pressure, high temperature, and reduced density of the electrode reactions destroy the passive layer on the electrode surface. On the other hand, cavitation reduces the particle size, which forms a high surface area, leading to more absorption of contaminants. Ultrasonic also facilitates the release of metal ions through the broken surface of the electrode and reducing the distribution of the deep double electrode layer formed on the metal surface, increasing the rate of floc formation. On the other hand, due to reaction with dissolved oxygen leads to the formation of adsorbent hydroxides of pollutants and oxidizing species helps improve the fluctuation and decomposition of pollutants Hence, sono-EC enhances radical production from ultrasound, therefore higher treatment rate and pollution degradation can be achieved. In [24]'s study, nanoparticles are used as one of the operating parameters in order to boost up the degradation efficiency. Nanoparticles refer to compounds that are less than 100 nm with a high specific surface area which, regarding activation conditions through the production of reactive radical species, increases the reaction rate of the contaminant removal. Ultrasound enables the continuous activity of the nanoparticles in oxidation processes by removing the deposited by-products of contaminants on the nanoparticle surface.

There are many more hybrid technologies which are to be studied and used practically in the wastewater treatment of the food industry. Nonetheless, all of the aforementioned applications and future applications require study of the optimal condition in order to ensure the feasibility at its economic and technical aspect and to tackle the arising complications in industrial scale up as well as environmental sustainability.

10.6 TRENDS, OPPORTUNITIES, CHALLENGES

10.6.1 ECONOMIC ASPECTS

Based on study from [4], the main cost components of the EC process are the cost of replacing the dissolved metal material and the cost of electricity. The operating costs (OC) of the EC process may be calculated according to Equation (6). In Equation (7), the formula for calculating the electrical energy consumption (EEC) is given. However, the economic values of a particular EC process calculated using equations are only theoretical.

where

$$OC = aEEC + bEMC + cSEC \qquad (6)$$

$$EEC = \frac{UIt}{60V} \qquad (7)$$

OC = operating cost [$ m^{-3}]
a, b, c = the current market prices of electricity [$ kWh^{-1}], electrode materials [$ kg^{-1}], and supporting electrolyte [$ kg^{-1}], respectively;
EMC = electrode material consumption [kgm^{-3}];
SEC = supporting electrolyte consumption [kgm^{-3}];
EEC = electrical energy consumption [kWh m^{-3}];
U = applied voltage [V];
I = applied Current [A];
t = treatment time [min]; and
V = volume of the treated water [dm^3].

To facilitate additional economic analysis and comparison with other treatment methods, EEC and OC per kg of pollutant (e.g., chemical oxygen demand, COD) may also be readily evaluated. These calculations are made based on the initial or final concentrations of the pollutants in the solution.

EC systems have low running costs as well as investment costs because of their small size and no large mixing or sedimentation pools are needed. They are also rather simple in design and do not contain moving parts, therefore, their maintenance costs are also low. It has significantly lower volumes of sludge produced as compared with coagulation. Moreover, the sludge generated by the EC system has lower water content, larger and more stable flocs with better settleability. The sludge can be further treated and possibly add value economically.

With electricity being the driving force of metal coagulant generation, the need for chemical addition in EC is reduced. However, for the less economically developed countries, electricity becomes a limiting factor for carrying out EC. By reducing current density, the power consumption costs can be tackled while obtaining the same EC efficiency by modifying the other key parameters of EC. For example, the costs can be minimized by reducing the power input by diminishing the inter electrode distance, electrode surface area and raising electrolyte conductivity. In addition, by incorporating renewable energy as power supply, EC can therefore further contribute to both economic and environmental sustainability in wastewater [9]. For example, windmills, fuel cells, or solar electricity, the price of which is currently in decline globally which could reduce the cost of EC application. Since EC systems are also compact in size, decentralization of EC treatment may be viable on a case-specific basis, even comprising application of EC in rural areas, where there is no access to power grids [4].

10.6.2 ECOLOGICAL ASPECTS

The EC process does not usually require other chemicals, thus preventing secondary pollution. However, case-specific optimization may be needed to ensure that residual anode metal concentrations remain low in the EC-treated water. The neutralization effect of water occurring in EC makes it more beneficial in comparison to the use of conventional coagulation chemicals which may significantly lower the pH of water due to hydrolysis of Al/Fe.

Certain types of EC sludge, such as nutrient-containing sludges may be utilized as additives in various ecological fertilizer products in future based on some studies, subject to the need of pre-drying. Similarly, some researchers have suggested based on thermal gravimetric analysis (TGA) and differential thermal analysis (DTA) that some types of EC sludge could be dried and then used as fuel in incinerators or in fuel briquette production.

On the other hand, some toxic chlorinated organic compounds may be formed in case of the removal of organic compounds. Wastewater with high humic and fulvic acid content may be amenable to the formation of trihalomethanes. It is a very harmful and carcinogenic agent. If phenols and algal metabolic and decomposition products are present, chlorine may lead to odor [8]. Moreover, the production of metallic aluminum and iron is very energy-consuming and leads to the formation of large amounts of emissions. Undesirable solids, liquids, and gases are generated both directly and indirectly during mining and processing. Beside the high energy intensity of their production, the metallic aluminum and iron are more costly (per kg) than the corresponding metal salts, because they are purely metal and do not contain crystallization water. This problem can be minimized by metal recycling. Recycling will substantially reduce energy demand, pollution, and environmental impact resulting from, for example, aluminum production as against the use of virgin metal materials. This is very important because in the overall supply chain of material needed in metal production, mineral resource extraction and processing are particularly critical stages in terms of potential emissions release.

10.6.3 Technical Aspects

There is no dominant reactor design and the number of electrodes may vary. Typically, an electrode with a gap of 5–20 mm between the anodes and the cathodes is used. Raising the gap will result in highly increased voltage and, thereby, energy consumption. On the other hand, if a too small gap is used, there may be a risk of clogging. Nonetheless, EC systems are readily automated and may be easily hybridized with other methods. EC has been successfully used in conjunction with other treatment such as electro-oxidation (EO), electromagnetic treatment (proposing slight changes to some of the physicochemical properties of water) prior to EC, hydrogen peroxide (as a modified electro-Fenton process based on iron-EC), ultraviolet (UV) light, ultrasound (sonolysis), aeration (to promote Fe^{2+} conversion to Fe^{3+}), as well as addition of chemicals, bentonite and PACl to enhance coagulation. Innovations such as collecting the hydrogen gas produced during EC using inexpensive and simple equipment, applying APC (alternating pulse current) instead of DC power and partial recirculation of supernatant from EC sludge drying have been studied, with promising results. Converting the recovered electrogenerated hydrogen gas to electricity for the EC process itself could reduce its EEC value up to about 10%.

According to [4], water and wastewater treatment by EC may be considered rapid compared with other methods. Optimal treatment time may typically be less than 60 min excluding sedimentation time. EC processes can be applicable to treating cold wastewater, which is meaningful especially to Nordic countries. EC processes have relatively low electricity consumption and may be designed to be run by green processes, for instance, windmills, fuel cells, or solar cells. Because EC systems are compact in size, deployment of EC treatment may be feasible in rural areas, where there is no access to power grids. Response surface methodology (RSM), along with analysis of variance (ANOVA) has been shown to be feasible in several EC studies, which may significantly help the design and optimization of a given EC application. Typically, high levels of significance and very low percentages of experimental error (related to theoretical models) may be achieved. This applies also to real wastewater instead of only synthetic model waters. It may also be possible to apply artificial intelligence (e.g., neural networks and adaptive neuro-fuzzy inference systems) to optimize EC processes on a case-specific basis.

The disadvantages of EC systems are the need to replace the electrodes periodically and also sufficient water conductivity for the EC to function. Sodium chloride (NaCl) is the most common supporting electrolyte used to raise wastewater conductivity in an EC process because of its low price, nontoxicity and availability. The usage of NaCl may cause the formation of toxic chlorinated organic compounds to form. The efficiency of EC treatment may be reduced by uneven anode dissolution. The cathode passivation may occur due to slow impermeable oxide film development on electrode surfaces, especially for aluminium electrodes. The oxide film will hinder electron exchange between the anode and the cathode as well as metal dissolution, thus increasing consumption of electrical energy. To solve this problem, using alternating current (AC) in EC systems is proposed. The electrical polarity is changed periodically and the cyclic energization between the anode or cathode delays the cathode passivation. Periodic cleaning of the electrodes might also prevent or minimize electrode passivation. Excessive residual concentrations of the anode metal in the environment should be prevented. Iron, used as soluble anode material, may cause coloration of water if overdosing. Hence, optimization of the EC process is needed to ensure proper dosing of the anode metal chosen for the process.

10.7 CONCLUSION

Recent advancements on the role of EC on food wastewater treatments has proven the potential of EC application in food industry wastewater. The characteristics of EC distinctively makes it an attractive option from its ability to treat a wide range of wastewater types, simplicity in setup, cost effectiveness and above all, environmental sustainability by not adding or generating hazardous chemicals. The important aspect of EC that requires consideration is the operational variables which are various with respect to the type of food wastewater. Optimization and modelling are required to ensure its feasibility in real-life application. Moreover, hybrid treatment processes with EC and other methods is encouraged with its promising results in order to reach higher treatment efficiency, to meet the environmental discharge standards and for water reclamation from wastewater and recycling in the industry. Replacing conventional power supply with solar power or other green powers for the EC process is the trend in achieving sustainability goals in many countries and at the same time, saving cost of the EC process. More research needs to be undertaken in terms of parameter optimization, system design and economic feasibility to investigate the potential of EC technology in industrial level specific to the food wastewater type.

REFERENCES

[1] Liu SX. *Food and Agricultural Wastewater Utilization and Treatment*. 2nd Edition. Wiley; 2014.
[2] Overview of the Malaysian oil palm industry 2021. Malaysian Palm Oil Board (MPOB). https://bepi.mpob.gov.my/index.php/en/summary-2/2021/summary-of-the-malaysian-palm-oil-industry-2021.
[3] Latifahmad A, Ismail S, Bhatia S. *Water Recycling from Palm Oil Mill Effluent (POME) using Membrane Technology*. vol. 157. John Wiley & Sons, Ltd; 2003.
[4] Kuokkanen VV. *Water Treatment by Electrocoagulation. Encyclopedia of Inorganic and Bioinorganic Chemistry*. John Wiley & Sons, Ltd; 2016, pp. 1–12. https://doi.org/10.1002/9781119951438.eibc2452.
[5] Holt PK, Barton GW, Mitchell CA. The future for electrocoagulation as a localised water treatment technology. Chemosphere 2005; 59: 355–67. https://doi.org/10.1016/j.chemosphere.2004.10.023.

[6] Reilly M, Cooley AP, Tito D, Tassou SA, Theodorou MK. Electrocoagulation treatment of dairy processing and slaughterhouse wastewaters. Energy Procedia 2019; 161: 343–51. https://doi.org/10.1016/j.egypro.2019.02.106.

[7] Bouhezila F, Hariti M, Lounici H, Mameri N. Treatment of the OUED SMAR town landfill leachate by an electrochemical reactor. Desalination 2011; 280: 347–53. https://doi.org/10.1016/j.desal.2011.07.032.

[8] Islam SMDU. Electrocoagulation (EC) technology for wastewater treatment and pollutants removal. Sustainable Water Resources Management 2019; 5: 359–80. https://doi.org/10.1007/s40899-017-0152-1.

[9] Tahreen A, Jami MS, Ali F. Role of electrocoagulation in wastewater treatment: A developmental review. Journal of Water Process Engineering 2020; 37. https://doi.org/10.1016/j.jwpe.2020.101440.

[10] Sahu O. Significance of iron compounds in chemical and electro-oxidation treatment of sugar industry wastewater: Batch reaction. Environmental Quality Management 2019; 29: 113–23. https://doi.org/10.1002/tqem.21654.

[11] Rakhmania, Kamyab H, Yuzir MA, Abdullah N, Quan LM, Riyadi FA, et al. Recent applications of the electrocoagulation process on agro-based industrial wastewater: A review. Sustainability (Switzerland) 2022; 14. https://doi.org/10.3390/su14041985.

[12] Moussa DT, El-Naas MH, Nasser M, Al-Marri MJ. A comprehensive review of electrocoagulation for water treatment: Potentials and challenges. Journal of Environmental Management 2017; 186: 24–41. https://doi.org/10.1016/J.JENVMAN.2016.10.032.

[13] Dizge N, Akarsu C, Ozay Y, Gulsen HE, Adiguzel SK, Mazmanci MA. Sono-assisted electrocoagulation and cross-flow membrane processes for brewery wastewater treatment. Journal of Water Process Engineering 2018; 21: 52–60. https://doi.org/10.1016/j.jwpe.2017.11.016.

[14] Sahana M, Srikantha H, Mahesh S, Mahadeva SM. Coffee processing industrial wastewater treatment using batch electrochemical coagulation with stainless steel and Fe electrodes and their combinations, and recovery and reuse of sludge. Water Science and Technology 2018; 78: 279–89. https://doi.org/10.2166/wst.2018.297.

[15] Ngobeni PV, Basitere M, Thole A. Treatment of poultry slaughterhouse wastewater using electrocoagulation: A review. Water Practice and Technology 2022; 17: 38–59. https://doi.org/10.2166/wpt.2021.108.

[16] Eryuruk K, Tezcan Un U, Bakır Ogutveren U. Electrochemical treatment of wastewaters from poultry slaughtering and processing by using iron electrodes. Journal of Cleaner Production 2018; 172: 1089–95. https://doi.org/10.1016/j.jclepro.2017.10.254.

[17] Nasrullah M, Ansar S, Krishnan S, Singh L, Peera SG, Zularisam AW. Electrocoagulation treatment of raw palm oil mill effluent: Optimization process using high current application. Chemosphere 2022; 299: 134387. https://doi.org/10.1016/j.chemosphere.2022.134387.

[18] Tezcan Ün Ü, Uğur S, Koparal AS, Bakir Öğütveren Ü. Electrocoagulation of olive mill wastewaters. Separation and Purification Technology 2006; 52: 136–41. https://doi.org/10.1016/J.SEPPUR.2006.03.029.

[19] Bazrafshan E, Moein H, Kord Mostafapour F, Nakhaie S. Application of electrocoagulation process for dairy wastewater treatment. Journal of Chemistry 2013. https://doi.org/10.1155/2013/640139.

[20] Khanitchaidecha W, Ratananikom K, Yangklang B, Intanoo S, Sing-Aed K, Nakaruk A. Application of electrocoagulation in street food wastewater. Water (Switzerland) 2022; 14. https://doi.org/10.3390/w14040655.

[21] Valero D, Ortiz JM, García V, Expósito E, Montiel V, Aldaz A. Electrocoagulation of wastewater from almond industry. Chemosphere 2011; 84: 1290–5. https://doi.org/10.1016/J.CHEMOSPHERE.2011.05.032.

[22] Bashir MJK, Hong Lim J, Abu Amr SS, Peng Wong L, Leng Sim Y. Post treatment of palm oil mill effluent using electro-coagulation-peroxidation (ECP) technique 2018. https://doi.org/10.1016/j.jclepro.2018.10.073.

[23] Asaithambi P, Sajjadi B, Abdul Aziz AR, Wan Daud WMA bin. Performance evaluation of hybrid electrocoagulation process parameters for the treatment of distillery industrial effluent. Process Safety and Environmental Protection 2016; 104: 406–12. https://doi.org/10.1016/j.psep.2016.09.023.

[24] Moradi M, Vasseghian Y, Arabzade H, Mousavi Khaneghah A. Various wastewaters treatment by sono-electrocoagulation process: A comprehensive review of operational parameters and future outlook. Chemosphere 2021; 263. https://doi.org/10.1016/j.chemosphere.2020.128314.

[25] Duan J, Gregory J. Coagulation by hydrolysing metal salts. Advances in Colloid and Interface Science 2003; 100–102: 475–502. https://doi.org/10.1016/S0001-8686(02)00067-2.

[26] Mohammad S, Baidurah S, Kobayashi T, Ismail N, Leh CP. Palm oil mill effluent treatment processes—A review. Processes 2021; 9. https://doi.org/10.3390/pr9050739.

[27] Lee ZS, Chin SY, Lim JW, Witoon T, Cheng CK. Treatment technologies of palm oil mill effluent (POME) and olive mill wastewater (OMW): A brief review. Environmental Technology and Innovation 2019; 15. https://doi.org/10.1016/j.eti.2019.100377.

[28] Chowdhury P, Viraraghavan T, Srinivasan A. Biological treatment processes for fish processing wastewater—A review. Bioresource Technology 2010; 101: 439–49. https://doi.org/10.1016/j.biortech.2009.08.065.

[29] Prajapati AK, Chaudhari PK. Physicochemical Treatment of Distillery Wastewater—A Review. Chemical Engineering Communications 2015; 202: 1098–117. https://doi.org/10.1080/00986445.2014.1002560.

[30] Kolev Slavov A. General characteristics and treatment possibilities of dairy wastewater – A review. Food Technology and Biotechnology 2017; 55: 14–8. https://doi.org/10.17113/ft b.55.01.17.4520.

[31] Bassala HD, Kenne Dedzo G, Njine Bememba CB, Tchekwagep Seumo PM, Donkeng Dazie J, Nanseu-Njiki CP, et al. Investigation of the efficiency of a designed electrocoagulation reactor: Application for dairy effluent treatment. Process Safety and Environmental Protection 2017; 111: 122–7. https://doi.org/10.1016/j.psep.2017.07.002.

[32] Sharma S, Simsek H. Treatment of canola-oil refinery effluent using electrochemical methods: A comparison between combined electrocoagulation + electrooxidation and electrochemical peroxidation methods. Chemosphere 2019; 221: 630–9. https://doi.org/10.1016/j.chemosphere.2019.01.066.

[33] Sharma S, Aygun A, Simsek H. Electrochemical treatment of sunflower oil refinery wastewater and optimization of the parameters using response surface methodology. Chemosphere 2020; 249. https://doi.org/10.1016/j.chemosphere.2020.126511.

[34] da Costa PRF, Emily ECT, Castro SSL, Fajardo AS, Martínez-Huitle CA. A sequential process to treat a cashew-nut effluent: Electrocoagulation plus electrochemical oxidation. Journal of Electroanalytical Chemistry 2019; 834: 79–85. https://doi.org/10.1016/J.JELECHEM.2018.12.035.

[35] Can OT. COD removal from fruit-juice production wastewater by electrooxidation electrocoagulation and electro-Fenton processes. Desalination and Water Treatment 2014; 52: 65–73. https://doi.org/10.1080/19443994.2013.781545.

11 Advanced Oxidation Processes (AOPs) on the Removal of Different Per- and Polyfluoroalkyl Substances (PFAS) Types in Wastewater

Z. Y. Yong, H. Y. Tey, E. L. Yong,
M. H. D. Othman, and H. H. See

CONTENTS

11.1 Introduction 203
11.2 Fenton-Based Reaction 206
11.3 Ozonation 209
11.4 Photocatalytic Process 211
11.5 Comparisons between Fenton, Ozonation and Photocatalysis Processes 213
11.6 Conclusions 214
References 215

11.1 INTRODUCTION

Per- and polyfluoroalkyl substances (PFAS) are a group of emerging anthropogenic chemical compounds that have been extremely persistent in the environment owing to their low molecular polarity and strong carbon-fluorine (C-F) bond making it highly resistant to change either chemically, physically or biologically [1–3]. Their vast applications in a variety of industries as well as in daily consumer products such as fast-food containers, water-proof clothing, firefighting foams, painting materials, cookware, papers, surface cleaners, cosmetics, *etc*. [4–10] led to their ubiquitous distribution in the environment [8, 11, 12]. Nevertheless, it is evident from the FDA's testing of foods cultivated or produced in regions known to have PFAS pollution that PFAS in the ecosystem, such as soil, water, or air may be absorbed by plants and animals and result in contaminated meals for humans. PFAS are able

DOI: 10.1201/9781003260738-11

to bioaccumulate through the food chain [13–15] and both occupationally exposed and non-occupationally exposed communities have these substances reported to be present in their blood, tissues, and breast milk [16–19]. The common exposure pathways of PFAS into the human body are through the consumption of food, drink, and inhaling of dust [20, 21]. However, the FDA has discovered that, generally, very few samples had detectable PFAS and those that do, have low levels after evaluating a wide variety of food samples collected from the general food supply for the Total Diet Study (TDS). In fact, the main source of PFAS exposure is drinking water, particularly when it is contaminated by industrial discharges, municipal wastewater treatment plant effluent, or even leaching from contaminated soil [22–25]. PFAS could be detected in drinking water worldwide from a concentration of pg/L to ng/L [22]. Great concerns have been raised regarding their possible adverse effects on human health especially on PFAS with a longer carbon chain, for instance, perfluoro-octane sulfonate (PFOS) and perfluorooctanoic acid (PFOA) since they are linked to high cholesterol, increased liver enzymes, impairment of the immune system, thyroid disorders, pregnancy-induced hypertension and preeclampsia, cancer (testicular and kidney) and adverse reproductive and development effects [26, 27]. Hence, Stockholm Convention on Persistent Organic Pollutants placed restrictions and regulations on the most widely used PFAS in production, PFOS and PFOA [28, 29]. However, short-chain PFAS (C4-C6) as well as some other newly developing PFAS substitutes, such F-53B, ADONA, and Gen-X, have now taken the place of those prohibited PFAS [26, 30–32]. Although the data on toxicity of these substitutes have not yet been sufficiently investigated, similar properties can be expected due to their analogous structure [26] and in recent years, both short-chain PFAS and its substitutes have frequently been found throughout surface and drinking water [25].

The ripple effect of PFAS usage could substantially challenge risk assessors and regulators once they have entered the environment, especially in water bodies due to the fact that conventional wastewater treatment plants (WWTPs) cannot remove them, posing a severe threat to a clean water supply [32–35]. The treatment methods being applied on standard WWTPs such as sorption and filtration were non-destructive enough to just shift the pollutants from one medium to another [36], while both membrane bioreactor (MBR) and biodegradation systems demonstrated to be ineffective for PFAS remediation in many WWTPs [37, 38]. This is due to PFAS cannot be broken down by the majority of microorganisms in water [39]. Some anaerobic or aerobic biological treatment processes were only able to break the carbon chains (C-C), resulting in the generation of short-chain PFAS and other PFAS transformation products [40]. This in turn results in an increased concentration of PFAS in treated water [41]. Granular activated carbon (GAC) and anion exchange resins that employed adsorption approaches are recently studied for the reduction of PFAS from environmental wastewater samples. GAC is a suitable adsorbent for the removal of hydrophobic pollutants due to its extremely porous nature while anion exchange resins often outperform GAC in terms of capacity yet with higher cost. Although activated carbon, which is commonly used in tertiary treatment, was found out to be able to remove 64 ± 11 and 45 ± 19% of PFOS and PFOA, respectively [42], any PFAS, particularly short-chain PFAS, could not be eliminated by activated carbon and anion exchange resins that had been utilised for more than a year [43]. Nevertheless, the synthesis of GAC and other treatment processes require higher pressure and temperatures, which are more costly. Other conventional WWTPs treatment methods such as chlorination, flocculation or precipitation are also less efficient and may produce toxic by-products [44–46].

Therefore, alternatives to PFAS mitigation such as oxidation processes are employed due to their destructive nature through breaking of the chemical bonds of the contaminants [47]. On that account, for the treatment of PFAS, advanced oxidation processes (AOPs), which were first introduced in the 1980s for drinking water treatment, have been developed as straightforward, clean, and effective alternatives and were widely studied for treatment of different wastewater [40]. This is due to their capability to degrade organic pollutants through the highly oxidizing agents, such as hydroxyl radicals ($\cdot$OH) or sulphate radicals (SO4$\cdot$), being produced in situ which in turn, transformed the contaminants into harmless products [48]. For instance, sulfate radical (SO4$\cdot$) could destroy the C-F bonds and degrade the PFOA compound to form F- ions and CO_2 [49]. Recent studies have shown an intense progress in AOPs, especially in water treatment where there are different ways, modifications and improvement on the decompositions of contaminants by free radicals [50]. Despite this, there are also arguments that the utilization of AOPs in the remediation of PFAS are not feasible as PFAS are extremely prone to degrade due to the poor defluorination caused by the C-F bond's chemical inertness and fluorine's strong electronegativity [51].

In the previous book chapter from this publisher, Kempisty et al. (2018) and Lester et al. (2021) have comprehensively discussed activated persulfate, electrochemical and ultrasonication oxidation methods for the PFAS degradation [52, 53]. Despite the fact that these methods ultimately generate hydroxyl radicals, the route to achieve the formation of this radical is quite complex. Activated persulfate entails the activation of persulfate into free radicals before it can degrade PFAS. Although direct reaction of persulfate with PFAS may occur, the process is slow and reaction specific. Moreover, activation of persulfate takes a long time, making it only suitable for groundwater treatment as its interaction with natural organic matter is slow, thus providing ample time for persulfate to reach the targeted contaminant. For electrochemical oxidation, PFAS are degraded either through its adsorption on the electrode surface or by the oxidizing agents released from the electrode into the bulk liquid. Here, the formation of PFAS and peroxyl radicals are crucial in achieving complete mineralization. Additionally, the effectiveness of this method is highly dependent on the electrolyte type, current density, electrode distance and pH. Meanwhile, ultrasonication would require extreme conditions to homogeneously dissociate water to produce hydroxyl radical. Another possible pathway of degradation in ultrasonication is the occurrence of pyrolysis in the gas bubble where PFAS accumulated. This, again, depends on the ultrasonic frequency, the type of gas employed during the sparging, temperature as well as the density of energy used. Comparatively, Fenton, ozonation and photocatalysis treatment methods are much more straightforward and do not demand for any sophisticated condition to attain the desired removal. Topping up with only a few operational parameters to perform optimization, it offers attractive alternatives for the degradation of PFAS. This chapter, therefore, sheds some light on the recent studies and developments of other AOPs methods available for the remediation of PFAS compounds, the degradation mechanism and their efficiency levels obtained from the degradation of target PFAS in water media.

11.2 FENTON-BASED REACTION

Iron is the most commonly used metal in activating hydrogen peroxide to produce hydroxyl radicals (OH·) in water. The chemical reaction that involves the interaction of hydrogen peroxide and Fe(II) for the generation of strong reactive species, for example the (OH·), is known as the Fenton reaction. Figure 11.1 shows the occurring reactions of the Fenton Process. OH· is generated through electron transfer between hydrogen peroxide and Fe(II). The generated OH· will then decompose the organic pollutants into non-harmful products [54]. Due to this principle, the Fenton process had been widely used in organic wastewater treatment [55]. Nevertheless, the generated OH· will be scavenged by excess amounts of either hydrogen peroxide or Fe(II), which may hinder the process efficiency. Hence, it is necessary to experimentally determine the optimal molar ratio of hydrogen peroxide and Fe(II) to reduce the undesired scavenging phenomenon.

On the other hand, Fenton reactions need to be conducted at an acidic pH condition (pH < 4) to ensure the interconversion of Fe(II) and Fe(III) that maximizes

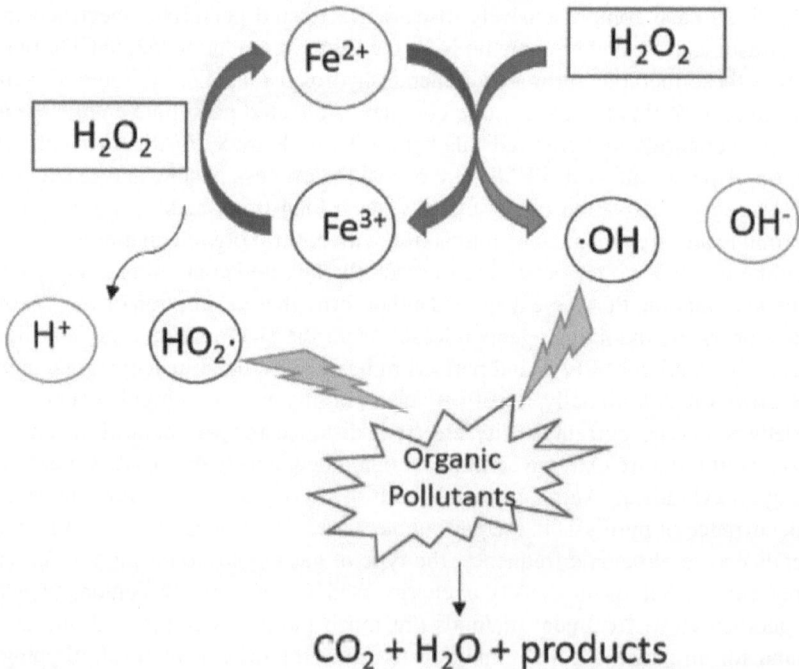

FIGURE 11.1   Reaction mechanism of Fenton process.

the effectiveness and efficiencies in the generation of hydroxyl radicals. When pH is above 4, Fe(III) is not recommended to be used as a catalyst in the Fenton system even though it can be reduced to Fe(II). It is because the rate constant in the reduction reaction of Fe(III) at higher pH is many orders of magnitude less than that in the oxidation reaction of Fe(II). As a result, Fe(III) produces iron hydroxide sludge and part of the catalyst properties is lost, hence reducing the efficacy of the Fenton reaction under conventional water and wastewater treatment settings. Moreover, the unwanted iron sludge needs to be disposed of separately in the wastewater treatment process as it serves as an adsorbent for the pollutants in the wastewater. As recycling of the iron sludge is not feasible, the handling and disposal of the iron sludge inadvertently increases the complexity of the treatment procedure and the operational expenses. Consequently, the use of Fenton reaction for wastewater treatment is constrained in practice.

To overcome the limitation in Fenton reactions in terms of cost and processing, including catalyst loss, high input of chemicals and the formation of sludge, Fenton-like reactions were developed. In Fenton-like reaction, Fe(II) is substituted with Fe(III) or other transition metal ions in the Fenton system. The Fenton-like reaction composed of three steps: adsorption of organics on the catalyst surface, in situ formation and attack of OH· radicals on organic pollutants and desorption of oxidation products from catalyst surface. The Fenton-like reaction is limited due to the narrow pH range, thus leading to the development of photo-Fenton reaction and electro-Fenton reaction.

By using 0.5 mM Fe(III), 0.25, 0.5, and 1 M hydrogen peroxide, and pH 3.5, Mitchell et al. [56] performed the Fenton PFOA oxidation process. The results obtained showed that PFOA was degraded to around 68, 85, and 89% over 150 minutes at initial hydrogen peroxide concentrations of 0.25, 0.5, and 1 M, respectively. However, on the contrary, Schröder and Meester [57] performed standard Fenton treatment on a fluorinated surfactant spiked in Milli-Q-water at pH 3.5. The authors found out that oxidative destruction of fluorinated compounds through Fenton's reagent was ineffective. A review of the literature on the Fenton process was done by Lutze et al. [58]. They claimed that the basic Fenton process is insufficient for PFAS breakdown.

Santos and coworkers [59] investigated the degradation of PFOA using the standard Fenton method and the modified Fenton method at room temperature. The standard Fenton method was deployed by using hydrogen peroxide at concentration of 165 mM with an addition of 3 mM of Fe(III) while a modified Fenton process was deployed by combining humic acid with the Fenton's like reagent (165 mM H_2O_2 and 3 mM Fe(III)) in a hybrid process. The results yielded show that no defluorination of PFOA was observed using the standard Fenton method while complete PFOA removal was achieved before 30 minutes using the modified Fenton method. This is due to the humic acid in the modified Fenton method which were oxidized by Fenton's reagent, modifying their chemical structure that enables them to entrap PFOA in an irreversible way similar to chemisorption.

The mechanism of the Photo-Fenton process in PFOA breakdown is depicted in Figure 11.2. In the photo-Fenton process, the traditional Fenton process is performed by means of UV light, which boosts the generation of hydroxyl radicals and,

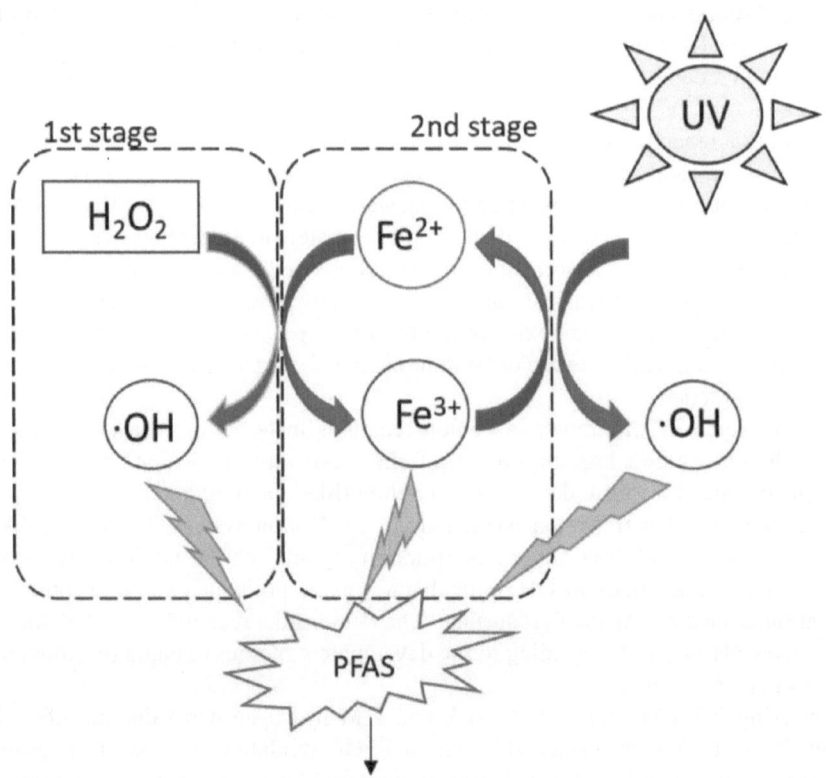

$$CO_2 + 2F^- + H^+ + \text{shorter chain PFAS}$$

FIGURE 11.2   The reaction mechanism of the Photo-Fenton process on PFAS.

as a result, speeds up the pace at which pollutants decompose [60]. With the aid of external high energies (UV), the breaking of the C-F bond can be enhanced [61]. Tang and coworkers [61] investigated the decomposition of PFOA in the UV-Fenton system. A pH 3.0 PFOA solution was used as the experimental medium, and a UV lamp served as the light source. The experiment was carried out in a cylindrical quartz photoreactor with a water jacket. The deterioration was started by turning on the UV light before ferrous sulphate, H_2O_2, and other ingredients were introduced. The results suggest that the UV–Fenton system might effectively decompose PFOA. Within an hour of reaction, the PFOA removal rate was around 87.9 %, as well as the defluorination efficacy was 35.8%. After an hour, the PFOA removal rate increased further with a defluorination efficiency of 53.2% and an efficiency of 95% at a 5-hour reaction time. This complemented the two-stage mechanism for the PFOA degradation process. The highly oxidizing •OH radicals that have been formed as a result of the first stage's rapid disintegration of PFOA sped up the process of decarboxylation. At the second step, most of the H_2O_2 had already been consumed, and interactions between PFOA and Fe^{3+} ions were primarily responsible for the defluorination.

Therefore, it is noteworthy that the modified Fenton process delivered more promising results in the decomposition of PFAS and investigations regarding them are on the rise [51]. Despite this, studies regarding the Fenton reaction on PFAS are still scarce [50].

11.3 OZONATION

For decades, the ozonation procedure has been used extensively to remove organic contaminants in wastewater [62]. Figure 11.3 depicts the classical ozonation process. The decomposition of organic pollutants takes place when the pollutants directly reacted with O_3 molecules or indirectly with ·OH produced from the decomposition of ozone [50]. In the presence of the right catalyst or in conjunction with other techniques, the effectiveness of ozone, one of the strongest oxidants, can be increased further [63]. For instance, Jiménez and coworkers [64] tested the efficiencies of classical ozonation and the combinations of ozone with H_2O_2 and/or Fe^{2+}. It was observed that the addition of Fe to the ozonation process was detrimental as O_3 was used up in the oxidation of Fe^{2+} but improved performance was observed in presence of H_2O_2. Rosenfeldt et al. [65] made a comparison between several AOPs including ozone, ozone+H_2O_2, low pressure UV (LP)+H_2O_2, and medium pressure UV (MP)+H_2O_2. It was found that ozone-based AOPs were more energy-efficient.

In comparison to all other AOPs used in water treatment, the usage of AOP with ozonation has shown greater industrial uses. Utilizing ozone technology, contaminated effluent from electroplating wastes, recycling, the production of electronic chips, marine aquaria, textiles, and oil refineries has been well managed and treated accordingly. The food industry has successfully used ozonation to treat wastewater from distilleries, olive mills, the meat industry, and molasses. Ozone has also been used in municipal treatment facilities to handle landfill leachates, wastewater from rubber additives, and detergents.

Schröder et al. [57] was the first to investigate the decomposition of different fluorinated compounds using ozonation, O_3 in conjunction with UV radiation (O_3/UV), O_3 in conjunction with hydrogen peroxide (O_3/H_2O_2), and H_2O_2 in conjunction with ferrous ions, sometimes referred to as Fenton's reagent (H_2O_2/Fe^{2+}). The experiment took place over the course of 120 minutes in a three-necked cylindrical glass reactor. The monitoring of decomposition was carried out by mass spectrometry with solid-phase pre-concentration.

It can be observed that PFOS resisted all the treatments applied. The O_3/UV irradiation to HFOSA-glycinic acid reduced it by 80% within 120 minutes while other perfluorinated compounds such as NEtFASE-PEG as well as NEtFASE-PEG methyl ether were converted to shorter chain fluorinated compounds. Studies of PFAS detection and removal efficiencies conducted in several countries such as Spain, Australia and Japan on WWTPs that deployed the ozonation process also showed that the PFAS compounds were not removed by the process [42, 43, 66].

On the other hand, modification of the classic ozonation process shows promising results in the removal of PFAS. For instance, Lin and coworkers [67] investigated the elimination of PFOA and PFOS in wastewater samples using ozonation

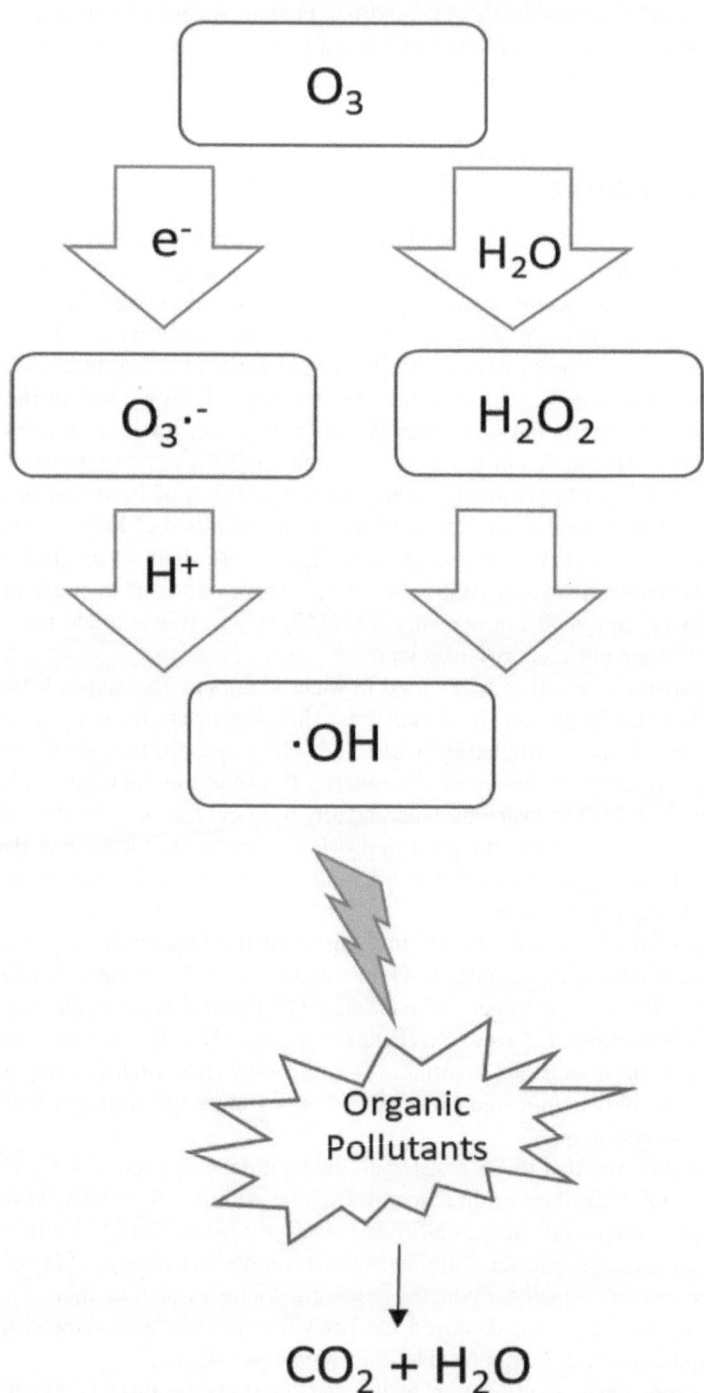

FIGURE 11.3 The mechanism of the ozonation process.

treatment under alkaline conditions. A 15-minutes-ozonation pretreatment was first run on the sample solution at ambient pH (pH 4–5). Following a pH adjustment to 11 using NaOH solution, the ozonation procedure was continued for an additional 6 hours. 92 percent of the PFOA and all the PFOS were completely removed from the wastewater samples, according to the data. Interestingly, Huang et al. [68] managed to combine photocatalyst technology with the ozone treatment process to degrade PFOA. Ozone was continually bubbled through during the experiment, which was carried out in a glass tube reactor at 25°C. TiO_2, which serves as a photocatalyst and is exposed to UV light, was mixed with PFOA solution inside the reaction solution system. The findings suggest that PFOA can be degraded effectively, generating intermediates including fluoride ions as well as short-chain perfluorocarboxylic acids (PFCAs). Its degrading efficiency was 99.1 % after around 4 hours of reaction time.

11.4 PHOTOCATALYTIC PROCESS

Figure 11.4 illustrates the PFOA degradation approach through photocatalysis. In the photocatalysis process, the photocatalyst absorbs light (*hv*), which excites its electrons and moves them from the valence band through into the conduction band. The wavelength required for photoexcitation is specific to the catalyst material because the energy absorbed must be sufficient to overcome the band gap, where larger band gaps require lower wavelengths than catalysts with smaller band gaps. An electron-hole pair is created when a photoexcited electron is present in the conduction band and a positively charged hole is produced in the valence orbital. The electron-hole pairs (e- and h^+) can either interact with molecules adsorbed upon the catalyst's surface or recombine, bringing the catalyst back to the unexcited ground state [39]. The photocatalytic degradation of PFAS can undergo two kinds of mechanisms namely direct and indirect mechanisms. The direct mechanism includes the reduction or oxidation of the contaminant by the e- and h^+ on the catalyst surface.

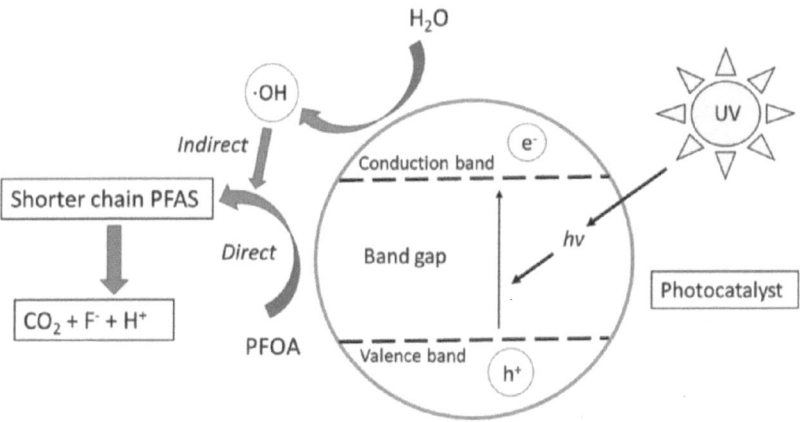

FIGURE 11.4 The mechanism of heterogenous photocatalysis processes for PFAS decomposition.

As for the indirect mechanism, degradation of the contaminant was done by the reactive species which are the hydroxyl radicals that were produced from the reaction of h^+ and H_2O [69]. In the photocatalytic process of degrading PFAS, the C-F bonds of the contaminant were destroyed consecutively forming shorter chain PFAS compounds and eventually into non-harmful products such as CO_2, F- and H^+ [51].

Titanium dioxide (TiO_2) was frequently utilised as a photocatalyst for the elimination of organic pollutants due to its chemical and physical stability, affordability, availability, and non-toxicity [70]. Under UV-Vis light irradiation, the photogenerated holes of photocatalysts such as TiO_2 react with hydroxide anion (OH^-) to produce surface adsorbed hydroxyl radicals, and these hydroxyl radicals engage in the photocatalytic breakdown of organics in the wastewater treatment process. Pure TiO_2 appears to be ineffective in the mineralization of PFAS. Organic contaminant adsorption on the catalytic surface is critical for the photocatalytic response. Improving sorption of pollutants by altering photocatalytic surfaces is one strategy for improving its performance and efficacy. Hence, other metal-doped or modified TiO_2 nanoparticles have been investigated and used to accelerate the degradation of PFAS.

The photocatalytic activity of TiO_2 as well as modified $Cu-TiO_2$ and $Fe-TiO_2$ for PFOA breakdown in aqueous solutions were assessed by Chen et al. [71]. A UV reactor was utilised to accelerate the breakdown of PFOA in aqueous solutions using Fe and Cu altered TiO_2 catalysts that were made using a photo deposition technique. From the results obtained, the degradation efficiency using TiO_2 at a reaction period of 720 minutes was only 15%. However, both $Cu-TiO_2$ and $Fe-TiO_2$ exhibit higher catalytic activity with degradation efficiencies up to 80% and 60% at 480 minutes, respectively. This results from the high rate of recombination of the electron-hole pairs of TiO_2 and the additional Fe and Cu provided traps to catch photo-induced electrons or holes, hence lowering electron–hole recombination during photocatalytic processes which in turn, enhanced its capability to decompose PFOA.

Due to its greater oxidation-reduction potential than TiO_2, gallium oxide (Ga_2O_3) was discovered to have more effective photocatalytic capabilities [72]. Shao et al. [73] studied the breakdown of PFOA using a Ga_2O_3 photocatalyst. Commercial Ga_2O_3 was being synthesized through a microwave irradiation hydrothermal method and the experiment was conducted in a customized photoreactor with a UV light tube and a cold-water recycling system. It was shown that PFOA degrades at rates of 15% at 180 minutes. On top of that, Ga_2O_3 also faced the same problem of fast recombination of electron-hole pairs which limit the performance [39]. Hence, studies on the improvement for Ga_2O_3 photocatalyst should be conducted to increase the degradation efficiency of PFAS compounds.

Li and coworkers [74] conducted a study on heterogeneous photocatalytic decomposition of PFOA by indium oxide (In_2O_3). The results showed that the degradation efficiencies of pure In_2O_3 photocatalyst on PFOA were up to 75% in 240 minutes. This result proved that In_2O_3 possessed 59 times higher photocatalytic activity in terms of PFAS compared to TiO_2. This is due to the higher absorption capacity for PFOA by In_2O_3 compared to TiO_2 with stronger electron-hole pairs created during the photocatalysis process, boosting the PFOA photocatalytic oxidation [74]. However, due to the similar disadvantages of fast recombination of electron-hole pairs, improvements in photocatalytic performance for PFOA breakdown should be addressed in a significant way [39].

Despite the immense potential of photocatalysts, several technological shortcomings prevent PFAS treatment from being used in a practical and scalable manner. For instance, the majority of the investigated photocatalysts for the removal of PFAS are found in nature as nanoparticles that are suspended in water. This would lead to the secondary contamination due to improper handling and improper separation, which is considered to be a major concern in the use of photocatalysts in the PFAS treatment. Furthermore, the significant number of co-contaminants and trace concentration of PFAS in wastewater samples make it difficult to treat directly and hence require further pretreatment techniques.

11.5 COMPARISONS BETWEEN FENTON, OZONATION AND PHOTOCATALYSIS PROCESSES

The comparison for the three AOPs approaches (Fenton-based reaction, ozonation-based process and photocatalytic process) are shown in Table 11.1. In all the AOPs approaches discussed, it can be seen that the degradation of PFOA was greatly improved by doing modification from their standard form.

Among all the modified AOPs, modified Fenton reactions with humic acid could degrade 100% of PFOA in the shortest amount of time. On the other hand, among all the standard AOPs approaches, In_2O_3 photocatalyst shows promising results with 75% PFOA degradation in 240 minutes compared to other photocatalysts, whereas the standard ozone process does not show any sign of removal. There might be a claim that Fenton-based processes were also able to degrade PFOA at a high rate (89%) [56] but a few studies have proven otherwise [57, 58]. Despite that, studies regarding degradation efficiencies of modified In_2O_3 photocatalyst onto PFAS is still scarce and no study has been conducted yet on the effect of immobilized photocatalyst towards PFAS compounds.

TABLE 11.1

Comparisons of the Three AOPs Approaches With Different Modifications. Data From [56, 57, 59, 61, 67, 68, 71, 74, 75]

AOPs type	Modification approach	Reaction period	Yield
Fenton-based reaction	+ humic acid	30 minutes	degrade 100% PFOA
	+ UV	60 minutes	degrade 87.9% PFOA
Ozone-based process	+ UV	120 minutes	0%
	+ H_2O_2	120 minutes	0%
	+ Fenton's reagent	120 minutes	0%
	alkaline conditions (pH11)	360 minutes	degrade 92% PFOA; degrade 100% PFOS
	+ photocatalyst (TiO_2)	240 minutes	degrade 99.1% PFOA
Photocatalytic Process	TiO_2	720 minutes	degrade 15% PFOA
	Cu⁻ TiO_2	480 minutes	degrade 80% PFOA
	Fe⁻ TiO_2	480 minutes	degrade 60% PFOA
	Ga_2O_3	180 minutes	degrade 15% PFOA
	In_2O_3	240 minutes	degrade 75% PFOA

11.6   CONCLUSIONS

The use of AOP is expanding in the wastewater treatment sector. To date, Fenton's reactions, ozonation, and the photocatalytic process have all been extensively studied and used to effectively remove organic pollutants and toxic compounds and subsequently to reduce these compounds that are frequently found in industrial and municipal wastewater. Numerous variables, including the starting concentration of the targeted pollutant, the quantity of oxidising agents and catalysts, light intensity, irradiation period, and the composition of the wastewater's solution, have a significant impact on these AOP processes. In addition, it is important to carry out experimental investigations in order to create a technique appropriate for the particular wastewater. On the other hand, it is also crucial to estimate the capital costs, and overhead and management costs of the AOP in water treatment. This could be accurately estimated by conducting pilot studies that could provide closer conditions to estimate accurate costs.

Due to their persistence in the environment and resistance to removal by standard WWTP methods, PFAS are known as "forever chemicals" and require the adoption of more cutting-edge technology. This chapter, focused on the Fenton-based reaction, Ozone-based process and Photocatalytic process as part of the AOPs in the decomposition of PFAS in water media. Although the approach of each method is different, their pollutant removal approaches are the same, which is to produce hydroxyl radicals and the hydroxyl radicals will decompose the target pollutant. Despite that, the classical methods for the three approaches mentioned are deemed to be ineffective in the treatment of PFAS compounds. This is due to the principal oxidant of AOPs in which the hydroxyl radicals are unable to break the strong C-F bonds of PFAS. However, the modified version of those classical methods gives promising results for PFAS decomposition. For instance, the modified version of the Fenton process either through altering the original surrounding pH to a lower pH or the incorporation of UV light into the system has achieved more than an 85% PFOA removal rate. Besides this the ozonation process incorporated with UV, changing surrounding pH to higher pH value or combining with photocatalyst technology produces a more than 80% of PFOA removal rate. Although a standard photocatalyst could remove PFAS to a certain amount within a period of time, the modified photocatalyst gives a higher degradation efficiency and is able to prevent the drawbacks of a classical photocatalyst which is the fast recombination of an electron-hole pair. Nonetheless, studies on the removal of PFAS and other fluorinated alternatives using the Fenton-based reaction, Ozone-based process and the Photocatalytic process are still scarce and more research on their modified version to improve their functions is recommended.

Although proof-of-concepts for AOP have been widely reported, effective ways to upscale AOP from laboratory scale to full operational scale remain limited. The main obstacle is the operating costs associated with AOP compared to other traditional wastewater treatment techniques. For future consideration in the development of new AOPs, the new approach should demonstrate its cost effectiveness as well as with wider applicability at the same time. The utilisation of renewable energy integrated AOPs could be an essential new element to significantly reduce the cost of energy used by self-producing energy needed during treatment processes. Moreover, a hybrid water treatment system could be designed whereby a cost-effective conventional water treatment system is available for untargeted water treatment while an AOP option is integrated and activated only when a definite treatment target is present. This represents a new trend of SMART treatment systems needed for the future developments, by combining fully automated AOPs with the ancient treatment technologies. Furthermore, these destructive AOP technologies require thorough investigations on the toxicity profile of the numerous PFAS breakdown products to minimise unforeseen repercussions.

REFERENCES

[1] Ahrens L, Bundschuh M. Fate and effects of poly- and perfluoroalkyl substances in the aquatic environment: A review. Environ Toxicol Chem 2014; 33:1921–9. https://doi.org/10.1002/etc.2663.

[2] O'Hagan D. Understanding organofluorine chemistry. An introduction to the C–F bond. Chem Soc Rev 2008; 37:308–19. https://doi.org/10.1039/b711844a.

[3] Vierke L, Möller A, Klitzke S. Transport of perfluoroalkyl acids in a water-saturated sediment column investigated under near-natural conditions. Environ Pollut 2014; 186:7–13. https://doi.org/10.1016/j.envpol.2013.11.011.

[4] Buck RC, Franklin J, Berger U, Conder JM, Cousins IT, De Voogt P, et al. Perfluoroalkyl and polyfluoroalkyl substances in the environment: Terminology, classification, and origins. Integr Environ Assess Manag 2011; 7:513–41. https://doi.org/10.1002/ieam.258.

[5] Jahnke A, Berger U. Trace analysis of per- and polyfluorinated alkyl substances in various matrices-How do current methods perform? J Chromatogr A 2009; 1216:410–21. https://doi.org/10.1016/j.chroma.2008.08.098.

[6] Paul AG, Jones KC, Sweetman AJ. A first global production, emission, and environmental inventory for perfluorooctane sulfonate. Environ Sci Technol 2009; 43:386–92. https://doi.org/10.1021/es802216n.

[7] Post GB, Cohn PD, Cooper KR. Perfluorooctanoic acid (PFOA), an emerging drinking water contaminant: A critical review of recent literature. Environ Res 2012; 116:93–117. https://doi.org/10.1016/j.envres.2012.03.007.

[8] Prevedouros K, Cousins I T, Buck RC, Korzeniowski SH. Critical Review Sources, Fate and Transport of Perfluorocarboxylates. Environ Sci Technol 2006; 40:32–40.

[9] Wang Z, MacLeod M, Cousins IT, Scheringer M, Hungerbühler K. Using COSMOtherm to predict physicochemical properties of poly- and perfluorinated alkyl substances (PFASs). Environ Chem 2011; 8:389–98. https://doi.org/10.1071/EN10143.

[10] Wang Z, Cousins IT, Scheringer M, Buck RC, Hungerbühler K. Global emission inventories for C4-C14 perfluoroalkyl carboxylic acid (PFCA) homologues from 1951 to 2030, Part I: Production and emissions from quantifiable sources. Environ Int 2014; 70:62–75. https://doi.org/10.1016/j.envint.2014.04.013.

[11] Chen S, Jiao XC, Gai N, Li XJ, Wang XC, Lu GH, et al. Perfluorinated compounds in soil, surface water, and groundwater from rural areas in eastern China. Environ Pollut 2016; 211:124–31. https://doi.org/10.1016/j.envpol.2015.12.024.

[12] Giesy JP, Kannan K. Global distribution of perfluorooctane sulfonate in wildlife. Environ Sci Technol 2001; 35:1339–42. https://doi.org/10.1021/es001834k.

[13] Haukås M, Berger U, Hop H, Gulliksen B, Gabrielsen GW. Bioaccumulation of per- and polyfluorinated alkyl substances (PFAS) in selected species from the Barents Sea food web. Environ Pollut 2007; 148:360–71. https://doi.org/10.1016/j.envpol.2006.09.021.

[14] Martin JW, Whittle DM, Muir DCG, Mabury SA. Perfluoroalkyl contaminants in a food web from Lake Ontario. Environ Sci Technol 2004; 38:5379–85. https://doi.org/10.1021/es049331s.

[15] Taniyasu S, Kannan K, Man KS, Gulkowska A, Sinclair E, Okazawa T, et al. Analysis of fluorotelomer alcohols, fluorotelomer acids, and short- and long-chain perfluorinated acids in water and biota. J Chromatogr A 2005; 1093:89–97. https://doi.org/10.1016/j.chroma.2005.07.053.

[16] Kannan K, Corsolini S, Falandysz J, Fillmann G, Kumar KS, Loganathan BG, et al. Perfluorooctanesulfonate and related fluorochemicals in human blood from several countries. Environ Sci Technol 2004; 38:4489–95. https://doi.org/10.1021/es0493446.

[17] Kärrman A, van Bavel B, Järnberg U, Hardell L, Lindström G. Perfluorinated chemicals in relation to other persistent organic pollutants in human blood. Chemosphere 2006; 64:1582–91. https://doi.org/10.1016/j.chemosphere.2005.11.040.

[18] Kärrman A, Domingo JL, Llebaria X, Nadal M, Bigas E, van Bavel B, et al. Biomonitoring perfluorinated compounds in Catalonia, Spain: Concentrations and trends in human liver and milk samples. Environ Sci Pollut Res 2010; 17:750–8. https://doi.org/10.1007/s11356-009-0178-5.

[19] Llorca M, Farré M, Picó Y, Teijón ML, Álvarez JG, Barceló D. Infant exposure of perfluorinated compounds: Levels in breast milk and commercial baby food. Environ Int 2010; 36:584–92. https://doi.org/10.1016/j.envint.2010.04.016.

[20] Björklund JA, Thuresson K, De Wit CA. Perfluoroalkyl compounds (PFCs) in indoor dust: Concentrations, human exposure estimates, and sources. Environ Sci Technol 2009; 43:2276–81. https://doi.org/10.1021/es803201a.

[21] Ericson I, Martí-Cid R, Nadal M, Van Bavel B, Lindström G, Domingo JL. Human exposure to perfluorinated chemicals through the diet: Intake of perfluorinated compounds in foods from the Catalan (Spain) market. J Agric Food Chem 2008; 56:1787–94. https://doi.org/10.1021/jf0732408.

[22] Domingo JL, Nadal M. Human exposure to per- and polyfluoroalkyl substances (PFAS) through drinking water: A review of the recent scientific literature. Environ Res 2019; 177. https://doi.org/10.1016/j.envres.2019.108648.

[23] Bach C, Dauchy X, Boiteux V, Colin A, Hemard J, Sagres V, et al. The impact of two fluoropolymer manufacturing facilities on downstream contamination of a river and drinking water resources with per- and polyfluoroalkyl substances. Environ Sci Pollut Res 2017; 24:4916–25. https://doi.org/10.1007/s11356-016-8243-3.

[24] Coggan TL, Moodie D, Kolobaric A, Szabo D, Shimeta J, Crosbie ND, et al. An investigation into per- and polyfluoroalkyl substances (PFAS) in nineteen Australian wastewater treatment plants (WWTPs). Heliyon 2019; 5:e02316. https://doi.org/10.1016/j.heliyon.2019.e02316.

[25] Yong ZY, Kim KY, Oh JE. The occurrence and distributions of per- and polyfluoroalkyl substances (PFAS) in groundwater after a PFAS leakage incident in 2018. Environ Pollut 2021; 268:115395. https://doi.org/10.1016/j.envpol.2020.115395.

[26] Wang Y, Chang W, Wang L, Zhang Y, Zhang Y, Wang M, et al. A review of sources, multimedia distribution and health risks of novel fluorinated alternatives. Ecotoxicol Environ Saf 2019; 182:109402. https://doi.org/10.1016/j.ecoenv.2019.109402.

[27] Bartell SM, Vieira VM. Critical review on PFOA, kidney cancer, and testicular cancer. J Air Waste Manag Assoc 2021; 71:663–79. https://doi.org/10.1080/10962247.2021.1909668.

[28] Suthersan SS, Horst J, Ross I, Kalve E, Quinnan J, Houtz E, et al. Responding to Emerging Contaminant Impacts: Situational Management. Groundw Monit Remediat 2016; 36:22–32. https://doi.org/10.1111/gwmr.12172.

[29] UNEP, Convention S. Stockholm Convention: General Guidance on POPs Inventory Development, New York: UNEP, 2020.

[30] Munoz G, Labadie P, Botta F, Lestremau F, Lopez B, Geneste E, et al. Occurrence survey and spatial distribution of perfluoroalkyl and polyfluoroalkyl surfactants in groundwater, surface water, and sediments from tropical environments. Sci Total Environ 2017; 607–608:243–52. https://doi.org/10.1016/j.scitotenv.2017.06.146.

[31] Gomis MI. From emission sources to human tissues: Modelling the exposure to per- and polyfluoroalkyl substances. Environ Sci 2017; 91.

[32] Yao Y, Zhu H, Li B, Hu H, Zhang T, Yamazaki E, et al. Distribution and primary source analysis of per- and poly-fluoroalkyl substances with different chain lengths in surface and groundwater in two cities, North China. Ecotoxicol Environ Saf 2014; 108:318–28. https://doi.org/10.1016/j.ecoenv.2014.07.021.

[33] Arvaniti OS, Stasinakis AS. Review on the occurrence, fate and removal of perfluorinated compounds during wastewater treatment. Sci Total Environ 2015; 524–525:81–92. https://doi.org/https://doi.org/10.1016/j.scitotenv.2015.04.023.

[34] Filipovic M, Berger U. Are perfluoroalkyl acids in waste water treatment plant effluents the result of primary emissions from the technosphere or of environmental recirculation? Chemosphere 2015; 129:74–80. https://doi.org/https://doi.org/10.1016/j.chemosphere.2014.07.082.

[35] Chen H, Yao Y, Zhao Z, Wang Y, Wang Q, Ren C, et al. Multimedia Distribution and Transfer of Per- and Polyfluoroalkyl Substances (PFASs) Surrounding Two Fluorochemical Manufacturing Facilities in Fuxin, China. Environ Sci Technol 2018; 52:8263–71. https://doi.org/10.1021/acs.est.8b00544.

[36] Urtiaga A, Fernández-González C, Gómez-Lavín S, Ortiz I. Kinetics of the electrochemical mineralization of perfluorooctanoic acid on ultrananocrystalline boron doped conductive diamond electrodes. Chemosphere 2015; 129:20–6. https://doi.org/https://doi.org/10.1016/j.chemosphere.2014.05.090.

[37] Yu J, Hu J, Tanaka S, Fujii S. Perfluorooctane sulfonate (PFOS) and perfluorooctanoic acid (PFOA) in sewage treatment plants. Water Res 2009; 43:2399–408. https://doi.org/https://doi.org/10.1016/j.watres.2009.03.009.

[38] Kwon H-O, Kim H-Y, Park Y-M, Seok K-S, Oh J-E, Choi S-D. Updated national emission of perfluoroalkyl substances (PFASs) from wastewater treatment plants in South Korea. Environ Pollut 2017; 220:298–306. https://doi.org/https://doi.org/10.1016/j.envpol.2016.09.063.

[39] Xu B, Ahmed MB, Zhou JL, Altaee A, Wu M, Xu G. Photocatalytic removal of perfluoroalkyl substances from water and wastewater: Mechanism, kinetics and controlling factors. Chemosphere 2017; 189:717–29. https://doi.org/10.1016/J.CHEMOSPHERE.2017.09.110.

[40] Araújo RG, Rodríguez-Hernandéz JA, González-González RB, Macias-Garbett R, Martínez-Ruiz M, Reyes-Pardo H, et al. Detection and Tertiary Treatment Technologies of Poly-and Perfluoroalkyl Substances in Wastewater Treatment Plants. Front Environ Sci 2022; 10.

[41] Gagliano E, Sgroi M, Falciglia PP, Vagliasindi FGA, Roccaro P. Removal of poly- and perfluoroalkyl substances (PFAS) from water by adsorption: Role of PFAS chain length, effect of organic matter and challenges in adsorbent regeneration. Water Res 2020; 171:115381. https://doi.org/10.1016/j.watres.2019.115381.

[42] Flores C, Ventura F, Martin-Alonso J, Caixach J. Occurrence of perfluorooctane sulfonate (PFOS) and perfluorooctanoate (PFOA) in N.E. Spanish surface waters and their removal in a drinking water treatment plant that combines conventional and advanced treatments in parallel lines. Sci Total Environ 2013; 461–462:618–26. https://doi.org/10.1016/J.SCITOTENV.2013.05.026.

[43] Takagi S, Adachi F, Miyano K, Koizumi Y, Tanaka H, Watanabe I, et al. Fate of Perfluorooctanesulfonate and perfluorooctanoate in drinking water treatment processes. Water Res 2011; 45:3925–32. https://doi.org/10.1016/J.WATRES.2011.04.052.

[44] Rout PR, Zhang TC, Bhunia P, Surampalli RY. Treatment technologies for emerging contaminants in wastewater treatment plants: A review. Sci Total Environ 2021; 753:141990. https://doi.org/10.1016/j.scitotenv.2020.141990.

[45] McCord JP, Strynar MJ, Washington JW, Bergman EL, Goodrow SM. Emerging Chlorinated polyfluorinated polyether compounds impacting the waters of southwestern New Jersey Identified by use of nontargeted analysis. Environ Sci Technol Lett 2020; 7:903–8. https://doi.org/10.1021/ACS.ESTLETT.0C00640/SUPPL_FILE/EZ0C00640_SI_001.PDF.

[46] Mazhar MA, Khan NA, Ahmed S, Khan AH, Hussain A, Rahisuddin, et al. Chlorination disinfection by-products in municipal drinking water – A review. J Clean Prod 2020; 273:123159. https://doi.org/10.1016/J.JCLEPRO.2020.123159.

[47] Franke V, Schäfers MD, Lindberg JJ, Ahrens L. Removal of per- and polyfluoroalkyl substances (PFASs) from tap water using heterogeneously catalyzed ozonation. Environ Sci Water Res Technol 2019; 5:1887–96. https://doi.org/10.1039/C9EW00339H.

[48] Yang L, He L, Xue J, Ma Y, Xie Z, Wu L, et al. Persulfate-based degradation of perfluorooctanoic acid (PFOA) and perfluorooctane sulfonate (PFOS) in aqueous solution: Review on influences, mechanisms and prospective. J Hazard Mater 2020; 393:122405. https://doi.org/10.1016/J.JHAZMAT.2020.122405.

[49] Hori H, Yamamoto A, Hayakawa E, Taniyasu S, Yamashita N, Kutsuna S, et al. Efficient Decomposition of Environmentally Persistent Perfluorocarboxylic Acids by Use of Persulfate as a Photochemical Oxidant. Environ Sci Technol 2005; 39:2383–8. https://doi.org/10.1021/ES0484754.

[50] Trojanowicz M, Bojanowska-Czajka A, Bartosiewicz I, Kulisa K. Advanced Oxidation/Reduction Processes treatment for aqueous perfluorooctanoate (PFOA) and perfluorooctanesulfonate (PFOS) – A review of recent advances. Chem Eng J 2018; 336:170–99. https://doi.org/10.1016/J.CEJ.2017.10.153.

[51] Ahmed MB, Alam MM, Zhou JL, Xu B, Johir MAH, Karmakar AK, et al. Advanced treatment technologies efficacies and mechanism of per- and poly-fluoroalkyl substances removal from water. Process Saf Environ Prot 2020; 136:1–14. https://doi.org/10.1016/J.PSEP.2020.01.005.

[52] Kempisty DM, Xing Y, Racz L. *Perfluoroalkyl Substances in the Environment: Theory, Practice, and Innovation*, Boca Raton: CRC Press, 2018.

[53] Lester Y. Recent Advances in oxidation and reduction processes for treatment of PFAS in Water. Forever Chem 2021:271–90. https://doi.org/10.1201/9781003024521-16.

[54] Pignatello JJ, Oliveros E, MacKay A. Advanced oxidation processes for organic contaminant destruction based on the fenton reaction and related chemistry 2007; 36:1–84. https://doi.org/10.1080/10643380500326564.

[55] Zhang M hui, Dong H, Zhao L, Wang D xi, Meng D. A review on Fenton process for organic wastewater treatment based on optimization perspective. Sci Total Environ 2019; 670:110–21. https://doi.org/10.1016/J.SCITOTENV.2019.03.180.

[56] Mitchell SM, Ahmad M, Teel AL, Watts RJ. Degradation of perfluorooctanoic acid by reactive species generated through catalyzed H 2 O 2 propagation reactions. Environ Sci Technol Lett 2013; 1:117–21. https://doi.org/10.1021/EZ4000862/SUPPL_FILE/EZ4000862_SI_001.PDF.

[57] Schröder HF, Meesters RJW. Stability of fluorinated surfactants in advanced oxidation processes—A follow up of degradation products using flow injection–mass spectrometry, liquid chromatography–mass spectrometry and liquid chromatography–multiple stage mass spectrometry. J Chromatogr A 2005; 1082:110–9. https://doi.org/10.1016/J.CHROMA.2005.02.070.

[58] Lutze H, Panglisch S, Bergmann A, Schmidt TC. Treatment options for the removal and degradation of polyfluorinated chemicals. Handb Environ Chem 2012; 17:103–25. https://doi.org/10.1007/978-3-642-21872-9_6/TABLES/3.

[59] Santos A, Rodríguez S, Pardo F, Romero A. Use of Fenton reagent combined with humic acids for the removal of PFOA from contaminated water. Sci Total Environ 2016; 563–564:657–63. https://doi.org/10.1016/J.SCITOTENV.2015.09.044.

[60] Ameta R, Chohadia AK, Jain A, Punjabi PB. Fenton and photo-fenton processes. Adv Oxid Process Wastewater Treat Emerg Green Chem Technol 2018:49–87. https://doi.org/10.1016/B978-0-12-810499-6.00003-6.

[61] Tang H, Xiang Q, Lei M, Yan J, Zhu L, Zou J. Efficient degradation of perfluorooctanoic acid by UV–Fenton process. Chem Eng J 2012; 184:156–62. https://doi.org/10.1016/J.CEJ.2012.01.020.

[62] Langlais B, Reckhow DA, Brink DR. *Ozone in Water Treatment: Application and Engineering*. London: Lewis Publishers, 1991.

[63] Nawrocki J, Kasprzyk-Hordern B. The efficiency and mechanisms of catalytic ozonation. Appl Catal B Environ 2010; 99:27–42. https://doi.org/10.1016/J.APCATB.2010.06.033.

[64] Jiménez S, Andreozzi M, Micó MM, Álvarez MG, Contreras S. Produced water treatment by advanced oxidation processes. Sci Total Environ 2019; 666:12–21. https://doi.org/10.1016/J.SCITOTENV.2019.02.128.

[65] Rosenfeldt EJ, Linden KG, Canonica S, von Gunten U. Comparison of the efficiency of OH radical formation during ozonation and the advanced oxidation processes O_3/H_2O_2 and UV/H_2O_2. Water Res 2006; 40:3695–704. https://doi.org/10.1016/J.WATRES.2006.09.008.

[66] Thompson J, Eaglesham G, Reungoat J, Poussade Y, Bartkow M, Lawrence M, et al. Removal of PFOS, PFOA and other perfluoroalkyl acids at water reclamation plants in South East Queensland Australia. Chemosphere 2011; 82:9–17. https://doi.org/10.1016/J.CHEMOSPHERE.2010.10.040.

[67] Lin AYC, Panchangam SC, Chang CY, Hong PKA, Hsueh HF. Removal of perfluorooctanoic acid and perfluorooctane sulfonate via ozonation under alkaline condition. J Hazard Mater 2012; 243:272–7. https://doi.org/10.1016/J.JHAZMAT.2012.10.029.

[68] Huang J, Wang X, Pan Z, Li X, Ling Y, Li L. Efficient degradation of perfluorooctanoic acid (PFOA) by photocatalytic ozonation. Chem Eng J 2016; 296:329–34. https://doi.org/10.1016/J.CEJ.2016.03.116.

[69] Sun Q, Zhao C, Frankcombe TJ, Liu H, Liu Y. Heterogeneous photocatalytic decomposition of per- and poly-fluoroalkyl substances: A review. Https://DoiOrg/101080/1064338920191631988 2019; 50:523–47. https://doi.org/10.1080/10643389.2019.1631988.

[70] Sornalingam K, McDonagh A, Zhou JL. Photodegradation of estrogenic endocrine disrupting steroidal hormones in aqueous systems: Progress and future challenges. Sci Total Environ 2016; 550:209–24. https://doi.org/10.1016/J.SCITOTENV.2016.01.086.

[71] Chen MJ, Lo SL, Lee YC, Huang CC. Photocatalytic decomposition of perfluorooctanoic acid by transition-metal modified titanium dioxide. J Hazard Mater 2015; 288:168–75. https://doi.org/10.1016/J.JHAZMAT.2015.02.004.

[72] Zhao B, Zhang P. Photocatalytic decomposition of perfluorooctanoic acid with β-Ga2O3 wide bandgap photocatalyst. Catal Commun 2009; 10:1184–7. https://doi.org/10.1016/J.CATCOM.2009.01.017.

[73] Shao T, Zhang P, Jin L, Li Z. Photocatalytic decomposition of perfluorooctanoic acid in pure water and sewage water by nanostructured gallium oxide. Appl Catal B Environ 2013; 142–143:654–61. https://doi.org/10.1016/J.APCATB.2013.05.074.

[74] Li X, Zhang P, Jin L, Shao T, Li Z, Cao J. Efficient photocatalytic decomposition of perfluorooctanoic acid by indium oxide and its mechanism. Environ Sci Technol 2012; 46:5528–34. https://doi.org/10.1021/ES204279U/SUPPL_FILE/ES204279U_SI_001.PDF.

[75] Zhao P, Xia X, Dong J, Xia N, Jiang X, Li Y, et al. Short- and long-chain perfluoroalkyl substances in the water, suspended particulate matter, and surface sediment of a turbid river. Sci Total Environ 2016; 568:57–65. https://doi.org/10.1016/J.SCITOTENV.2016.05.221.

12 Photocatalysis for Oil Water Treatment

Baskaran Sivaprakash and N. Rajamohan

CONTENTS

12.1 Introduction 221
12.2 Principle of Photocatalysis 223
12.3 Application of Photocatalyst in the Treatment of Oily Wastewater 225
    12.3.1 TiO$_2$ Catalyst 229
    12.3.2 ZnO Catalyst 232
    12.3.3 Miscellaneous Catalysts...233
12.4 Restrictions in Photocatalysis 234
12.5 Conclusion 234
References...235

12.1 INTRODUCTION

In the year 1972, Kenichi Honda and Akira Fujishima studied the technique of photocatalysis of splitting up water by utilizing titanium dioxide (TiO$_2$) as electrode under the influence of ultraviolet light [1], stating that photocatalysis is the procedure which involves photon (light) induced reaction takes place in the presence of a catalyst which greatly accelerates the chemical reaction. The catalyst has a significant impact in lowering the activation energy by accelerating the reaction, whereas illuminators like UV light, solar radiation, visible light, and light-emitting diodes enhance the photons required [2]. Multiple factors including the addition of catalyst concentration, pH value, temperature range, intensity, and wavelength of light, influence the photocatalysis phenomenon. The photocatalysis proposed a good photochemical stability and better removal efficiency [3].

Photocatalysis can be done by utilizing two types of catalyst such as, homogeneous and heterogeneous. Heterogeneous photocatalysis is superior to homogeneous since heterogeneous catalysts involve advanced oxidation processes, which can be applied in the treatment of air and water, splitting of water, reduction of CO$_2$ and fixation of N$_2$ [4, 5]. The heterogeneous reaction is more advantageous than homogeneous because it gives complete mineralization, cheap cost, there is no disposal of waste and it does not require a high amount of temperature and pressure conditions [6].

The designing of photocatalytic reactors is used in the treatment of wastewater since it increases the reaction rate and area of reaction which tremendously enhances

DOI: 10.1201/9781003260738-12

the photocatalytic reaction. The reactors are categorized according to the light availability and based on the system and parameters such as concentration of light and the catalyst used, that is, the suspended and immobilized reactor, packed bed reactor and fluidized bed reactor [5–7]. The photocatalytic reactor can be integrated with a membrane which is termed as the membrane photocatalytic reactor. This synergistic membrane photocatalytic reactor also exhibits some disadvantages like fouling, maintenance difficulty, UV light induced membrane damage and it requires high maintenance cost. There are photo electrocatalytic reactors which are constructed using catalysts and copper as a cathode.

Photocatalysis can be used in the application of treatment of wastewaters like oily wastewater. Oily wastewaters discharged from petrochemical, drilling sight, vegetable oil industries, refineries, and oil and gas industries cause a great impact on human health and are capable of causing biological oxygen demand and chemical oxygen demand. Oil refineries contribute significantly to the oily wastewater with more aliphatics and phenols which are extremely toxic and unstable that leads to endocrine disruption and carcinogenic hazards. The quantum of wastewater released from crude oil processing ranges from 0.4 to 1.6 times the quantity of oil handled. Ineffective discharge of this creates a very high imbalance in the ecosystem and diminishes the profile of water bodies and soil. Unlike urban wastewater, oily wastewaters are highly complex and indefinite in their constituents and composition which depend on the type of crude oil, operational parameters, plant designs, etc. A typical oil refinery wastewater comprises hydrocarbons as high as 600 to 1500 ppm in suspended and dissolved forms in addition to 20 to 300 ppm of phenolic ingredients and COD level of 200 to 950 ppm [8]. Being categorized under emerging contaminants, there is an urgent demand to deal with this oily wastewater treatment. Mechanical, physicochemical and biochemical treatment methods including flotation, adsorption, hydrocyclones, membrane separation, plate separation and hydrothermal oxidation methods suffer from many limitations like sludge formation, limited applicability, deterioration of materials used, lesser efficiency, etc [8]. Advanced oxidation induced by photocatalysis has proved to be a more efficient and predominant method.

In naturally occurring self- purification photocatalytic processes for the cleansing of water, transition metals that emanate from rocks, minerals, steel wastes, etc. play a crucial role. Advanced oxidative degradation supported by photocatalysis can be considered as a sustainable and eco-friendly method. This is owing to the fact that the environmental oriented photochemical reactions exist forever due to the UV intensification from solar sources to act on the surface waters' photic layers. Oxidation of oily contaminants is accomplished by hydrogen peroxide, molecular oxygen and ozone influenced by light and photocatalysts. Heterogeneous photocatalysis with semiconductors rich in valence and conductance zones and with high dispersive character has good prospects in oily waste treatment [9]. This chapter presents an outlook on the application of the widely used catalysts TiO_2 and ZnO in the treatment of oily wastewaters. The essential characteristic feature of a material to be applied as a photocatalyst is the potential to get excited by light and release electron – hole pairs. The required energy should be higher than the band gap. Among the various semiconductor catalysts, TiO_2 and ZnO find wide applicability as they possess relatively higher ability to absorb visible and UV light, high stability towards chemical, biological and photo agents and are also cheaper when compared to other heterogeneous catalysts. Very importantly they exhibit promising degradation ability in a variety of organic wastewaters including oily contaminants [3–10]. This chapter presents the principles of heterogeneous photocatalysis, application of photocatalysts in oil contaminated wastewater treatment with special focus on TiO_2 and ZnO based solid semiconductor materials in different forms, their mechanism of action on target pollutants, comparative studies on their performance and limitations in commercialization.

12.2 PRINCIPLE OF PHOTOCATALYSIS

The primary principle of the photocatalytic process is to produce CO_2, H_2O, and mineral salts by decomposing and mineralization of organic compounds [11]. On a broader context, a photocatalyst produces reactive oxygen species like •OH, O_2^-, H_2O_2 and $^1O_2$ that can degrade organic contaminants to harmless intermediates and carbon dioxide and water. This happens by three aspects which includes: (a) heat/light/ultrasonic induced energy sources, (b) oxidants and (c) catalysts/semiconductors [12]. In the photocatalytic process, the wavelength of the photon, catalyst, and bandgap energy needs to be cautiously chosen so that we get the best out of the process. In wastewater treatment of photocatalysis, the process is taken in steps, the first step is excitation, in which under the providence of adequate energy the photon generated electrons—hole pairs, as the means of excited electrons leaving the valence band and jump to the conduction band of a semiconductor producing voids in the valence band. This mechanism offers the photocatalyst with a small bandgap to capture more photons. The voids in the valence band oxidize the donor molecule and combine with water molecules to generate hydroxyl radicals, which acts as a powerful oxidizing agent capable of removing a large amount of effluent. The next step is the separation of photon-generated electrons from holes, where the dissociated holes react with hydroxyl ions or water molecules [13]. The hydroxyl radicals directly engage in oxidative degradation due to their strong oxidative nature. The electron in the conduction band reacts with oxygen species in the formation of superoxide ions; they induce a redox reaction by increasing the mobility of the charge. The holes and electrons experience consecutive oxidation and redox reaction in the presence of the catalyst, the contaminants are adsorbed on the surface of the catalyst. The overall principle and mechanism of photocatalysis are portrayed in Figure 12.1.

A typical photocatalysis comprises several steps in its mineralization of toxic organic contaminants [14].

1. Synthesis of charge carriers facilitated by absorption of photons
2. Recombination of the charge carriers
3. Trapping of electrons and holes
4. Evolution of reductive and oxidative pathways by electrons and holes respectively
5. Mineralization of complex compounds by photocatalytic and thermal reactions

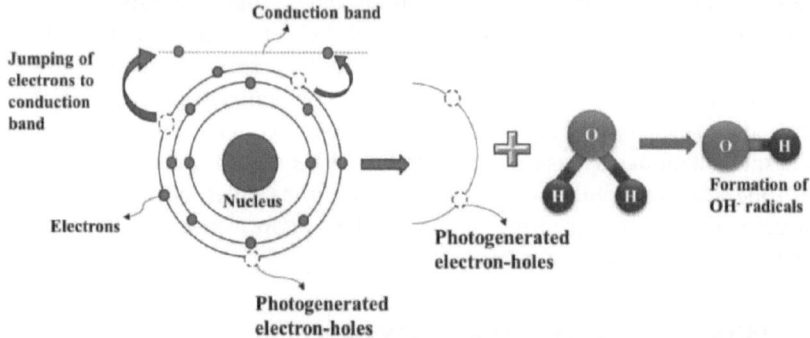

FIGURE 12.1   Mechanism of photocatalysis achieved by generating electron hole pairs with hydroxyl radicals under the influence of catalyst and illuminators for the degradation of pollutants.

The action of holes and electrons on photocatalysis is a key factor in carrying out the degradation. They occupy the surface of the active sites due to two reasons: 1. Rather than the bulk, relaxation of lattice for trapping of electrons and holes has lesser constraints in the surface, 2. The energy levels of the self-trapped carriers in the surface are higher than that of the bulk, thereby enabling the migration of the electrons and holes towards the surface with higher driving force. There is yet another interesting mechanistic fact in photocatalysis called the antenna effect. The excitons liberated by absorption of light in nano catalyst material keep migrating till they get trapped on the active site. The trapping is facilitated by adsorption of the electron receivers or donors in the aggregate particles. The charge carrier trapped gets involved in redox reactions, whereas the other finds its way along the particle network assembly. Thus, the cooperative effect of the charge carriers within the network enables improvised photocatalytic activity [14].

Depending on the usage of the catalyst, the process is categorized as homogeneous and heterogeneous photocatalysis. The homogeneous photocatalytic process is where metal complexes like iron, copper, chromium, etc. are mostly used as catalysts, in this process hydroxyl radicals of high metal ion complexes are generated under photon and thermal conditions with higher oxidation state, these hydroxyl radicals remove toxic substance when reacted with organic substance. In heterogeneous process semiconducting materials like TiO_2, ZnO, SnO_2, and CeO_2 act as heterogenous photocatalysts due to their good electronic structures having an occupied valence band and a vacant conduction band, which makes them have good light adsorption properties, good transport of charge, a long lifetime in an excited state. This makes the heterogeneous photocatalytic process more advantageous than the homogeneous process [15–17].

In the photocatalytic reaction, the catalyst plays an important role, utilizing semiconductors with lower energy bandgap in aqueous solution that undergoes photo corrosion by rapid recombination and redox reaction in the process of photogenerated electrons-holes pair. TiO_2 and ZnO are the most stable catalysts as these catalysts provide better thermal, sufficient chemical strength with strong mechanical properties, when exposed to acidic and basic compounds. TiO_2 exists in crystalline form as anatase, brookite, and rutile where anatase exhibit high photocatalytic activity, TiO_2 is a highly oxidative compound that can degrade both organic and inorganic substance at a low concentration and it has a high pigmentary character which enhances the adsorption of ultraviolet and it is very stable in aqueous solution at any pH conditions. The catalyst ZnO has a good photocatalytic activity due to its good charge transport characteristics, long-term photon stability and it is non-toxic in nature. Hexagonal wurtzite, cubic rock salt, and cubic zinc mixture are the three types of ZnO crystal form. A semiconducting metal Bi_2MoO_6 shows an excellent photocatalytic activity because of its high stability, nontoxic in nature, low cost, and it is an n-type semiconductor. To enrich the photocatalytic activity of Bi_2MoO_6 many methods of semiconductor combination, and elemental doping is used. Pure Bi_2MoO_6 is not good in photocatalytic activity due to its bad quantum yield and fast recombination of photo-induced charge carriers. In the photocatalytic reaction, various modification methods, such as doping and simultaneous co-doping with metal or non-metal catalysts were widely investigated [18–20].

12.3 APPLICATION OF PHOTOCATALYST IN THE TREATMENT OF OILY WASTEWATER

The combined contribution of oil, greases, salts, surfactants, fats discharge and hydrocarbons deposition on the water surface at higher concentration is termed as oily waste. These oily wastes can be recognized or recollected as either unstable or high stable, free-floating, or as a suspension in the aquatic environment. The surmounted chemical and biological oxygen demand, along with total suspended solids, total hydrocarbons, and other toxic organic pollutants were categorized under oily wastewater. The oil from palm, engine, olive, vegetables, and oilfields together

TABLE 12.1
Pros and Cons of TiO_2 and ZnO in Photocatalysis Applications

	TiO_2	ZnO
Nature	Inorganic toxic heavy metal origin	Inorganic non-toxic heavy metal origin
Crystalline structure	Anatase, Brookite and Rutile	Hexagonal wurtzite, cubic zinc blende
Bandwidth	3–3.2 eV	3.3 eV
Photon adsorption under UV	Has good photon reception for wider range of wavelength	Lesser than that of TiO_2
Minimum radiance of wavelength	387.5 nm	425 nm
COD removal potential	Low	High
Limitations	Has disadvantages in doping with non-metallic ions	Has wide band gap rising quick recombination causing less efficiency
Economy	Expensive	Cheaper
Safety	Toxic	Non – toxic

contributes to the oily waste generation [3]. Some important sources of oily waste discharges were restaurants, fertilizers, beverage industries, slaughterhouses, petrochemical industries, offshore oil leakage, pharmaceutical industries, refineries, dairy industries, metallurgical, leather industries, and so on. These discharges cause adverse and detrimental effects to both environments as well as the living forms. The direct and indirect consumption of oily waste from contaminated water sources will lead to carcinogenic and mutagenic effects on human health. Also, the oil ejection affects the soil, and marine ecosystem, suppresses superficial and groundwater quality, and also emits unwanted volatile contaminants into the biosphere.

The presence of biological oxygen demand (BOD) and chemical oxygen demand (COD) on the aquatic surfaces will restrict the capacity of sunlight to penetrate deep sea level. The COD is the oxygen requirement to oxidize the organic substance present, whereas the requirement of the oxygen supply needed by bacterial species to break down the organic waste is termed as BOD. These oil wastes deplete the total dissolved oxygen content in water, leading to insufficiency of oxygen for aquatic species and even leading to eutrophication. These effects will pose a serious threat to aquatic flora, fauna, and the micro-organism community. The possessed gelatinous nature of oil waste blocks the sewage and drainage holes and eventually affects the agricultural sector, thereby washing out the physico-chemical terminologies in both land and water surface. So there is an urgent need to treat the oil wastes in the biosphere. Based on the size of oil droplets and their nature of availability in water, the oily waste can be typically classified. The oil droplet size greater than 150 micron is termed as free-floating oily waste, between 20 and 150 micron is termed as dispersed oily waste, size less than 20 micron as emulsified oily waste, and dissolved oil ranges below 5 microns. The regulatory limit of oil discharge in wastewater should be within 5 to 100 mg/L [10].

There are many advanced and traditional treatment technologies, including oxidation, membrane technology, floatation, chemical coagulation, electrochemical method, adsorption, bioremediation, electron-Fenton process, ultrasonic process, gravity separation, sedimentation, and supercritical water technology prevailed for oily wastewater treatment. Though there are several techniques, many methods remained ineffective in terms of showing improper removal potential. Among them few treatments recorded a sufficient decrease in oil content on the water surface, yet failed to satisfy the cost feasibility, intermediate formation and by-product (pollutant) discharge, and high regulatory limit. These treatments pose supreme removal for only the oily wastes, which were floating or dispersed at the surface. But for emulsified, minutely dispersed or dissolved oily contaminants, the treatment procedure should be more relevant with respect to oil discharges. So, there is an urgent need for some surpassing technology for oily wastewater treatment [21].

Photocatalysis, being an extraordinary treatment technique, washed out all the practical barriers and provided progressive oil removal on the aquatic surfaces. One of the astonishing treatments for oily wastewater is photocatalysis, where the process takes place with the action of various catalysts equipped. The process of photocatalysis gives rise to the formation of hydroxyl radicals and superoxide radicals, thereby mineralizing the pollutant. This radical formation plays a pivotal role in determining the mechanism under which photocatalysis takes place. The integrated system of photocatalyst and membrane technology also provides pre-dominant results, which prevents the formation of any other intermediates and secondary pollutants. The apparent rate constant was found to be greater for photocatalysis than the membrane technology [22].

The hydrocarbons from petroleum refinery associated with oily wastewater were detected and exposed to photocatalytic activity with the help of various catalysts used. These petroleum oily wastes contain more recalcitrant compounds, and many hydrocarbon groups viz, paraffin, naphthene, and aromatics. The carbon numbers between 12 and 39 (C_{12}-C_{39}) are responsible for about 90 % of oily waste [13]. The superposed selective photocatalyst breaks down the long-chain molecules into smaller organic fractions (C_{12}-C_{16}). In treating this oily (petroleum refinery) waste, the generated hydroxyl radicals, as a result of photon acceleration, attack the double bond of alkenes or aromatic rings. This mechanism happens by means of radical attachment to the aromatic ring, radical replacement to the aryl ring, and hydrogen atom abstraction. Among the three listed mechanisms, the radical attachment to the aromatic ring or on alkenes was the most dominant mechanism, due to the empowered photon acceleration even at room temperature. This mechanism contributes to the majority of hydrocarbon degradation using photocatalysis [23, 24].

As same as petroleum (hydrocarbons) waste, the waste generated from palm oil mill effluent also plays a major role in oily wastewater discharge. It is a brown high viscous liquid, available at a temperature range between 80 and 100 °C. It is usually available as an acidic nature with the pH ranging from 3.4 to 5.2. Generally, this waste is estimated to contain about COD and BOD ranging from 16000 mg/L to 100000 mg/L and between 10,000 mg/L and 44,000 mg/L, respectively. The safer regulatory limit issued for BOD content in treated oily waste discharge is 100 mg/L and for COD it was issued as 350 mg/L. Initially, the regulatory limit of BOD and COD ranged higher (250 mg/L and 500 mg/L) and later issues decreased the range of regulatory limit. This palm oil mill effluent contains phenolic compounds, available at concentrations between 33 and 630 mg/L. Palm oil mill effluent contaminated with phenolic compound is more dreadful to human health as it causes severe inertness, mutagenic and teratogenic effects, and possesses bioaccumulation potential and endocrine-disrupting capability. The scope of photocatalysis on the treatment of oily waste treatment is presented in Figure 12.2.

Some of the listed catalyst types used in photocatalysis of oily waste treatment are, titanium oxide (TiO_2), tungsten trioxide (WO_3), cadmium sulfide (CdS) Bismuth vanadate ($BiVO_4$), zinc oxide (ZnO), tin oxide (SnO_2), zinc sulphide (ZnS), niobium trioxide (Nb_2O_3), iron oxide (Fe_2O_3), vanadium pentoxide (V_2O_5), copper oxide (CuO_2), antimony oxide (Sb_2O_4), cerium oxide (CeO_2), graphitic carbon nitride and so on. Among these, the TiO_2 catalyst shows more effective oil degradation followed by ZnO catalyst. To obtain effective removal of oily (petroleum waste), the adaptation of multifunctional photocatalyst is appreciable. The multifunctional catalysts can be obtained by doping or impregnating the pure catalyst with metallic compounds viz, silver (Ag), platinum (Pt), cerium (Ce), calcium (Ca), tungsten (W). This doping imparts the formation of active sites with a higher affinity towards oily waste. The

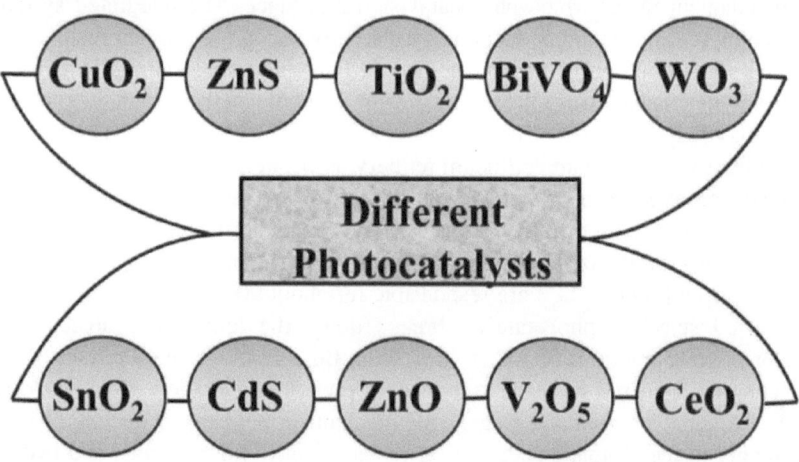

FIGURE 12.2 Different types of catalysts used in the application of treatment of oily wastewater by photocatalysis.

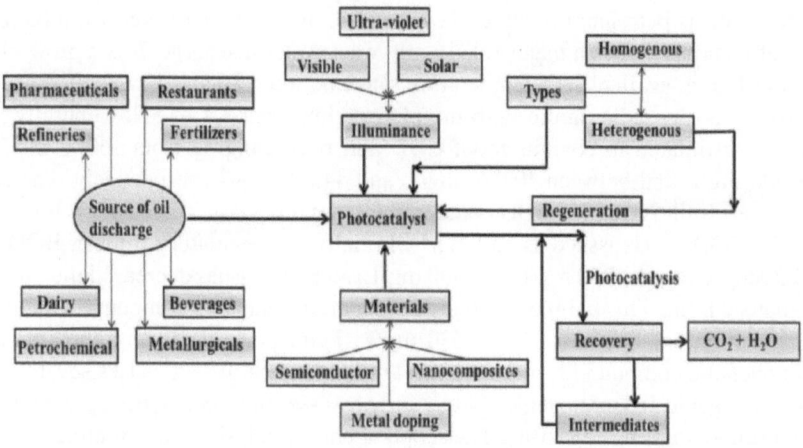

FIGURE 12.3 Details of sources of oil waste discharge with photocatalysis mechanism and simultaneous recovery of products and intermediates were figured out.

treatment of hydrocarbons using a catalyst is a time-consuming process and even prolonged time of exposure is required for treating contaminants at higher concentrations. Though doping using metals, non-metals, or co-doping and dye sensitization induces the property of photocatalytic property, it reduces the stability of photocatalysts in aqueous solutions, leaches out the doped metal from the reactor and causes dreadful disease formation. Though the role of surface ions in low coordination is of higher significance in photocatalysis, the exact role is yet to be explored. Chemical treatment methods were found to give promising results in functionalizing the surface to control the coordinative unsaturation degree of the surface ions. This was able

TABLE 12.2

Performance of TiO$_2$ and ZnO Catalysts in Oily Wastewater Treatment

Wastewater	Catalyst	Performance	Reference
Synthetic oily wastewater	TiO$_2$/γ-Al$_2$O$_3$-modified ceramic membrane	Separation—90%	[11]
Synthetic oily wastewater	PVDF-TiO$_2$ membrane	TOC removal – 80% Separation – 90%	[22]
Oily wastewater	ZnO	Separation—96 %	[25]
Oily wastewater	ZnO	Separation—90%	[26]
Oil/water separation	ZnO	Separation—95%	[27]
Carcinogenic benz[a] anthracene and benzo[a] pyrene	ZnO	Separation—90%	[28]
Oil-water separation	ZnO	Separation—98%	[29]
Oil/water separation	ZnO	Separation – 97%	[30]
Oily wastewater in extraction of solvent	ZrO$_2$-TiO$_2$	TOC removal—86.20%	[30]
Oily wastewater	ZnO	Separation – 99.9%	[31]
Restaurant oily wastewater	TiO$_2$	BOD removal- 43±2 %, COD removal—63±3 %, Selectivity—70±3%	[32]
Metal-processing industry	TiO$_2$	COD removal- 70%	[33]
Petroleum refinery wastewater	TiO$_2$	COD removal—92%, Separation—93%	[34]
Industrial oily wastewater	TiO$_2$	Separation – 99%	[35]

to yield high reactivity on single site catalysts owing to the influence of highly active electron-hole pairs [14]. The quantitative data of the performance on the catalysts are presented in Table 12.2 and also elaborated in sections 12.1.3.1 and 12.1.3.2.

12.3.1 TiO$_2$ CATALYST

The TiO$_2$ is the natural oxide from the titanium metal, termed as titania. The TiO$_2$ catalyst is a remarkable nanoparticle for its eminent traits including impressive thermal and physical stability, self-cleaning, low toxic, cheapest, excellent oxidation properties, and anti-bacterial and antifouling properties. Also, it is easy to operate at ambient temperature and pressure with no compromise in degradation potential [36]. Among the three metastable phases, namely, rutile, brookite and anatase, the anatase form was witnessed to give maximum performance under the influence of ultraviolet radiation. The higher energy acquired from the ultraviolet rays overtake the bandgap energy of TiO$_2$ stimulating the excitation of electrons to leave the valence band and reach the conductance band and causing holes with positive charges in the former band. The pairs of electrons and holes reach the surface of the catalyst to induce the series of redox reactions. While the electrons convert Ti(IV) to Ti(III) and react with oxygen on catalyst surface in aqueous system to yield superoxide radical ions (O$_2$.$^-$), the holes contribute in generating hydroxyl radicals (OH.) from water molecules adsorbed on the surface. The oily contaminants thus get oxidized to carbon dioxide and water by the influence of oxygen reactive radicals [37]. The TiO_2 photocatalyst possesses the photoinduced-hydrophilicity which explores the acclaimed photocatalysis. The application of TiO_2 catalyst was widely applied in various fields including physicochemical, toxicological, and biocompatibility. The TiO_2 catalyst better suits for photons illumination using UV light, rather than solar and visible light. Its efficiency as a catalyst depends on the pollutant concentration, intensity and duration of irradiation and the light penetration level onto the surface. The TiO_2 catalyst requires irradiance at the wavelength range of 387.5 nm to activate the catalyst. Exclusively, the vacuum ultraviolet illuminated catalyst behaves as an additive-free process, which can be utilized for pretreatment of oil recovery as an intensified process [32].

Depending upon the crystalline structure of the TiO_2 catalyst, it possesses a large bandgap ranging between 3 and 3.2 eV. The TiO_2 catalyst exhibits three main crystal structures based on the nature of stability. Of the three forms, namely anatase, rutile, and brookite, rutile contributes to the most stable phase and remaining metastable and each has a bandgap of 3.03 eV, 3.23 eV, and 3.14 eV, respectively. The combined effect of rutile and anatase phase in TiO_2 catalyst provides slow recombination and possesses a large surface area attributing to absorb a higher level of radicals. Though the combined effect provides better results, considering the bandgap of rutile (3.03 eV) is lower than anatase (3.2 eV), considering the difference in the bandgap of crystal structure, the amount of rutile phase contribution should be lower in order to obtain higher catalytic activity.

Compared to white TiO_2 catalyst, the colored catalyst attracted the research area in oily waste treatment, since it neglects certain limitations by providing sufficient visible light absorption, narrow bandgap, and delayed recombination. Few surface defects, including surface oxygen vacancies, additional Ti species, and core-shell structure change the white TiO_2 catalyst to a colored TiO_2 catalyst. The color of the catalyst might vary like, black, blue, red, grey, yellow, or brown. Also, the TiO_2 nanotubes could provide more promising results than powder TiO_2 catalysts [38].

In addition to all these characteristics, the withheld pigmentary property, stability at any pH, and ultraviolet absorption provide prolonged stability even in an acidic medium, attributing excellent catalytic activity to petroleum (oily) refinery wastewater treatment. The TiO_2 catalyst can be doped with any other metal or non-metallic ions, as a means to enhance the structure and electronic property of the catalyst. Generally, the improved characteristics can be exposed while doping TiO_2 catalyst with Ag, Cu, and Pt. The process of doping influences the transfer of charge and thus aids in the segregation of photo-generated charges. Additionally, the doping provides the mesoporous structure with a high surface area, thereby increasing the sites for substrate attachment [39].

However, the TiO_2 catalyst can exhibit some disadvantageous conditions towards doping with non-metallic ions, since it provides a way for the formation of surface oxygen vacancies, quick recombination of electron-hole pair, less thermal stability, and further leach out of doped substance. Also, the non-metallic substance doped could get deactivated while being exposed to visible light in an aqueous medium. Similarly, in the case of dye sensitization upon TiO_2 catalyst, the formation of a physical bond rather than a chemical bond will lead to leaching out of dyes used which could get dissolved in water and cause the formation of dye-based pollutants. Also, the dye-sensitized TiO_2 catalyst decreases the electron transfer potential.

The three-step procedure of real rolling wastewater containing oily waste utilizes TiO_2 catalyst for photodegradation after the process of acidification and adsorption. The utilization of TiO_2 catalyst decreases the COD content after the oil recovery. Under ultraviolet illumination, the photocatalytic reactor coupled with the membrane filtration technique provided a synergetic photocatalytic membrane effect for the treatment of total organic carbon in the oily wastewater. Many polymeric and ceramic membranes containing typical permeability and selectivity were utilized for synthetic oily waste treatment [11]. A ceramic membrane was coated using γ-Al_2O_3 and coupled with a titanium dioxide catalyst (TiO_2) equipped reactor. Comparatively the anatase phase TiO_2 nanoparticles were witnessed to outperform the rutile phase, which dominated the entire photocatalysis process. Additionally, the UV illuminance exercised the strong oxidant accumulation.

The integrated hybrid system of polyvinylidene fluoride membrane submerged with TiO_2 photocatalyst dealt with the degradation of oily wastewater under the irradiance effect [13]. The catalysts used in the integrated system of photocatalytic membrane reactors can be either coated or blended with the membranes used. While comparing TiO_2 catalyst being coated and blended upon polyethersulfone membrane, the TiO_2 coated catalyst provided enhanced porous structure, thereby increasing the degradation capacity of oily waste. While investigating with the same TiO_2 catalyst being coated and blended upon three different types of membranes viz, polysulfone, polyacrylonitrile, and polyvinylidene, the system proposed better efficiency for TiO_2 coated catalyst. These disparate results were obtained due to fouling mitigation increment, dense particle accommodation upon membrane surface, and mainly due to membrane selectivity. Also, the initial and steady-state flux maintenance throughout the entire filtration process also supported the catalytic activity of TiO_2. The prolonged solar irradiance or ultraviolet exposure on the TiO_2 catalyst coated membrane provides an additional advantageous note [40].

The TiO_2 coated ceramic membrane was also attributed to about 150 % increment in flux, which induced the hydrophilic property of the membrane. While investigating engine oil treatment using TiO_2 catalyst synergized with ceramic membrane, the result suggested the improvised antifouling property. The oil oil-emulsion system carried out with glass fiber coated TiO_2 nanostructured catalyst, provides many recycles. Further, the regeneration of catalysts was also studied.

The petroleum refinery wastes were removed using nano-TiO_2/Fe composites, which provided excellent removal of chemical oxygen demand obtained at specified optimal parameters. Similarly, the oily wastes in an aqueous solution were treated by TiO_2/SiO_2 photocatalytic performance. A similar oily compound formaldehyde degradation was also made under the activity of TiO_2 nanowires upon Ti substrate as TiO_2/TiO_2 foil catalyst where the regeneration capacity was also evaluated [5]. The lipid regulators like metformin and amoxicillin were treated using pure TiO_2 and synthetic TiO_2-ZnO_2 coupled catalysts, respectively.

A major limitation of TiO_2 catalyst is the challenge in separation from the treated water. This has been overcome by introduction of immobilized nano-sized TiO_2 onto different supporting materials like glass, clay, polymer films and sand. The immobilization techniques include chemical vapour deposition, filtration, activation reaction, blending, Sol-gel dip coat method, atmospheric plasma spraying, hydrothermal and self-assembly methods. The photocatalytic membrane reactor concept has been introduced by combination of photocatalytic oxidation and membrane filtration. Four distinct configurations of such integrated approaches as mentioned below have been applied [37].

1. Slurry photocatalytic reactor and a membrane filtration unit in series
2. Slurry photocatalytic reactor with a submerged membrane made by polymer or inorganic material
3. A photoreactor with a membrane coated with photocatalysts in the inner walls
4. Photocatalytic membranes made of pure or composite materials of TiO_2

Among these four integrated approaches, the photocatalytic membranes are more preferred owing to the benefits of both physical separation and degradation of organic pollutants. Moreover, they outpace conventional types of membranes on the aspects of fouling and permeate qualities. These membranes were witnessed to remove recalcitrant organic compounds of not only oil wastes but also several other pollutants like creatinine, chlorophenols, Acid Red, Congo Red, methyl orange, methylene blue, Rhodamine B, etc.

12.3.2   ZnO Catalyst

ZnO is a white powdery inorganic compound possessing a non-polar trait, which exhibits two crystalline structures viz, hexagonal wurtzite and cubic zinc blende. The wurtzite structure was identified to pose high stability even at room temperature. ZnO catalyst is cheaper than TiO_2 catalyst but provides better degradation results. ZnO catalyst also owes to impressive photocatalytic behavior including excellent charge transport property, chemical stability, long-term photostability and non-toxic effects. It includes many exclusive traits including an almost stable level of excitons and large bond strength, which were demonstrated by high binding energy, and cohesive energy and melting point, respectively. The bandwidth of the ZnO catalyst ranged at 3.3 eV and excitation binding energy at 60 meV. The amount of threshold energy required for ZnO electrons to obtain excitation was 425 nm. The ZnO catalyst also exhibits significant absorption towards ultraviolet light than the visible and solar irradiance [41].

Based upon preparatory methods, ZnO catalyst exhibits different morphological structures, including rod, spherical, hexagonal disk, rice husk, dumbbell, and pompon-like shapes. All these morphologies provided varied results corresponding to appropriate pH levels. However, the three-dimensional nanostructure provided by the pompon structure exhibited closely packed petals budding from the center of pompons [38]. This pompons structure of ZnO catalyst will come up with extreme COD removal at specified pH. This structure issues high specific surface area and wide pore size distribution, in order to provide many active sites for oily contaminants to get attached to the catalyst. This specialized structure of ZnO catalyst exposes some inhibited traits, including oil adsorption, efficient light absorption as photons to accelerate electrons, and photo-generated charge carriers transferred to the surface of the photocatalyst. Also, the pompon structured ZnO nanostructure contributed to providing better results even after five recycles, and even sustainable results were recorded.

This ZnO catalyst contains some limitations to get operating for photocatalytic degradation. The wide bandgap of the catalyst gives rise to quick recombination and provides less efficiency for electrons to absorb the UV light. Also, the ZnO catalyst is a photo-corrosion material. Exclusively, the ZnO catalyst boasts an amphoteric effect, which provides an advantageous note to survive both in an acidic and alkaline medium. The zero-point charge of the ZnO catalyst aids to alter the reaction mechanism of hydroxyl radicals.

The synergistic effect of photocatalysis upon membrane technology using ZnO catalyst upon polydopamine functionalized nanofibres was investigated. This polydopamine contains a catechol group, which provides adhesive force to get attached to ZnO nanostructure, the capacity of being an electron trap, slow recombination of electron-hole pairs, thereby providing effective removal.

Due to the unique electronic, mechanical, and optical parameters possessed by ZnO, they are widely distributed among various fields viz, light-emitting diodes, biosensors, solar cells, electronic, UV detectors, and gas sensors, and so on. These ZnO catalysts were biocompatible, biodegradable, and bi-safe catalysts. Also, it is widely distributed in the medical field. Further, the development of auto-dispersion of non-aggregated catalysts is appreciated. Additionally, for its ferromagnetic property, its application in spintronic was promising [42].

12.3.3 MISCELLANEOUS CATALYSTS

Other catalysts that were utilized for oily waste next to TiO_2 catalyst and ZnO catalyst also exhibited decent degradation capacity for oily waste treatment. The bismuth-based catalyst is an n-type semiconductor used in photocatalysis of oily wastewater treatment. The bismuth-based catalyst like Bi_2WO_6, $BiVO_4$, and $BiMoO_6$ were generally produced in many morphological structures, including spherical, T-shape, and rod-shaped. This bismuth-based catalyst was collectively termed as Bi_aAO_b. The bandgap of the catalyst ranged between 2.5 eV and 2.8 eV. The advantages of this catalyst base include high chemical stability, separation of photogenerated charge carriers, non-toxic, and so on [43]. However, the disadvantages were more inappropriate which include, poor quantum yield, quick recombination, chemical corrosion, and so on. The bismuth-based catalyst used the visible light irradiance for photons to accelerate the electrons. The Bi_2MoO_6 catalyst coupled with the $CoMoO_4$ membrane possesses super hydrophilicity and underwater superoleophobicity property, which was better suited in treating oil in water emulsions. Exclusively, the Bi_2MoO_6 catalyst exhibits high intrinsic anisotropic properties.

The tungsten trioxide (WO_3) catalyst is a yellow-colored compound widely distributed in the photocatalysis of oily wastewater. Other applications, including usage in paints, pigments, and ceramics, were widely distributed [41]. The bandgap of the catalyst ranges between 2.7 and 2.8 eV. This WO_3 catalyst better absorbs visible light rather than ultraviolet light. They exhibit a monoclinic crystalline structure at room temperature. However, the advantages and disadvantages were also to be examined.

Some advantages include narrow bandgap, high thermal stability, and physic-chemical stability. Due to the scarcity of this catalyst, it is somewhat costlier to proceed with oily wastewater treatment.

Carbon-based substances like carbon quantum dots, graphene, and carbon nanotubes can behave as characterized catalysts for oily wastewater treatment. They contain adequate thermal stability, high absorptive rate of light irradiance and a large surface area. The graphene can be used either as a pure catalyst or as a composite with TiO_2, which enhances the hydrophilicity nature of the membrane at which the catalyst is immobilized. Also, it contains better stability and conduction capacity under suitable optical parameters [40].

12.4   RESTRICTIONS IN PHOTOCATALYSIS

The heterogeneous method of photocatalysis in wastewater treatment has been a promising area of research for several decades. But the feasibility in real time applications on industrial scale has not been reported due to technological and process limitations. The low photo-conversion efficiency yielded during the advanced oxidation process for treating oil content in the wastewater metrics is a major process-oriented disadvantage. Photonic efficiency is a key term in photocatalysis which is defined as the rate of formation of products from reaction divided by the photon flow fed. Unfortunately, most of the semiconductor photocatalysis exhibits less than 10 % of photonic efficiency which is mainly attributed to the rapid recombination of electron – hole pairs generated [14]. However higher photonic efficiencies can be attained by illuminating with laser pulses of higher intensity in sequence instead of supplying an identical quantum of photons continuously. The technological limitations of photocatalysis include the fragile method of synthesis of catalysts, design of large-scale reactors, high energy costs and slurry management [44]. The other major restriction of photocatalysis is that it cannot be applied to the treatment of wastewaters contaminated with heavy metals. Photocatalysis are least susceptible to thermodynamic observations as the Gibbs energy calculations of photon related operations haven't been explored to correlate the activation energy and equilibrium constants directly to the photolysis.

12.5   CONCLUSION

Oily wastewater is a major class of pollutant originating from industries petroleum refineries, petrochemicals, pharmaceuticals, fertilizers, etc. The toxic hydrocarbons emanated from these sources occur in the concentration of 50–1000 ppm. Since this range is too much higher than the permissible level of 10–15 ppm, they pose severe damages to soil, water bodies and marine environment [12]. Advanced oxidation using heterogeneous photocatalysis plays a significant role in abatement of oil contamination from wastewaters using solid semiconductors. The widely used TiO_2 and ZnO catalytic materials proved to yield promising results in this domain owing to their versatility, availability, stability and reusability. Heterogeneous photocatalysis has been used in integrated technologies and tertiary treatment steps by several researchers on lab scale in treatment of oily wastewaters. However, the challenges that incur in commercialization of these approaches on industrial scale applications haven't been overcome yet due to process difficulties in photocatalyst synthesis to handle a large throughput of complex matrices of real-time wastewater in continuous mode. The technological limitations are the design of large-scale reactors, low efficiency and high energy consumption.

REFERENCES

[1] Wu J, Ren J, Pan W, Lu P, Qi Y. *Photo-catalytic Control Technologies of Flue Gas Pollutants*. Singapore: Springer; 2019.

[2] Li J, Sun Y, Zhang L, Xiao X, Yang N, Zhang L, Yang X, Peng F, Jiang B. Visible-light induced CoMoO4@ Bi2MoO6 heterojunction membrane with attractive photocatalytic property and high precision separation toward oil-in-water emulsion. Separation and Purification Technology. 2021; 277:119568.

[3] Adetunji AI, Olaniran AO. Treatment of industrial oily wastewater by advanced technologies: A review. Applied Water Science. 2021; 11(6):1–9.

[4] Ani IJ, Akpan UG, Olutoye MA, Hameed BH. Photocatalytic degradation of pollutants in petroleum refinery wastewater by TiO_2 and ZnO-based photocatalysts: Recent development. Journal of Cleaner Production. 2018; 205:930–54.

[5] Ren G, Han H, Wang Y, Liu S, Zhao J, Meng X, Li Z. Recent advances of photocatalytic application in water treatment: A review. Nanomaterials. 2021; 11(7):1804.

[6] Saravanan R, Gracia F, Stephen A. Basic principles, mechanism, and challenges of photocatalysis. In Nanocomposites for Visible Light-induced Photocatalysis. New York: Springer; 2017.

[7] Sundar KP, Kanmani S. Progression of photocatalytic reactors and its comparison: A review. Chemical Engineering Research and Design. 2020; 154:135–50.

[8] Prihod'ko RV, Soboleva NM. Photocatalysis: Oxidative processes in water treatment. Journal of Chemistry. 2013; 2013.

[9] Tetteh EK, Rathilal S, Naidoo DB. Photocatalytic degradation of oily waste and phenol from a local South Africa oil refinery wastewater using response methodology. Scientific Reports. 2020; 10(1):1–2.

[10] Abuhasel K, Kchaou M, Alquraish M, Munusamy Y, Jeng YT. Oily wastewater treatment: Overview of conventional and modern methods, challenges, and future opportunities. Water. 2021; 13(7):980.

[11] Golshenas A, Sadeghian Z, Ashrafizadeh SN. Performance evaluation of a ceramic-based photocatalytic membrane reactor for treatment of oily wastewater. Journal of Water Process Engineering. 2020; 36:101186.

[12] Nasir AM, Awang N, Jaafar J, Ismail AF, Othman MH, Rahman MA, Aziz F, Yajid MA. Recent progress on fabrication and application of electrospun nanofibrous photocatalytic membranes for wastewater treatment: A review. Journal of Water Process Engineering. 2021; 40:101878.

[13] Zhong Q, Shi G, Sun Q, Mu P, Li J. Robust PVA-GO-TiO_2 composite membrane for efficient separation oil-in-water emulsions with stable high flux. Journal of Membrane Science. 2021; 640:119836.

[14] Schneider J, Matsuoka M, Takeuchi M, Zhang J, Horiuchi Y, Anpo M, Bahnemann DW. Understanding TiO_2 photocatalysis: Mechanisms and materials. Chemical Reviews. 2014; 114(19):9919–86.

[15] Gopinath KP, Madhav NV, Krishnan A, Malolan R, Rangarajan G. Present applications of titanium dioxide for the photocatalytic removal of pollutants from water: A review. Journal of Environmental Management. 2020; 270:110906.

[16] Li ZK, Liu Y, Li L, Wei Y, Caro J, Wang H. Ultra-thin titanium carbide (MXene) sheet membranes for high-efficient oil/water emulsions separation. Journal of Membrane Science. 2019; 592:117361.

[17] Lou L, Kendall RJ, Smith E, Ramkumar SS. Functional PVDF/rGO/TiO$_2$ nanofiber webs for the removal of oil from water. Polymer. 2020; 186:122028.

[18] Lorwanishpaisarn N, Kasemsiri P, Srikhao N, Jetsrisuparb K, Knijnenburg JT, Hiziroglu S, Pongsa U, Chindaprasirt P. Fabrication of durable superhydrophobic epoxy/cashew nut shell liquid based coating containing flower-like zinc oxide for continuous oil/water separation. Surface and Coatings Technology. 2019; 366:106–13.

[19] Venkatesh K, Arthanareeswaran G, Bose AC, Kumar PS, Kweon J. Diethylenetriaminepentaacetic acid-functionalized multi-walled carbon nanotubes/titanium oxide-PVDF nanofiber membrane for effective separation of oil/water emulsion. Separation and Purification Technology. 2021; 257:117926.

[20] Qu M, Liu Q, Yuan S, Yang X, Yang C, Li J, Liu L, Peng L, He J. Facile fabrication of TiO$_2$-functionalized material with tunable superwettability for continuous and controllable oil/water separation, emulsified oil purification, and hazardous organics photodegradation. Colloids and Surfaces A: Physicochemical and Engineering Aspects. 2021; 610:125942.

[21] He T, Zhao H, Liu Y, Zhao C, Wang L, Wang H, Zhao Y, Wang H. Facile fabrication of superhydrophobic Titanium dioxide-composited cotton fabrics to realize oil-water separation with efficiently photocatalytic degradation for water-soluble pollutants. Colloids and Surfaces A: Physicochemical and Engineering Aspects. 2020; 585:124080.

[22] Ong CS, Lau WJ, Goh PS, Ng BC, Ismail AF. Investigation of submerged membrane photocatalytic reactor (sMPR) operating parameters during oily wastewater treatment process. Desalination. 2014; 353:48–56.

[23] Hassan AA, Al-zobai KM. Chemical oxidation for oil separation from oilfield produced water under UV irradiation using Titanium dioxide as a nano-photocatalyst by batch and continuous techniques. International Journal of Chemical Engineering. 2019; 9810728.

[24] Wei Q, Oribayo O, Feng X, Rempel GL, Pan Q. Synthesis of polyurethane foams loaded with TiO$_2$ nanoparticles and their modification for enhanced performance in oil spill cleanup. Industrial & Engineering Chemistry Research. 2018; 57(27):8918–26.

[25] Zhang Y, Wang X, Wang C, Liu J, Zhai H, Liu B, Zhao X, Fang D. Facile fabrication of zinc oxide coated superhydrophobic and superoleophilic meshes for efficient oil/water separation. RSC advances. 2018; 8:35150–6.

[26] Cao W, Ma W, Lu T, Jiang Z, Xiong R, Huang C. Multifunctional nanofibrous membranes with sunlight-driven self-cleaning performance for complex oily wastewater remediation. Journal of Colloid and Interface Science. 2022; 608:164–74.

[27] Velayi E, Norouzbeigi R. A mesh membrane coated with dual-scale superhydrophobic nano zinc oxide: Efficient oil-water separation. Surface and Coatings Technology. 2020; 385:125394.

[28] Rani M, Shanker U. Sunlight mediated improved photocatalytic degradation of carcinogenic benz [a] anthracene and benzo [a] pyrene by zinc oxide encapsulated hexacyanoferrate nanocomposite. Journal of Photochemistry and Photobiology A: Chemistry. 2019; 381:111861.

[29] Huang A, Chen LH, Kan CC, Hsu TY, Wu SE, Jana KK, Tung KL. Fabrication of zinc oxide nanostructure coated membranes for efficient oil/water separation. Journal of Membrane Science. 2018; 566:249–57.

[30] Yaacob N, Sean GP, Nazri NA, Ismail AF, Abidin MN, Subramaniam MN. Simultaneous oily wastewater adsorption and photodegradation by ZrO$_2$–TiO$_2$ heterojunction photocatalysts. Journal of Water Process Engineering. 2021; 39:101644.

[31] Velayi E, Norouzbeigi R. Synthesis of hierarchical superhydrophobic zinc oxide nano-structures for oil/water separation. Ceramics International. 2018; 44:14202–8.

[32] Kang JX, Lu L, Zhan W, Li B, Li DS, Ren YZ, Liu DQ. Photocatalytic pretreatment of oily wastewater from the restaurant by a vacuum ultraviolet/TiO$_2$ system. Journal of Hazardous Materials. 2011; 186:849–54.

[33] Yang Q, Xu R, Wu P, He J, Liu C, Jiang W. Three-step treatment of real complex, variable high-COD rolling wastewater by rational adjustment of acidification, adsorption, and photocatalysis using big data analysis. Separation and Purification Technology. 2021; 270:118865.

[34] Ulhaq I, Ahmad W, Ahmad I, Yaseen M, Ilyas M. Engineering TiO$_2$ supported CTAB modified bentonite for treatment of refinery wastewater through simultaneous photocatalytic oxidation and adsorption. Journal of Water Process Engineering. 2021; 43:102239.

[35] Sun F, Li TT, Ren HT, Shiu BC, Peng HK, Lin JH, Lou CW. Multi-scaled, hierarchical nanofibrous membrane for oil/water separation and photocatalysis: Preparation, characterization and properties evaluation. Progress in Organic Coatings. 2021; 152:106125.

[36] Hadjltaief HB, Zina MB, Galvez ME, Da Costa P. Photocatalytic degradation of methyl green dye in aqueous solution over natural clay-supported ZnO–TiO$_2$ catalysts. Journal of Photochemistry and Photobiology A: Chemistry. 2016; 315:25–33.

[37] Leong S, Razmjou A, Wang K, Hapgood K, Zhang X, Wang H. TiO$_2$ based photocatalytic membranes: A review. Journal of Membrane Science. 2014; 472:167–84.

[38] Nawaz R, Kait CF, Chia HY, Isa MH, Huei LW, Sahrin NT, Khan N. Countering major challenges confronting photocatalytic technology for the remediation of treated palm oil mill effluent: A review. Environmental Technology & Innovation. 2021; 23:101764.

[39] Avilés-García O, Espino-Valencia J, Romero-Romero R, Rico-Cerda JL, Arroyo-Albiter M, Solís-Casados DA, Natividad-Rangel R. Enhanced photocatalytic activity of titania by co-doping with Mo and W. Catalysts. 2018; 8(12):631.

[40] Nascimben Santos E, László Z, Hodúr C, Arthanareeswaran G, Veréb G. Photocatalytic membrane filtration and its advantages over conventional approaches in the treatment of oily wastewater: A review. Asia-Pacific Journal of Chemical Engineering. 2020; 15:2533.

[41] Saputera WH, Amri AF, Daiyan R, Sasongko D. Photocatalytic technology for palm oil mill effluent (POME) wastewater treatment: Current progress and future perspective. Materials. 2021; 14:2846.

[42] Ng KH, Yuan LS, Cheng CK, Chen K, Fang C. TiO$_2$ and ZnO photocatalytic treatment of palm oil mill effluent (POME) and feasibility of renewable energy generation: A short review. Journal of cleaner production. 2019; 233:209–25.

[43] Monisha RS, Mani RL, Sivaprakash B, Rajamohan N, Vo DV. Green remediation of pharmaceutical wastes using biochar: A review. Environmental Chemistry Letters. 2021:1–24.

[44] Iervolino G, Zammit I, Vaiano V, Rizzo L. Limitations and prospects for wastewater treatment by UV and visible-light-active heterogeneous photocatalysis: A critical review. Heterogeneous photocatalysis. 2020:225–64.

13 Integrated Treatment Process for Industrial Gas Effluent

Daniel Dobslaw

CONTENTS

13.1 Background and Necessity of New Exhaust Air Purification Concepts 239
13.2 Conventional Waste Air Treatment Techniques 240
    13.2.1 Condensation 241
    13.2.2 Absorption 241
    13.2.3 Membrane Processes 242
    13.2.4 Thermal Combustion Processes 242
    13.2.5 Oxidative Catalysis 243
    13.2.6 Non-Thermal Plasma (NTP) 243
    13.2.7 UV Oxidation..244
    13.2.8 Thermal Plasma 244
    13.2.9 Adsorption 245
    13.2.10 Biological Processes 246
13.3 Parameters for Suitability of the Processes...248
13.4 Combined Treatment Processes 253
    13.4.1 High Contaminated Waste Air Conditions.....................................254
    13.4.2 Low Contaminated Waste Air Conditions 256
13.5 Conclusion 259
References...259

13.1 BACKGROUND AND NECESSITY OF NEW EXHAUST AIR PURIFICATION CONCEPTS

Emissions from industrial production processes are usually characterized by a complex mixture of waste air components with potential health and environmental hazards. The toxicological spectrum of these anthropogenic waste air components ranges from harmful and corrosive to mutagenic, teratogenic, and chronically toxic to carcinogenic or acutely toxic. The World Health Organization (WHO) currently classifies anthropogenically induced air pollution as the greatest source of risk to human health. In 2017, a medical study of the Institute for Health Metrics and Evaluation of solely dust and ozone exposures showed that approximately 4.3 million people worldwide suffer from early deaths induced by outdoor dust and ozone exposures,

DOI: 10.1201/9781003260738-13

and 2.6 million people die prematurely by indoor air exposures [1]. Since emissions of inorganic components such as NH_3, N_2O or H_2S as well as organic components (VOC) have not yet been considered in comparable studies, a much more serious impact of anthropogenic emissions on human life expectancy can be assumed. It is firstly necessary to minimize these emissions by means of suitable waste air purification concepts.

Secondly, the recent study of the International Energy Agency IEA [2] on world energy consumption and related greenhouse gas emissions shows that for the presented 'Sustainable Development Scenario', which predicts a global warming of 1.7 °C by 2050, global CO_2 emissions of 33.7 Gt CO_2/a must be reduced by 12 Gt CO_2/a by 2030 and stabilized at about 8 Gt CO_2/a by 2050. This ultimately corresponds to a CO_2 footprint of approx. 1 t CO_2/(inhabitant · a). The path taken by individual countries or world regions varies greatly. Developed countries and emerging economies have significantly higher emissions because of higher living standards (EU: 5.1 t CO_2/(inhabitant · a); China: 9 t CO_2/(inhabitant · a); USA: 10.4 t CO_2/(inhabitant · a) [2]. If the target of max. 1 t CO_2/(inhabitant · a) is to be realized realistically, it will be necessary not only to make personal cuts in the standard of living, but also to develop new environmentally friendly technologies and mobility concepts. In the context of the evaluation of greenhouse gas emissions, existing waste air purification technologies will also have to be re-evaluated about their direct emissions (i.e., CO_2, CH_4 and N_2O) as well as indirect emissions (i.e., energy production and transportation), since in future it will no longer be possible to afford greenhouse gas emissions from these techniques.

Furthermore, considering current crises with dramatically increasing energy prices, there will be and must be an increasing focus on energy conservation actions in addition to the costs of energy-intensive waste air treatment technologies. Since the removal of contaminated waste air from production processes is inevitably linked to the supply of pre-heated fresh air during heating periods, concepts for air-reuse are gaining in importance. However, in addition to sufficient oxygen concentration and low CO_2 concentration, the air may not contain any toxicologically harmful concentrations of VOCs or transformation products.

In view of the need to purify waste air and the objective to achieve a technology that is as climate-neutral as possible, the following section will briefly describe existing technologies and present new approaches that allow energy-efficient and climate-friendly treatment of industrial waste air. Current trends in process combinations as well as processes for integrated air-reuse concepts will be presented. The possibilities and requirements of integrated processes will be illustrated by two examples.

13.2   CONVENTIONAL WASTE AIR TREATMENT TECHNIQUES

VOC abatement can be accomplished by the following processes, some of which are used in combination as waste air complexity increases. Especially the combination of scrubbers as a pretreatment stage with a subsequent cleaning stage are widely used.

13.2.1 CONDENSATION

During condensation, the partial pressure of the VOCs in the VOC-containing exhaust air is raised above the saturated vapour pressure of the compounds by cooling and optional compression of the air, and at concentrations of approx. 101.7–103.2% of saturation [3, 4] heterogeneous condensation takes place. Here, present dust particles and water droplets cause VOC precipitation in the form of liquid droplets, which are then separated mechanically. The effectiveness of the process strongly depends on temperature and cooling capacity, but also on levels of saturated vapour pressure, surface tension, individual gas constant, gas density and the concentration of accompanying components (dust, water droplets, ionised gases). It is only suitable from VOC concentrations of at least 25 g C · m^{-3} and has approximately twice the energy demand of post-combustion processes. The legal limit values (usually less than 100 mg C · m^{-3}) can only be achieved in individual cases. Solvent recovery from the condensate is possible, but requires extensive purification steps.

13.2.2 ABSORPTION

Absorption relies on the wash-out of gaseous hydrophilic VOCs with low Henry partition coefficients by an aqueous phase. Four process variants are used here:

1) Physical scrubber: water is used as the absorbent.
2) Acid or base scrubber: The absorbent is dilute acid or base and is used in particular for the scrubbing of basic (e.g., NH$_3$, amines) or acidic reacting compounds (e.g., H$_2$S, organic acids), which are converted into the complementary salts by the pH shift and thus exhibit extremely low residual vapor pressures.
3) Chemical oxidative scrubbers: These rely on oxidation of present VOCs to partially oxidized transformation products exhibiting higher water solubility and lower vapor pressure at the same time. Aqueous solutions of oxidants such as H$_2$O$_2$ or NaOCl are used as absorbents.
4) Scrubbers with (aqueous) organic phase: To enable absorption of VOCs with higher Henry partition coefficients and thus more lipophilic compounds, solubilizers can be added to the aqueous phase in concentrations of 0.5–30 vol% or an organic phase can be used directly as absorbent [5]. In the 1970s, especially water-silicone oil emulsions were used as organic phases. Modern solubilizers are based on polyethylene glycols and derivatives of phthalic acid esters. Due to the complex disposal situation of the loaded absorbent, these scrubbers are only operated in exceptional cases.

Absorbers can basically be used over the entire VOC concentration range, with physical scrubbers mainly used at high VOC waste air concentrations (>1 g C · m^{-3}) and chemical scrubbers at low VOC waste air concentrations (<1 g C · m^{-3}) due to the chemical demand correlating with the present VOC concentrations. The legal limit value can be safely complied with for scrubbers of type 1–3, especially for chemical scrubbers, as long as the waste air exclusively contains hydrophilic VOC components.

13.2.3 MEMBRANE PROCESSES

Membrane processes for VOC separation are based on the principle of cross-flow thin-film membranes, whereby the partial pressure of the VOCs on the feed side is increased by building up high pressure gradients between feed and permeate (p_{Feed}/ $p_{Permeate} \approx 20$–30) and thus getting a mass transport towards the permeate side. The mass transfer obeys Fick's diffusion law and, in addition to the pressure gradient, is particularly dependent on the membrane interface. Silicone membranes in particular are used as membrane materials, but these limit the application of the process to lipophilic VOCs. The permeate enriched in VOCs is regenerated via a condensation stage as a downstream step and residual gaseous permeate is returned into the feed, while the condensate is separated. Hence, solvent recovery is possible. In addition to the restriction to lipophilic waste air components, high requirements are placed on the upstream air treatment (no dust and condensing droplets). Furthermore, the process is characterized by high operating and investment costs and is only economically interesting at VOC concentrations higher than 10 g C $\cdot$ m^{-3}. Compliance with legal limit values is not possible.

13.2.4 THERMAL COMBUSTION PROCESSES

During combustion, total oxidation of the gaseous VOCs to CO_2 and H_2O takes place by heating the VOC-containing waste air in the combustion chamber to temperatures above the ignition temperature of the waste air components (750–1000 °C). The required heat supply is usually provided by the use of fossil energy (gas, oil) or by electrical heating (for small plants), where autothermal operating conditions cannot be achieved. Autothermal operation is possible from a substance-specific VOC concentration of 2–3 g C $\cdot$ m^{-3}. Heat recovery is based either on the principle of recuperators (so-called thermal post-combustion) or regenerators (so-called regenerative post-combustion). By installing additional heat exchangers downstream of the process, the differential temperature between the clean gas and the crude gas can be reduced to 55–70 K in the case of modern combustion systems. Nevertheless, this mode of operation also requires the use of approx. 2 m^3 of natural gas and 3 kWh of electricity per 1000 m^3 of treated waste air and thus causes direct and indirect CO_2 emissions of approx. 5.1 kg CO_2 $\cdot$ 1000 m^{-3}. Furthermore, thermally induced NO_x emissions of 10–200 mg NO_x $\cdot$ m^{-3} and CO emissions of 5–90 mg CO $\cdot$ m^{-3} [6] occur. Another approach in reduction of energy demand is the application of catalytically active heat beds, whereby the activation energy for oxidation and thus the combustion chamber temperature can be lowered (so-called catalytic post-combustion). Although the reduction of the temperature difference to 35–50 K and thus of the energy consumption is successful, the use of the catalysts requires in-depth knowledge of the waste air composition in order to prevent catalyst poisoning. The consumption costs of catalysts can be estimated at approx. 2000 € $\cdot$ 1000 m^{-3} waste air capacity $\cdot$ year^{-1} [7].

13.2.5   OXIDATIVE CATALYSIS

Oxidative catalysis exploits the catalytic-oxidative conversion of VOCs in hot waste air streams by supported noble metal catalysts (Pt, Rh, Pd, Ag, Au or Ru on supports of molsieves, ceramics, SiO_2, Al_2O_3, TiO_2 but also CeO_2, Co_3O_4, ZrO_2, MnOx or mixtures of the components), metal oxide catalysts (usually CeO_2, Fe_2O_3, Co_3O_4 and MnO_2 or mixtures of the components, possibly with doping of lanthanum and zirconium) and metal oxide catalysts with special morphology (tubes, needles, rods). components, optionally with doping of lanthanum and zirconium) and metal oxide catalysts with special morphology (tubes, needles, rods, platelets, spheres, flowers, cubes or general highly porous structures) [5] without separate supply of secondary energy. Hence, the process requires a sufficiently high waste air temperature (substance-specific; min. 150 °C) and its efficiency depends on a number of process and catalyst-specific parameters [5, 7]. The disadvantages of catalytic oxidation, apart from the presence of waste air components that can passivate or irreversibly inactivate the catalysts by forming a capping layer (e.g., dusts, siloxanes, sulfur-containing waste air substances; metallic dusts) [8–10], are mostly unknown reaction mechanisms that can occur, especially in the case of more complex waste air compositions.

13.2.6   NON-THERMAL PLASMA (NTP)

In NTP processes, strong electric fields are established by supplying electrical energy, in which secondary ionization and activation reactions are caused by primary electron emission (1–10 eV) [11, 12]. The resulting ionic, metastable as well as radical species are highly reactive and allow a variety of reactions under atmospheric pressure conditions in the presence of oxygen, leading to oxidation of VOCs. The process can be favored by moderate humidity, where hydroxide radicals are more readily formed. High humidity or even the presence of droplets or dust particles are detrimental to performance due to their strong radical scavenging properties [13]. Technically, ionization can be carried out in a variety of ways, with dielectric barrier discharge being particularly relevant for large-volumes of waste air. Here, an insulator is installed as a dielectric barrier between the two electrodes in the air gap, whereby electron exchange takes place only between the electrodes and the dielectric. Electrical power is transferred to the plasma by this electron transfer. Other designs do not use a dielectric, but have lower energy as well as cleaning efficiencies due to the larger electric current [14]. Both energy efficiency and conversion efficiency of the non-thermal plasma can be increased by combining the NTP stage with a catalytically active adsorber as an integrated (so-called 'inner plasma catalyst', IPC) or downstream (so-called 'post plasma catalyst') packing. Due to the short lifetime of the reactive intermediates, the IPC mode is basically characterized by higher efficiency and performance. As a result of the direct (linear) correlation between pollutant concentration and energy requirement, NTP processes are especially suitable for the treatment of odorous or low-loaded VOC waste air. With sufficient energy input, they enable compliance with legal limits even at higher VOC concentrations, but show unreasonably high energy consumption in this case. Therefore, recent developments also show the possibility of combining NTP processes with biological treatment stages to reduce energy requirements [15, 16].

13.2.7 UV OXIDATION

For UV oxidation processes, organic waste air compounds are directly decomposed by exposure to high-energy UV photons or oxidized by the formation of reactive oxygen species [17, 18]. The yield of reactive oxygen species can be increased by the addition of H_2O_2 or other oxidizing agents [19]. Processes that can be derived from this include photochemical dissociation, photolysis of hydrogen peroxide, Photo-Fenton processes (mostly in water phase), heterogeneous photocatalysis, and photolysis with ozone addition [20]. Although UV processes have also been combined with photocatalytically active catalysts in the recent years to improve their energy and purification efficiencies [6, 21–23], the energy efficiency of this process is significantly worse due to the limitation of emerging UV-C photons ($\lambda \leq 254$ nm; max. 15 % of the electrical input used). Furthermore, the photon yield is reduced over the lifetime of the lamps because they become dirty, salt deposits occur, or the glass body recrystallizes, which causes the photon yield to drop significantly after only 500 hours of operation [5]. Also, the lifetime of the lamps themselves is very limited with 1500–10,000 operating hours.

A hybrid between the conventional UV method and the NTP method is the excimer technology, in which light photons with $\lambda = 172$ nm (equivalent to 7.23 eV) are emitted by dissociation of high-energy Xe_2 molecules (so-called excimers) [24]. To date, excimer technology has been used industrially mainly for matting, hardening and increasing the chemical resistance of surfaces, as well as for curing paints and varnishes [24]. To date, it has only been used sporadically in waste air purification concepts on a laboratory scale for methanol, ethanol, isopropanol, 1,4-dioxane, cyclohexane and n-hexane [25] and for TCE on an industrial scale [26].

13.2.8 THERMAL PLASMA

Thermal plasmas are a waste air purification technology restricted to niche applications such as the treatment of CFCs, CHCs or PFCs in waste air, which has been in use since the 1980s. In contrast to NTP processes, by applying a DC voltage of up to 60 kV with over 95% efficiency, the electrical power is coupled in as heat, allowing flame temperatures of up to 12,000 K, or 14,000–22,000 K in the case of water vapor plasmas. The high temperature allows efficient conversion of very difficult-to-degrade components such as CF_4 or SF_6 with very short contact times and thus compact systems. Technically interesting plasma gases for the degradation of pollutants are helium, argon, nitrogen, water vapor, air and oxygen. In the case of the noble gases and nitrogen, the addition of liquid water, water vapor or air is required for oxidative degradation of the pollutants, although increased secondary emissions such as NO_x, COF_2, CO and similar compounds may occur. More recently, H_2 and CO_2 have also become increasingly important as plasma gases for syngas generation [27–29].

However, the efficiency of the process strongly correlates with the encoupled power and thus the plasma temperature, the plasma gas, the geometry of the plasma flame, the addition of additives, the mixing behavior of the waste gas into the plasma zone, and cooling-related energy losses. The formation of undesirable by-products also depends on these factors. Nevertheless, energetically optimized water vapor plasmas can have an SIE value of about 2000–2200 kWh $\cdot$ 1000 m^{-3} for the poorly degradable CF_4, for example [30–32].

13.2.9 ADSORPTION

During adsorption, VOCs accumulate at the valence-saturated solid interfaces of the adsorbent, whereby the maximum load capacity strongly depends on the properties of the adsorbent (vapor pressure, molar mass, molecular structure, specific adsorption enthalpy, chemical reactivity, surface tension, lipophilicity, etc.), on parameters of the process waste air (pressure, temperature, relative humidity, pollutant concentration, composition of the pollutant mixture) and on the adsorbent itself (specific inner surface area of the adsorbent). Activated carbon is mostly used as adsorbent, less frequently zeolites, since they have a high specific surface area and thus a high loading capacity.

As the operating time of the adsorber increases, the loading capacity is depleted, requiring regeneration of the loaded adsorbent. Regeneration of the adsorbent is possible in principle, but regenerated materials possess a lower loading capacity than virgin materials due to a partially irreversible chemical reaction of the adsorbent. Only physically adsorbed pollutants can be desorbed by regeneration (hot steam, hot dry air, electrothermal desorption, extraction, wet oxidation) [33]. The chemically sorbed compounds must be eliminated by pyrolysis as well as reactivation of the coal. Technically, the most common method is a combined regeneration and reactivation via thermal processes with dry air, where activated carbon is treated in a three-stage process in a furnace (e.g., rotary furnace, deck furnace, fluidized bed). After initial drying (T to 100 °C), the coal is regenerated by desorption and pyrolysis of the VOCs (100 °C < T < 700 °C), and finally the pyrolysis carbon covering the pores is removed by the so-called water gas reaction (C + H$_2$O → CO + H$_2$), thus reactivating the carbon [34]. The VOCs expelled in this process are captured in a highly concentrated waste gas and treated by thermal post-combustion at 1200 °C. Due to the high requirements in energy and equipment, the loaded adsorbent is therefore not regenerated on-site, but is instead treated by external service providers [35]. Furthermore, material losses due to attrition during both adsorption and desorption/ reactivation culminate to 5–19 % [36, 37], which means that activated carbon must be purchased on a recurring basis.

However, the stringent implementation of China's climate targets has led to a shutdown of many Chinese activated carbon producers in recent years, causing activated carbon prices to rise sharply. While in 2000 costs were still around 1500–1600 €/t (virgin) and 600–700 €/t (regenerated) [37], they rose to 2000–2200 €/t (virgin) by 2020 [38, 39] and currently to around 3000 €/t (virgin conventional coal) and 3400 €/t (virgin iodized coal) [7].

Various pollutants such as chlorine, ammonia, amines, sulfur dioxide, mercury, hydrogen sulfide, mercaptans, formaldehyde or carbonyl sulfide, but also polar pollutants, cannot be adsorbed or only insufficiently adsorbed due to their high vapor pressure and the lipophilic surface of the activated carbon. However, through suitable manufacturing processes, activated carbons can be provided with a more polar surface or can be specifically impregnated (e.g., with acids, alkali solution, or iodine), whereby the carbons also have a very high loading capacity with respect to the highly volatile pollutants mentioned.

However, the effectiveness of the activated carbon is severely limited above a humidity of 60 % due to the capillary condensation of water. Furthermore, polymerization reactions may occur for reactive VOCs, and adsorber fires for VOCs with a low ignition temperature.

13.2.10   Biological Processes

In addition to the described non-biological processes, the microbial degradation potential can be used for biological waste air treatment processes. The established processes can be reduced to the three basic techniques of bioscrubber, biotrickling filter and biofilter as shown in Figure 13.1.

Bioscrubbers represent a two-stage process consisting of an absorption column and a regeneration tank. The contaminated waste air is usually passed through the absorption column in countercurrent to the scrubbing liquid, allowing polar waste air components to be absorbed. To a limited extent, however, sufficient absorption of moderately polar compounds can be achieved by increasing the scrubbing liquid flow rate, but also by adding surfactant additives or by presenting a biphasic liquid

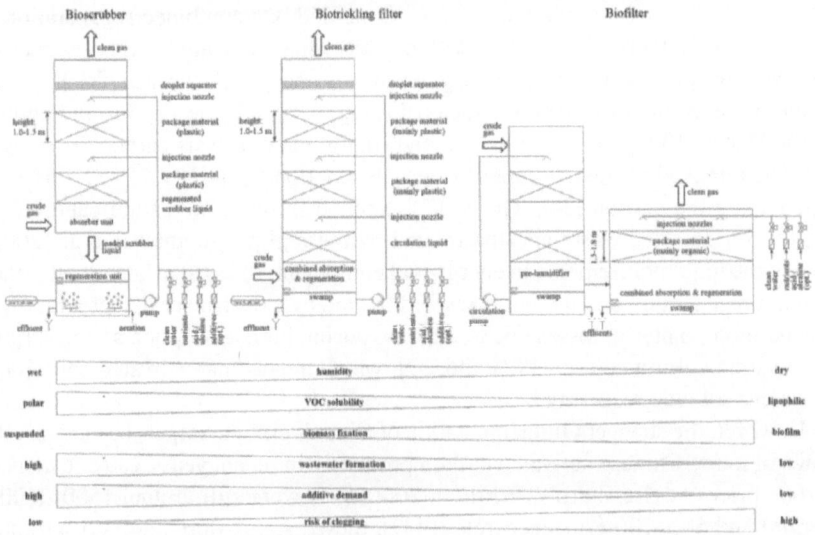

FIGURE 13.1   Functional diagram with components of an air conditioning system.

(*Source:* based on Greinacher 2019) [100].

phase, although the disposal of two-phase effluents in particular proves to be difficult [40–42]. The VOCs of the loaded absorption phase are biologically mineralized in the aerated regeneration tank, regenerating the liquid. Especially in the case of poorly biodegradable lipophilic compounds, biomass fixation on submerged carriers in the regeneration tank can also be targeted [43]. The liquid is then recirculated back into the absorber for repeated VOC absorption. The absorber is usually equipped with a structured packing or bulk material to improve the mass transfer from the gas phase to the liquid phase through a higher specific surface area and increased grade of turbulence.

If the share of poorly water-soluble and thus lipophilic VOCs in the waste air increases, a biotrickling filter can be applied. This is a single-stage process similar to the absorption stage of the scrubber. According to a rudimentary and unaerated swamp and significantly reduced irrigation densities, microbial growth as a biofilm is enforced on the carriers of the absorber. Packing materials with defined geometries are preferred as carriers, but lava gravel, hydrogranules and similar minerals are also suitable. Due to the immobilization of the biological phase, absorption of the VOCs and microbial regeneration simultaneously take place. The system can be adapted to dominantly more polar or lipophilic VOC components by selective control of the irrigation density and the packing used (type of material and structure). In industrial environments, intermittent humidification has become established if required, whereas this should be urgently avoided in the agricultural sector due to increasing N_2O emissions [44].

A further reduction of the sprinkling density and thus an even drier mode of operation is used with the biofilter. This represents a single-stage process where a mostly organic filter material is colonized with a biofilm for VOC degradation. As in the case of the biotrickling filter, simultaneous absorption and mineralization of the VOCs takes place, but irrigation density is dramatically reduced or sprinkling is dispensed with altogether (in which case a prewetting stage is mandatory), which means that this dry mode of operation is optimized in particular for lipophilic-dominated exhaust air streams. Compared with the BTF, the sprinkling density is again significantly reduced or even not used (in which case a pre-humidification is mandatory), which means that this dry mode of operation is optimized for lipophilic-dominated waste air. However, the lack of or low irrigation density severely restricts the supply of nutrients as well as the discharge of intermediates and pH-shifting degradation products.

Hence, the selection of the appropriate process especially depends on the polarity and concentration of the VOCs in the waste air. For example, as described in Figure 13.2, bioscrubbers are specifically used for waste air streams with high loadings of up to 2 g C $\cdot$ m^{-3} of polar VOCs, while biofilters are mainly used for lipophilic low-concentrated VOCs as well as odors. The biotrickling filter is a hybrid system of the above two processes and is used in continuous operation especially for VOC mixtures of a broad polarity spectrum at concentrations of several 100 mg C $\cdot$ m^{-3}, rarely up to 1 g C $\cdot$ m^{-3} (resulting volume-specific loads approx. 10–100 g C $\cdot$ m$^{-3} \cdot$ h^{-1}. In individual cases, volume-specific loads of up to 4000 g $\cdot$ m$^{-3} \cdot$ h^{-1} have also been realized under short-term conditions [45–47], but for long-term operation they lead to clogging of the bed by abundant biomass growth (so-called clogging).

All three biological processes are characterized by relatively low operating costs (i.e., electricity as well as additives) and have almost no aqueous or gaseous secondary emissions, which means that they can be seen as a sustainable waste air treatment process, especially in light of climate change. Besides existing physical limits (water solubility, mass transfer behavior), damage may occur in case of improper system design or lack of maintenance, especially in the aeration system (clogging of ceramic aerators, pore expansion or cracks in membrane aerators) with the consequence of pH drop and secondary odor emissions, increased foam formation due to cell lysis, as well as migration of the suspended biocenosis to the packing material of the absorber column with the formation of extensive clogging of the packing. Furthermore, a bioscrubber-specific problem may arise from microbial regulatory mechanisms and thus limited degradability of VOC mixtures [48–51]. Biofilters are particularly limited in their efficiency by improper humidification, as a lack of humidity leads to insufficient biofilm growth and thus a lack of degradation performance. Excessive humidification, on the other hand, leads to wetting, compaction of the packing material with a sharp increase in pressure loss and thus energy demand of the fan, as well as inhomogeneous gas distribution with the formation of anaerobic zones in the packing material. Besides secondary odor emissions, this is reflected in particular in a drop in cleaning efficiency. In long-term operation, both gradual microbial degradation of the filter material and excessive biomass formation as a result of high VOC loads can lead to a simultaneous drop in performance with increasing pressure drop. Clogging is also known for the biotrickling filter, but to a lesser extent, since mechanically stable packed beds or packing materials are used here instead of beds of organic materials. Finally, the use of biofilters for the degradation of organosulfur compounds and H_2S is particularly challenging, since the biodegradation leads to the formation of sulfuric acid as the final product and thus to a strong acid formation [52]. If a strong pH drop occurs, neutralization of biofilter materials during operation is almost impossible and forces shifting of the microbial flora to higher fungal proportions, which can lead to a fungus-typical inherent odor.

13.3   PARAMETERS FOR SUITABILITY OF THE PROCESSES

The suitability of these methods to different groups of VOCs depends primarily on the chemical properties and concentrations of the compounds to be removed as shown in Figure 13.2; i.e., depending on the treatment method, the elimination capability depends on the vapor pressure (condensation, adsorption), the water solubility (absorption), the adsorption capacity (adsorption), the diffusion constant (membrane process), the ignition temperature as well as the calorific value (combustion, thermal plasma), or the reactivity and degree of saturation (oxidative catalysis, NTP, UV).

An overview of the suitability of these processes for various substances/substance classes is given in Table 13.1.

In addition to VOC emissions, odor emissions attract increasing attention in waste air treatment because, on the one hand, odor emissions can be easily detected by everyone through the human nose and, on the other hand, legislative amendments to minimum distances towards buildings can lead to an increasing conflict potential.

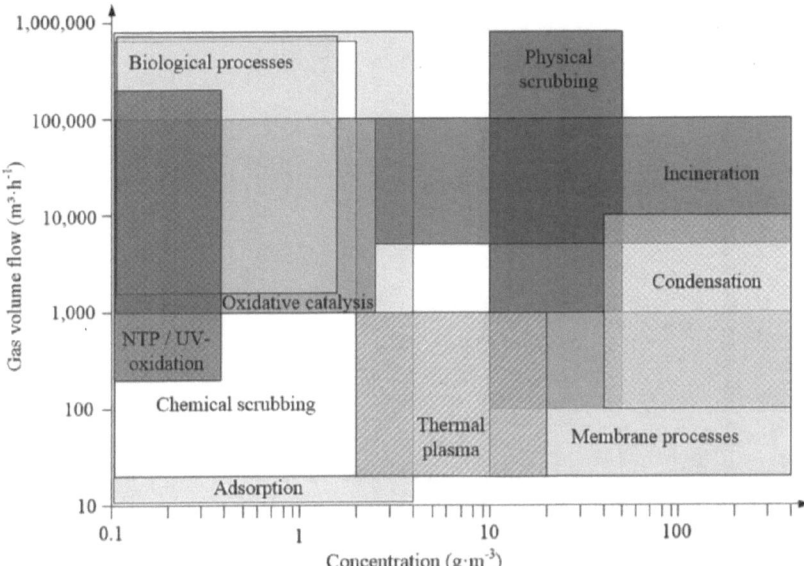

FIGURE 13.2 Overview of the fields of application of stand-alone waste air treatment processes.

(*Source:* adapted from Webster 1999) [103].

For example, in the present German TA-Luft (Technical Instructions on Air Quality Control), the hitherto rigid regulation of a minimum distance of 300 m from odor-intensive production sites towards residential areas was replaced by a more flexible solution requiring a mandatory odor emission forecast depending on the occurring or expected plant emission potential and the site-specific dispersion conditions, but makes the required distance to the nearest residential area more flexible and minimizes the minimum distance to 100 m [54]. In this context, odor emission measurements are carried out according to a specified grid or a defined dispersion plume with inspection dates spread over 6 months in accordance with relevant guidelines [55], while odor emission mass flows are quantified by sampling with sample bags and subsequent olfactometric analysis [55, 56]. In this case, odor concentration is determined by an olfactometer, where highly diluted sample air is offered to a test team of 4 persons and the dilution level is gradually reduced within three test runs until the participants perceive it as positive. The calculated 50th percentile then corresponds to the odorant concentration.

Since, according to Table 13.2, malodorous compounds have odor thresholds in the ppt to ppb range, oxidative processes such as NTP and UV oxidation, but also biofilters, are particularly suitable for odor elimination. In the latter case, the odorant concentrations present are primarily too low to induce biological pollutant degradation enzymatically; however, the adsorptive potential of the biofilter enables pollutant accumulation and thus subsequent enzyme induction [57].

TABLE 13.1

Suitability of Waste Air Treatment Processes to Different Substances and Substance Classes

Parameter	Condensation	Absorption	Membrane processes	Combustion	Catalytic oxidation	NTP	UV-Oxidation	Thermal plasma	Adsorption	Bioscrubber	Biotrickling filter	Biofilter
Aliphatics	+/-	-	+	+	-	+/-	+/-	+	+	-	+	+
Alcohols	+	+	-	+	+/-	+/-	+/-	+	+	+	+	+/-
Aldehydes	+/-	+	-	+	+	+	+	+	+/-	+	+	+
Ketones	+	+	-	+	+/-	+/-	+/-	+	+	+	+	+
Organic acids	+	+	-	+	+	+	+	+	+	+	+	+/-
Esters	+	+	-	+	+	+/-	+	+	+	+	+	+/-
Ethers	+/-	-	+/-	+	-	+/-	+/-	+	+	-	+	+
Aromatics	+	+/-	+	+	+	+/-	+/-	+	+	+	+	+/-
Sulphurorganics	-	+/-	+	+/-	+	+/-	+/-	+	+	+	+	+/-
Haloorganics	+/-	-	+/-	+/-	-	+/-	+/-	+/-	+	-	+	+/-
CH$_4$	-	-	-	+	-	-	-	+	-	-	-	+/-
NH$_3$	-	+	-	+/-	+/-	+	+	+	-	+	+	+
N$_2$O	-	-	-	+/-	+/-	-	-	+	-	-	-	+/-
H$_2$S	-	+	-	+/-	+	+/-	+/-	+	+/-	+	+	+/-
Odour	+/-	+	-	+	+	+	+	+	+	+/-	+/-	+/-
Germs	-	+/-	-	+	+	+	+	-	+	-	+	+
Dust	+/-	+	-	-	-	-	-	-	-	+	+/-	-
Humidity	+	+	-	+/-	-	-	-	-	-	+	+	+

(*Source*: adapted from Dobslaw and Ortlinghaus 2020) [53]. +: high suitability, +/-: moderate suitability, -: no/low suitability.

TABLE 13.2
Odor Thresholds in ppm of Odorous Substances

Substance	Threshold	Substance	Threshold
Alkanes		*Esters*	
Propane	1500	Methyl formate	130
n-Butane	1200	Ethyl formate	2.7
n-Pentane	1.4	n-Propyl formate	0.96
Isopentane	1.3	Isopropyl formate	0.29
n-Hexane	1.5	n-Butyl formate	0.087
2-Methylpentane	7.0	Isobutyl formate	0.49
3-Methylpentane	8.9	Methyl acetate	1.7
2,2-Dimethylbutane	20	Ethyl acetate	0.87
2,3-Dimethylbutane	0.42	n-Propyl acetate	0.24
n-Heptane	0.67	Isopropyl acetate	0.16
2-Methylhexane	0.42	n-Butyl acetate	0.016
3-Methylhexane	0.84	Isobutyl acetate	0.0080
3-Ethylpentane	0.37	38 sec. Butyl acetate	0.0024
2,2-Dimethylpentane	38	tert. Butyl acetate	0.071
2,3-Dimethylpentane	4.5	n-Hexyl acetate	0.0018
2,4-Dimethylpentane	0.94	Methyl propionate	0.098
n-Octane	1.7	Ethyl propionate	0.0070
2-Methylheptane	0.11	n-Propyl propionate	0.058
3-Methylheptane	1.5	Isopropyl propionate	0.0041
4-Methylheptane	1.7	n-Butyl propionate	0.036
2,2,4-Trimethylpentane	0.67	Isobutyl propionate	0.020
n-Nonane	2.2	Methyl n-butyrate	0.0071
2,2,5-Trimethylhexane	0.90	Methyl isobutyrate	0.0019
n-Undecane	0.87	Ethyl n-butyrate	0.000040
n-Decane	0.62	Ethyl isobutyrate	0.000022
n-Dodecane	0.11	n-Propyl n-butyrate	0.011
		Isopropyl n-butyrate	0.0062
Cyclic aliphatics		n-propyl isobutyrate	0.0020
Methylcyclopentane	0.15	Isopropyl isobutyrate	0.035
Cyclohexane	2.5	n-Butyl n-butyrate	0.0048
Methylcyclohexane	0.15	Isobutyl n-butyrate	0.0016
		n-Butyl isobutyrate	0.022
Alkenes		Isobutyl isobutyrate	0.075
Propylene	13	Methyl n-valerate	0.0022
1-Butene	0.36	Methyl isovalerate	0.0022
Isobutene	10	Ethyl n-valerate	0.00011
1-Pentene	0.10	Ethyl isovalerate	0.000013
1-Hexene	0.14	n-Propyl n-valerate	0.0033
1-Heptene	0.37	n-Propyl isovalerate	0.000056
1-Octene	0.0010	n-Butyl isovalerate	0.012
1-Nonene	0.00054	Isobutyl isovalerate	0.0052
1,3-Butadiene	0.23	Methyl acrylate	0.0035

(Continued)

TABLE 13.2
Continued

Substance	Threshold	Substance	Threshold
Isoprene	0.048	Ethyl acrylate	0.00026
		n-Butyl acrylate	0.00055
Alcohols		Isobutyl acrylate	0.00090
Methanol	33	Methyl methacrylate	0.21
Ethanol	0.52	2-Ethoxyethyl acetate	0.049
n-Propanol	0.094		
Isopropanol	26	*Aromatics*	
n-Butanol	0.038	Benzene	2.7
Isobutanol	0.011	Toluene	0.33
sec. Butanol	0.22	Styrene	0.035
tert. Butanol	4.5	Ethylbenzene	0.17
n-Pentanol	0.10	o-Xylene	0.38
Isopentanol	0.0017	m-Xylene	0.041
sec. Pentanol	0.29	p-Xylene	0.058
tert. Pentanol	0.088	n-Propylbenzene	0.0038
n-Hexanol	0.0060	Isopropylbenzene	0.0084
n-Heptanol	0.0048	1,2,4-Trimethylbenzene	0.12
n-Octanol	0.0027	1,3,5-Trimethylbenzene	0.17
Isooctanol	0.0093	o-Ethyltoluene	0.074
n-Nonanol	0.00090	m-Ethyltoluene	0.018
n-Decanol	0.00077	p-Ethyltoluene	0.0083
2-Ethoxyethanol	0.58	o-Diethylbenzene	0.0094
2-n-Buthoxyethanol	0.043	m-Diethylbenzene	0.070
1-Butoxy-2-propanol	0.16	p-Diethylbenzene	0.00039
		n-Butylbenzene	0.0085
Aldehydes		1,2,3,4-Tetramethylbenzene	0.011
Formaldehyde	0.50	1,2,3,4-Tetrahydronaphthalene	0.0093
Acetaldehyde	0.0015		
Propionaldehyde	0.0010	*Phenols*	
n-Butylaldehyde	0.00067	Phenol	0.0056
Isobutylaldehyde	0.00035	o-Cresol	0.00028
n-Valeraldehyde	0.00041	m-Cresol	0.00010
Isovaleraldehyde	0.00010	p-Cresol	0.000054
n-Hexylaldehyde	0.00028		
n-Heptylaldehyde	0.00018	*Sulfuric compounds*	
n-Octylaldehyde	0.000010	Sulfur dioxide	0.87
n-Nonylaldehyde	0.00034	Hydrogen sulfide	0.00041
n-Decylaldehyde	0.00040	Carbonyl sulfide	0.055
Acrolein	0.0036	Dimethyl sulfide	0.0030
Methacrolein	0.0085	Methyl allyl sulfide	0.00014
Crotonaldehyde	0.023	Diethyl sulfide	0.000033
		Allyl sulfide	0.00022
Ketones		Carbon disulfide	0.21
Acetone	42	Dimethyl disulfide	0.0022
Methyl ethyl ketone	0.44	Diethyl disulfide	0.0020

Substance	Threshold	Substance	Threshold
Methyl n-propyl ketone	0.028	Diallyl disulfide	0.00022
Methyl isopropyl ketone	0.50	Methyl mercaptane	0.000070
Methyl n-butyl ketone	0.024	Ethyl mercaptane	0.0000087
Methyl isobutyl ketone	0.17	n-Propyl mercaptane	0.000013
Methyl sec. butyl ketone	0.024	Isopropyl mercaptane	0.0000060
Methyl tert. butyl ketone	0.043	n-Butyl mercaptane	0.0000028
Methyl n-amyl ketone	0.0068	Isobutyl mercaptane	0.0000068
Methyl isoamyl ketone	0.0021	sec. Butyl mercaptane	0.000030
Diacetyl (Butane-2,3-dione)	0.000050	tert. Butyl mercaptane	0.000029
		n-Amyl mercaptane	0.00000078
Organic acids		Isoamyl mercaptane	0.00000077
Acetic acid	0.0060	n-Hexyl mercaptane	0.000015
Propionic acid	0.0057	Thiophene	0.00056
n-Butyric acid	0.00019	Tetrahydrothiophene	0.00062
Isobutyric acid	0.0015		
n-Valeric acid	0.000037	*Nitrogeneous compounds*	
Isovaleric acid	0.000078	Nitrogen dioxide	0.12
n-Hexanoic acid	0.00060	Ammonia	1.5
Isohexanoic acid	0.00040	Methylamine	0.035
		Ethylamine	0.046
Terpenes		n-Propylamine	0.061
α-Pinene	0.018	Isopropylamine	0.025
β-Pinene	0.033	n-Butylamine	0.17
Limonene	0.038	Isobutylamine	0.0015
		sec. Butylamine	0.17
Chlorinated compounds		tert. Butylamine	0.17
Chlorine	0.049	Dimethylamine	0.033
Dichloromethane	160	Diethylamine	0.048
Chloroform	3.8	Trimethylamine	0.000032
Trichloroethylene	3.9	Triethylamine	0.0054
Carbon tetrachloride	4.6	Acetonitrile	13
Tetrachloroethylene	0.77	Acrylonitrile	8.8
		Methacrylonitrile	3.0
Others		Pyridine	0.063
Ozone	0.0032	Indole	0.00030
Furane	9.9	Skatole (3-Methylindole)	0.0000056
2,5-Dihydrofurane	0.093	Ethyl-o-toluidine	0.026
Geosmin	0.0000065		

(*Source:* based on Yoshio and Nagate 2003) [58].

13.4   COMBINED TREATMENT PROCESSES

As shown in Table 13.1, none of the listed processes can efficiently reduce the entire spectrum of contaminants. Thus, with increasing complexity of the waste air situation, it is necessary to establish multi-stage waste air treatment concepts that are adapted to the actual waste air situation. Taking into account aspects such as energy and cleaning efficiencies, national and international political objectives for sustainability and climate protection, price increases and shortages of energy sources and raw materials, as well as the social acceptance of the installed treatment processes, biological treatment processes and integrated air handling concepts, on the one hand, and air-reuse concepts, on the other hand, are currently gaining dramatically in importance. The objectives of biological treatment and air-reuse initially appear to be incompatible, since biological treatment processes potentially emit bioaerosols whose pathogenicity is not known or whose apathogenicity cannot be guaranteed over the lifetime of the bioprocess. This conflict of interest can be solved by an appropriate combination of waste air treatment processes as well.

Integrative industrial waste gas treatment can be carried out either on an energetic or on a material level. In the case of a purely energetic approach, the waste heat in the clean gas is recovered and utilized by heat exchangers. Depending on the temperature level, it can be used either for thermal processes (e.g., drying processes; $T \approx 120–180 \, °C$), for steam production ($T > 180 \, °C$), for preheating of contaminated crude gas ($T \approx 40–120 \, °C$) or for preheating of fresh air for air-conditioning systems ($T \approx 20–40 \, °C$). In addition, warm waste gas can also be used for cooling purposes in air conditioning systems according to the principle of the cooling pump [59].

For material utilization of the waste gas, a distinction must be made between air for human presence and simple process air. In the latter case, for example, as waste gas from compost piles shows sufficient oxygen content (10–20%) [60] it can be directly recycled within the piles, which homogenizes the temperature distribution within the piles and increases the degradation kinetics by up to 40% and reduces greenhouse gas emissions by up to 25.7% [60, 61]. On the other hand, if cleaning of the waste air is required by process reasons or as otherwise people will be exposed to waste air contaminants, the installation of a combined waste air cleaning process is necessary. However, the techniques to be used depend primarily on the degree of contamination of the waste gas with VOCs, so that a fundamental distinction must be made between an industrial waste gas situation with high VOC content (e.g., from enclosed production processes) and a waste gas situation with low VOC content (e.g., workplace or building exhaust ventilation).

13.4.1   HIGH CONTAMINATED WASTE AIR CONDITIONS

Addressing the above-mentioned economic, political and social objectives, process combinations of non-biological and biological stages are particularly suitable for exhaust air situations with high contamination levels of VOCs, but also inorganic pollutants, whereby the non-biological stage is usually upstream, since these can either eliminate biological interfering substances or lead to partial oxidation of the pollutants and thus to increased bioavailability through oxidative mechanisms [23]. In the former case, process combinations of acid scrubber and biofilter, possibly supplemented by an integrated alkaline scrubber, are now widely used to remove NH_3 and H_2S from VOC emissions from biogenic sources before VOC mineralization takes place. For the oxidative reaction mechanisms, advanced oxidation technologies such as non-thermal plasmas (NTP), ultraviolet (UV) oxidation or process combinations of ionization and deposition via electric fields are used. Studies on these techniques are presented in detail in Dobslaw and Ortlinghaus 2020 [53]. Abundant ozone as more stable reactive oxygen species may react within the biological stage, where it can positively prevent clogging in long terms [62–70].

However, a disadvantage of this process combination is that bioaerosols can potentially be emitted and the treated waste gas has a high humidity. These disadvantages also occur in case of multiple biological stages, which is why these process combinations are not suitable for air-reuse concepts with subsequent long-lasting air exposure to humans.

If air-reuse concepts for VOC containing waste gases are targeted, the process sequence can be exchanged. In this case a bioprocess is located upstream of an oxidative treatment process, and quantitative VOC removal is performed by biodegradation processes, therefore representing an energy-efficient process. Hence, the oxidative process is solely used for air finishing and disinfection by inactivating or killing of any bioaerosols potentially discharged from the bioprocess. Abundant ozone may be depleted by a carbon fleece downstream of the process. This process combination enables formation of high-quality clean gas, which can be re-entered into the production halls and thus directly at the workplaces located there. In the past, with exception of expensively pre-treated air (e.g., process air in clean rooms), the relevance of material re-use of air was low due to the lack of CO_2 emission pricing and low heating costs. However, against the background of sharply rising energy prices and inefficient heat recovery in heat exchangers, it gets more and more important as a possibility to reduce operational costs.

UV process combinations have recently been used to purify chlorobenzene [71, 72], dichloromethane [73], ethylbenzene [74], n-hexane [75], PCE [76], α-pinene [65, 77, 78], styrene [21, 79], TCE [76], toluene [80, 81], and o-xylene [82] as single contaminants and of VOC mixtures containing of toluene and o-xylene [63], a mixture of ethyl acetate, toluene, ethyl benzene, xylene, ethyl toluene and trimethyl benzene [83] as well as a mixture of toluene, ethyl benzene and xylene isomers [23]. NTP-based combined processes were used in the treatment of chlorobenzene [64, 84], styrene [14], 1,2-DCE [85], dimethyl disulphide [86] and limonene [87] as single substances and in a various number of VOC mixtures [15, 85, 88–94]. In addition, the first experiments with magnetic fields as an activation step were successfully carried out [95, 96]. A successful demonstration for VOC abatement as well as disinfection of secondary bioaerosols from a styrene degrading biotrickling filter was provided by Helbich et al 2020 [14]. A detailed overview of the use of these processes in combination with biological treatment processes can be found in Dobslaw and Ortlinghaus 2022 [53]. In all studies, the process combination succeeded by significant increase of the purification efficiency compared to the summarised efficiency of the two individual plants (i.e., due to partial oxidation of the contaminants, the occurring transformation products can be degraded much more efficiently due to their higher hydrophilicity and the associated higher bioavailability) [5]. Quantitatively, an efficiency increase of 3–15% (magnetic field), 20–50% (UV oxidation) and 25–80% (NTP) occurs compared to the sole bioprocess [53]. The advantage of treatment performance is particularly evident in the degradation of biopersistent compounds, which can thus be activated for biodegradation. For example, in the case of Freon R-22, dehalogenised reaction products have been identified after exposure to NTP, which tend to exhibit higher biodegradability than the original substance [97].

In individual cases, a multi-stage bioprocess may be more effective than VOC degradation in a single biological stage, which is particularly useful for VOC mixtures with strongly diverging polarity and a high pollutant load and thus an increased tendency to clogging.

For example, a combination of BTF and BF was used to treat H_2S, dimethyl sulphide and dimethyl disulphide [98]. The water-soluble hydrogen sulphide was completely mineralised in the BTF, while the degradation of the two water-insoluble organosulphur compounds took place in the biofilter. Gerl et al [51] investigated the biodegradation of a 1:1 mixture of polar methoxypropyl acetate and lipophilic xylene isomers by a combination of bioscrubber and biotrickling filter compared to a biofilter as reference system. While the performance of the biofilter was around 20% and increasing pressure losses were observed, the combined process showed long-term operational stability with elimination efficiencies of more than 70% at the same EBRT. In combination with NTP or UV, these combinations are also suitable for air-reuse concepts.

13.4.2   LOW CONTAMINATED WASTE AIR CONDITIONS

In the industrial environment, low-contaminated waste gas particularly occurs in ventilation of factory halls and buildings, since low workplace limit values must be complied with for the workplaces located there. Since these limit values are lower than the emission limit values, no treatment usually takes place here, but the contaminated air is merely replaced proportionally by unpolluted fresh air. The possibilities for waste air purification especially depend on the type of air conditioning system (centralized/decentralized) installed.

In a central air-conditioning system, as shown in Figure 13.3, the fresh air is first passed through a supply air filter (F1) to eliminate dust and pollen (filter class G4-F9; usually up to F7) and then heated or cooled by a heat exchanger (HEX) to recover room heat (winter) or room coolness (summer). The air is then passed through a silencer (S) and a supply inlet air fan (IF) again via a second silencer and then conditioned. The conditioning consists of up to 4 stages, namely the air heater (H), air cooler (AC), air humidifier (AH) and a second air heater. After that, the air is filtered again (usually F7–F9) and supplied to the indoor space. The used indoor air is pre-filtered (F3, usually as F7) and cooled (winter) or heated (summer) via the heat exchanger in counterflow to the inlet air and discharged into the ambient air via an exhaust fan (EF) with silencers (S). In the case of an energetically favorable recirculation mode, the indoor air is fed in directly upstream of the silencer of the supply air system and further preconditioned. The recirculation rate is usually set to 80–90% in order to achieve a sufficiently high oxygen concentration in the room on the one hand and to reduce the CO_2 concentration and heating energy costs on the other. 10–20% of the total air flow is thus replaced by fresh air.

The primary purpose of preconditioning fresh air is to provide an ideal comfortable climate, which is defined by the room temperature and humidity, but not to treat

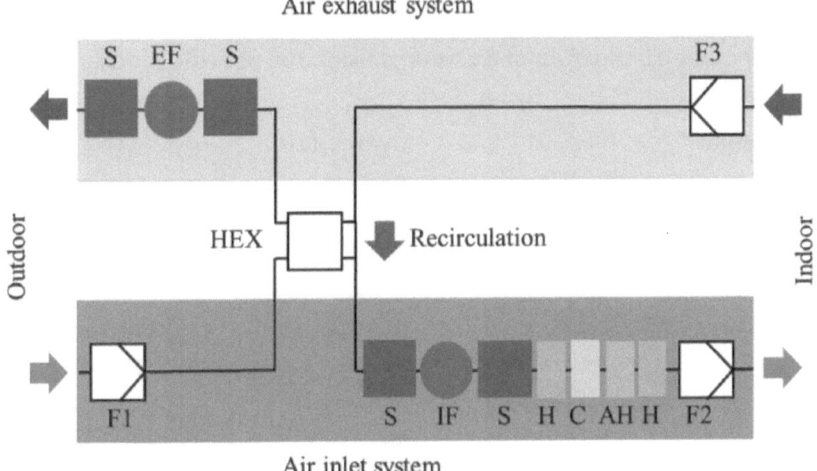

Air exhaust system

Air inlet system

FIGURE 13.3   Basic processes of biological waste air treatment with relevant key parameters.

gaseous waste gas compounds or pathogens. The installed filters are not suitable for eliminating gaseous emissions. According to DIN EN 779 [99], the installation of an F7 filter also filter no. 2 according to Figure 13.3 allows a minimum efficiency of 35% for pathogens. Since the installed air heater performs reheating to a maximum of 50 °C with very short contact times [101], its reduction efficiency with respect to pathogens is negligible. Centralized air-conditioning systems are therefore not suitable for significant disinfection or abatement of gaseous waste air compounds.

In contrast to central air-conditioning systems, decentralized systems enable rooms to be ventilated as needed. According to the place of installation, they are divided into ceiling ventilation systems, window ventilation systems, parapet ventilation systems and underfloor ventilation systems. As they are even more simple in design than centralized air-conditioning systems, they only show limited effect in removal of dust and pollen. Pre-conditioning or even sterilization of the air is therefore also not possible.

These existing deficiencies became apparent during the current SARS-CoV-2 pandemic, so that various stand-alone devices for the disinfection of indoor air were established on the market. The methods used are based in particular on medical disinfection methods of surfaces and liquids such as ionization, the use of high-energy γ-radiation, UV, thermal oxidation, chlorination (e.g., Cl_2, HOCl), ozonation, the use of oxidants (e.g., H_2O_2), special forms of chemical disinfection (e.g., I_2, Br_2, BrCl), physical processes such as filtration (e.g., sedimentation, cyclone, sand filter, activated carbon filter, ultrafiltration, reverse osmosis, ultrasound), pulsed lasers, and non-thermal plasmas (DBD, corona, arc, plasma torch) [102]. These processes have process-typical advantages and disadvantages, which are summarized for filtration, UV oxidation, and NTP processes as the processes with the highest efficiencies in Table 13.2. The other processes also have a low market relevance due to lower efficiencies, restrictions in application or difficulties in handling.

TABLE 13.3
Overview of the Elimination of Microorganisms and Specific Mechanisms of Action

Target parameter	Filtration	UV-Oxidation	NTP
Vegetative cells	High (>3 logs)	High (2.5–6.1 logs)	High (2.0–7.8 logs)
Gram-neg. bacteria	High (>3 logs)	High (3–6.1 logs)	High (3.0–7.8 logs)
Gram-pos. bacteria	High (>3 logs)	Medium (1–3 logs)	High (2.0–6.0 logs)
Bacterial spores	High (>3 logs)	Low (0.3–1 logs)	High (2.0–5.0 logs)
Yeasts	High (>3 logs)	High (2.5–6.1 logs)	High (2.0–7.8 logs)
Molds	High (>3 logs)	Medium (1.2–4.8 logs)	High (1.5–6.5 logs)
Viruses	High (2–3 logs)	High (>3 logs)	High (>3 logs)
Cell wall	No	Yes	Yes
Cell membrane	No	Partially	Yes
Cytosol	No	No	Yes
Membrane proteins	No	Yes	Yes
Cytosol proteins	No	No	Yes
Plasmids	No	Low (0–1.0 logs)	Yes
RNA	No	Yes	Yes
DNA	No	Yes	Yes
Ionization	No	No	Yes
Dimer formation	No	Yes	Yes
Cell perforation	No	No	Yes
Oxidative effect	No	Indirect	Yes
Effect of humidity	No effect	Negative	Positive
Denaturation	No	Yes	Yes
VOC removal	No	Yes	Yes
Inorganics removal	No	Low	Yes

To date, only individual solutions for air disinfection have been launched on the market as retrofit kits for air conditioning systems. Due to the high initial pressure drop of approx. 220 Pa (in continuous operation up to 450 Pa), HEPA filter solutions for existing systems have not become established due to the lack of fan power.

The existing solutions as retrofit kits are therefore exclusively UV-based systems. These can be differentiated between UV-LED processes, which are used for ceiling filter cassettes and split air conditioning units, and between processes with UV tube modules for centralized air-conditioning systems. Since the power consumption of UVC-LEDs (approx. 20–100 mW) is very low, only very slight disinfection is expected, at least at common air exchange rates.

Products for central air-conditioning systems, on the other hand, use UVC lamps because the absorption maxima of DNA (260 nm) and proteins (280 nm) and thus their efficient denaturation can be covered by the emitted wavelength spectrum. Since SIE values of approx. 0.05 kWh $\cdot$ 1000 m^{-3} are realized, a noteworthy disinfection effect can be assumed in the context of Table 13.2. Moderate VOC elimination can also be expected, which is why air-reuse concepts with appropriate retrofit kits are certainly feasible. However, the success of the action depends directly on the energy input.

NTP processes also appear to be suitable, but are not yet available as marketable products. However, since 274 patents for air purifiers based on NTP processes had been filed by October 2021, NTP-based disinfection modules for air-conditioning systems can also be expected in the near future.

13.5 CONCLUSION

Until now, processes for the treatment of industrial waste gas have almost exclusively been regarded as end-of-pipe technologies and the treated gas was discharged into the environment. Current aspects such as energy and cleaning efficiencies, national and international political efforts towards sustainability and climate protection, price increases and shortages of energy sources and raw materials as well as the social acceptance of the installed treatment processes lead to a strengthening of biological waste air treatment processes, which can be regarded as sustainable processes due to their low energy and resource requirements. In parallel, there is an increased desire to implement energy-efficient air-reuse concepts, which, however, may not lead to health-relevant secondary emissions (e.g., combustion gases, bioaerosols) during VOC abatement.

This requirement cannot be implemented by biological or non-biological processes as stand-alone processes, which is why there is an increased interest in integrated process combinations for the treatment of waste gases. The process combination to be used depends primarily on the type and concentration of the waste gas compounds. In case of air heavily contaminated with VOCs, integrated process combinations consisting of a biological treatment stage and oxidative post-treatment by UV or NTP can enable both VOC removal and disinfection of the waste gas, thus allowing air-reuse. In case of low polluted exhaust air (e.g., indoor air) the biological stage can be omitted and a VOC finishing as well as a disinfection of the indoor air can be realized energy-efficiently by air-conditioning systems with an oxidative stage. In the context of the current SARS-CoV-2 pandemic, existing technological gaps have been closed, so that integrated solution concepts for centralized and decentralized air-conditioning systems are available in addition to stand-alone units. Thus, the technological prerequisites have been created to increasingly implement air-reuse concepts in the future. From an economic point of view, these concepts become more and more interesting the more requirements are made and thus the higher the production costs of process air are (e.g., in clean room applications).

REFERENCES

[1] Institute for Health Metrics and Evaluation. *Global Burden of Disease (GBD) Results Tool/GHDx*. Seattle: Institute for Health Metrics and Evaluation; 2017.
[2] World Energy Outlook. International Energy Agency (IEA). Paris, France; 2021. https://www.iea.org/reports/world-energy-outlook-2021

[3] Niklas J, Haller F, Pitz M, Hellmann A, Ripperger S. Heterogeneous condensation of water vapor on nanoparticles in a membrane-based high-flow process monitored by a new in situ measuring cell. Aerosol Science and Technology. 2015; 49:950–8. doi:10.10 80/02786826.2015.1086480.

[4] Xu J, Yu Y, Zhang J. Heterogeneous condensation on fine particles of water vapor in a moderated growth tube. Energy Procedia. 2017; 118:201–9. doi:10.1016/j. egypro.2017.07.025.

[5] Dobslaw D. *Herausforderungen in der biologischen und nicht-biologischen Abluftreinigung*. Stuttgart: Universität Stuttgart; 2020.

[6] Verein Deutscher Ingenieure VDI, editor. *Waste Gas Cleaning—Methods of Thermal Waste Gas Cleaning: VDI-Guideline 2442*. Berlin: Beuth Verlag; 2014.

[7] Si, X., Lu, R., Zhao, Z. et al. Catalytic production of low-carbon footprint sustainable natural gas. Nat Commun. 2022; 13:258. https://doi.org/10.1038/s41467-021-27919-9

[8] Argyle M, Bartholomew C. Heterogeneous catalyst deactivation and regeneration: A review. Catalysts. 2015; 5:145–269. doi:10.3390/catal5010145.

[9] Marécot P, Paraiso E, Dumas JM, Barbier J. Deactivation of nickel catalysts by sulphur compounds. Applied Catalysis A: General. 1992; 80:79–88. doi:10.1016/0926-860X(92)85109-O.

[10] Verein Deutscher Ingenieure VDI, editor. *Waste Gas Cleaning—Methods of Thermal Waste Gas Cleaning—Selective catalytic reduction: VDI-Guideline 3476 Part 3*. Berlin: Beuth Verlag; 2012.

[11] Marotta E, Schiorlin M, Rea M, Paradisi C. Products and mechanisms of the oxidation of organic compounds in atmospheric air plasmas. J. Phys. D: Appl. Phys. 2010; 43:124011. doi:10.1088/0022-3727/43/12/124011.

[12] Marotta E, Scorrano G, Paradisi C. Ionic reactions of chlorinated volatile organic compounds in air plasma at atmospheric pressure. Plasma Process. Polym. 2005; 2:209–17. doi:10.1002/ppap.200400047.

[13] Cheng Z-W, Zhang L-L, Chen J-M, Yu J-M, Gao Z-L, Jiang Y-F. Treatment of gaseous alpha-pinene by a combined system containing photo oxidation and aerobic biotrickling filtration. J Hazard Mater. 2011; 192:1650–8. doi:10.1016/j.jhazmat.2011.06.092.

[14] Helbich S, Dobslaw D, Schulz A, Engesser K-H. Styrene and bioaerosol removal from waste air with a combined biotrickling filter and DBD–plasma system. Sustainability. 2020; 12:9240. doi:10.3390/su12219240.

[15] Dobslaw D, Schulz A, Helbich S, Dobslaw C, Engesser K-H. VOC removal and odor abatement by a low-cost plasma enhanced biotrickling filter process. Journal of Environmental Chemical Engineering. 2017; 5:5501–11. doi:10.1016/j.jece.2017.10.015.

[16] Sivachandiran L, Thevenet F, Rousseau A. Isopropanol removal using MnXOY packed bed non-thermal plasma reactor: Comparison between continuous treatment and sequential sorption/regeneration. Chemical Engineering Journal. 2015; 270:327–35. doi:10.1016/j.cej.2015.01.055.

[17] Colowick SP, Kaplan NO, Packer L, Sies H. *Methods in Enzymology*. San Diego, London: Academic Press; 2000.

[18] Manahan SE. *Environmental Chemistry*. 10th ed. Boca Raton: CRC Press; 2017.

[19] Buxton GV, Greenstock CL, Helman WP, Ross AB. Critical review of rate constants for reactions of hydrated electrons, hydrogen atoms and hydroxyl radicals ($\cdot$OH/$\cdot$O$^-$ in Aqueous Solution. Journal of Physical and Chemical Reference Data. 1988; 17:513–886. doi:10.1063/1.555805.

[20] Sher MSA, Oh C, Park H, Hwang YJ, Ma M, Park JH., Catalytic oxidation of methane to oxygenated products: Recent advancements and prospects for electrocatalytic and photocatalytic conversion at low temperatures. Adv. Sci. 2020; 7:2001946

[21] Runye Z, Christian K, Zhuowei C, Lichao L, Jianming Y, Jianmeng C. Styrene removal in a biotrickling filter and a combined UV–biotrickling filter: Steady- and transient-state performance and microbial analysis. Chemical Engineering Journal. 2015; 275:168–78. doi:10.1016/j.cej.2015.04.016.

[22] Wang C, Xi J-Y, Hu H-Y, Yao Y. Effects of UV pretreatment on microbial community structure and metabolic characteristics in a subsequent biofilter treating gaseous chlorobenzene. Bioresour Technol. 2009; 100:5581–7. doi:10.1016/j.biortech.2009.05.074.

[23] Zeng P, Li J, Liao D, Tu X, Xu M, Sun G. Performance of a combined system of biotrickling filter and photocatalytic reactor in treating waste gases from a paint-manufacturing plant. Environ Technol. 2016; 37:237–44. doi:10.1080/09593330.2015.1068375.

[24] Kogelschatz U. Atmospheric-pressure plasma technology. Plasma Phys. Control. Fusion. 2004; 46:B63–B75. doi:10.1088/0741-3335/46/12B/006.

[25] Oppenländer T. Mercury-free sources of VUV/UV radiation: Application of modern excimer lamps (excilamps) for water and air treatment. Journal of Environmental Engineering and Science. 2007; 6:253–64. doi:10.1139/s06-059.

[26] Schubert J, Mehnert R, Prager L, Langguth H. DE19712305A1 — Verfahren und Anlagen zum Abbau von Schadstoffen aus Abluft, insbesondere zur Grundwassersanierung.

[27] Sharma R, Singh G, Singh K. Modelling of the Thermophysical properties in Ar-He-H2 thermal plasmas with electronic excitation. J. Korean Phy. Soc. 2011; 58:1703–7. doi:10.3938/jkps.58.1703.

[28] Tao X, Bai M, Wu Q, Huang Z, Yin Y, Dai X. CO_2 reforming of CH_4 by binode thermal plasma. International Journal of Hydrogen Energy. 2009; 34:9373–8. doi:10.1016/j.ijhydene.2009.09.048.

[29] Tao X, Qi F, Yin Y, Dai X. CO_2 reforming of CH_4 by combination of thermal plasma and catalyst. International Journal of Hydrogen Energy. 2008; 33:1262–5. doi:10.1016/j.ijhydene.2007.12.057.

[30] Dobslaw D, Helbich S, Dobslaw C, Glocker B. Einsatz eines strahlungsgekühlten, thermischen Wasserdampfplasmas zur Behandlung von treibhausrelevanten, perfluorierten Abluftströmen: 5. VDI-Fachtagung Emissionsminderung 2018, Nürnberg, June 12th and 13th, 2018. Düsseldorf: VDI Verlag GmbH; 2018.

[31] Han S-H, Park H-W, Kim T-H, Park D-W. Large scale treatment of perfluorocompounds using a thermal plasma scrubber. Clean Technology. 2011; 17:250–8. doi:10.7464/ksct.2011.17.3.250.

[32] Lee CH, Chun YN. Reduction of Tetrafluoromethane using a waterjet gliding arc plasma. Korean Chemical Engineering Research. 2011; 49:485–90. doi:10.9713/kcer.2011.49.4.485.

[33] Verein Deutscher Ingenieure VDI, editor. *Waste Gas Cleaning by Adsorption—Process Gas and Waste Gas Cleaning—VDI 3674*. Berlin: Beuth Verlag; 2013.

[34] Anonymous. Neue Adsorptionsmaterialien und Regenerationsverfahren zur Elimination von Spurenstoffen in kommunalen und industriellen Kläranlagen: ZEROTRACE— Schlussbericht 03XP0098A-E. Oberhausen; 26.02.2021.

[35] European Parliament and Council of the European Union. EU-Directive 2009/28/EC— Promotion of the use of energy from renewable sources. Official Journal of the European Union. 2009.

[36] Weng C-H, Hsu M-C. Regeneration of granular activated carbon by an electrochemical process. Separation and Purification Technology. 2008; 64:227–36. doi:10.1016/j.seppur.2008.10.006.

[37] EPA United States Environmental Protection Agency. *Wastewater Technology Fact Sheet: Granular Activated Carbon Adsorption and Regeneration*. Washington, DC: EPA; 2000.

[38] IndexBox. Global Activated Carbon Market: Rising Demand for Water Purification Drives U.S. Trade—Index. 2021. www.globenewswire.com/news-release/2021/12/16/2353239/0/en/Global-Activated-Carbon-Market-Rising-Demand-for-Water-Purification-Drives-U-S-Trade-IndexBox.html.

[39] Mishra C. *On-site Regeneration of Granular Activated Carbon: A Literature Study, Comparison and Assessment of Different Regeneration Methods to Find Potential On-Site Regeneration Method in Sweden.* Stockholm: KTH Vetenskap och Konst; 2021.

[40] Zamir SM, Babatabar S, Shojaosadati SA. Styrene vapor biodegradation in single- and two-liquid phase biotrickling filters using Ralstonia eutropha. Chemical Engineering Journal. 2015; 268:21–7. doi:10.1016/j.cej.2015.01.040.

[41] Wang L, Yang C, Cheng Y, Huang J, Yang H, Zeng G, et al. Enhanced removal of ethylbenzene from gas streams in biotrickling filters by Tween-20 and Zn(II). J Environ Sci (China). 2014; 26:2500–7. doi:10.1016/j.jes.2014.04.011.

[42] San-Valero P, Álvarez-Hornos J, Ferrero P, Penya-Roja JM, Marzal P, Gabaldón C. Evaluation of Parallel-Series Configurations of Two-Phase Partitioning Biotrickling Filtration and Biotrickling Filtration for Treating Styrene Gas-Phase Emissions. Sustainability. 2020; 12:6740. doi:10.3390/su12176740.

[43] Leong YL, Krivak D, Kiel M, Laski E, González-Sánchez A, Dobslaw D. Triclosan biodegradation performance of adapted mixed cultures in batch and continuous operating systems at high-concentration levels. Cleaner Engineering and Technology. 2021; 5:100266. doi:10.1016/j.clet.2021.100266.

[44] Verein Deutscher Ingenieure VDI, editor. *Biological Waste Gas Purification—Biological Trickle Bed-Reactors: VDI-Guideline 3478 Part 2 — Expert Draft.* Berlin: Beuth Verlag; 2022.

[45] Kan E, Deshusses MA. Development of foamed emulsion bioreactor for air pollution control. Biotechnol Bioeng. 2003; 84:240–4. doi:10.1002/bit.10767.

[46] Wang L, He S, Xu J, Li J, Mao Z. Process performance of a biotrickling filter using a flow-directional-switching method. Clean Soil Air Water. 2013; 41:522–7. doi:10.1002/clen.201100574.

[47] Ryu HW, Kim SJ, Cho K-S, Lee TH. Toluene degradation in a polyurethane biofilter at high loading. Biotechnol Bioproc E. 2008; 13:360–5. doi:10.1007/s12257-008-0025-4.

[48] Liao D, Li E, Li J, Zeng P, Feng R, Xu M, Sun G. Removal of benzene, toluene, xylene and styrene by biotrickling filters and identification of their interactions. PLoS One. 2018; 13:e0189927. doi:10.1371/journal.pone.0189927.

[49] Dobslaw D. *Der bakterielle Abbau von halogen- und methylsubstituierten Aromatengemischen und dessen technische Anwendung in der biologischen Abluftreinigung.* München: Oldenbourg-Industrieverl; 2009.

[50] Dobslaw D, Schöller J, Krivak D, Helbich S, Engesser K-H. Performance of different biological waste air purification processes in treatment of a waste gas mix containing tert-butyl alcohol and acetone: A comparative study. Chemical Engineering Journal. 2019; 355:572–85. doi:10.1016/j.cej.2018.08.140.

[51] Gerl T, Engesser K-H, Fischer K, Dobslaw D. Biologische Abluftreinigung einer Lackierabluft im Kombinationsverfahren. Chemie Ingenieur Technik. 2016; 88:1145–50. doi:10.1002/cite.201500050.

[52] Schäfer H, Myronova N, Boden R. Microbial degradation of dimethylsulphide and related C1-sulphur compounds: Organisms and pathways controlling fluxes of sulphur in the biosphere. J Exp Bot. 2010; 61:315–34. doi:10.1093/jxb/erp355.

[53] Dobslaw D, Ortlinghaus O. Biological waste air and waste gas treatment: Overview, challenges, operational efficiency, and current trends. Sustainability. 2020; 12:8577. doi:10 3390/su12208577.

[54] TA-Luft. *Technical Instructions on Air Quality Control.* Berlin: TA-Luft 2021; 2021.

[55] DIN EN 13725. *Stationary Source Emissions—Determination of Odour Concentration by Dynamic Olfactometry and Odour Emission Rate*. Berlin: Beuth Verlag; 2022.

[56] Verein Deutscher Ingenieure VDI, editor. *Olfactometry—Determination of Odour Concentration by Dynamic Olfactometry—Supplementary Instructions for Application of DIN EN 13725: VDI-Guideline 3884 Part 1*. Berlin: Beuth Verlag; 2015.

[57] Márquez P, Siles JA, Gutiérrez MC, Alhama J, Michán C, Martín MA. A comparative study between the biofiltration for air contaminated with limonene or butyric acid using a combination of olfactometric, physico-chemical and genomic approaches. Process Safety and Environmental Protection. 2022; 160:362–75. doi:10.1016/j.psep.2022.02.024.

[58] Yoshio Y, Nagata E. Measurement of odor threshold by triangle odor bag method: Corpus ID: 4973091. Environmental Science. 2003; 4973091:118–127.

[59] Yang J, Zhao H, Li C, Li X. A direct energy reuse strategy for absorption air-conditioning system based on electrode regeneration method. Renewable Energy. 2021; 168:353–64. doi:10.1016/j.renene.2020.12.012.

[60] Zhang H, Li C, Li G, Zang B, Yang Q. Effect of Spent Air Reusing (SAR) on Maturity and Greenhouse Gas Emissions during Municipal Solid Waste (MSW) composting-with different pile height. Procedia Environmental Sciences. 2012; 16:59–69. doi:10.1016/j.proenv.2012.10.009.

[61] Bari QH, Koenig A. Effect of air recirculation and reuse on composting of organic solid waste. Resources, Conservation and Recycling. 2001; 33:93–111. doi:10.1016/S0921-3449(01)00076-3.

[62] Covarrubias-García I, Aizpuru A, Arriaga S. Effect of the continuous addition of ozone on biomass clogging control in a biofilter treating ethyl acetate vapors. Sci Total Environ. 2017; 584–585:469–75. doi:10.1016/j.scitotenv.2017.01.031.

[63] Covarrubias-García I, Jonge N de, Arriaga S, Nielsen JL. Effects of ozone treatment on performance and microbial community composition in biofiltration systems treating ethyl acetate vapours. Chemosphere. 2019; 233:67–75. doi:10.1016/j.chemosphere.2019.05.232.

[64] Saingam P, Baig Z, Xu Y, Xi J. Effect of ozone injection on the long-term performance and microbial community structure of a VOCs biofilter. J Environ Sci (China). 2018; 69:133–40. doi:10.1016/j.jes.2017.09.008.

[65] Saingam P, Xi J, Xu Y, Hu H-Y. Investigation of the characteristics of biofilms grown in gas-phase biofilters with and without ozone injection by CLSM technique. Appl Microbiol Biotechnol. 2016; 100:2023–31. doi:10.1007/s00253-015-7100-5.

[66] Dobslaw D, Woiski C, Winkler F, Engesser K-H, Dobslaw C. Prevention of clogging in a polyurethane foam packed biotrickling filter treating emissions of 2-butoxyethanol. Journal of Cleaner Production. 2018; 200:609–21. doi:10.1016/j.jclepro.2018.07.248.

[67] Cheng Z-W, Zhang L-L, Xi J-Y, Chen J-M, Hu H-Y, Jiang Y-F. Recovery of biological removal of gaseous alpha-pinene in long-term vapor-phase bioreactors by UV photodegradation. Chemical Engineering Journal 2011. doi:10.1016/j.cej.2011.09.113.

[68] Jiang L, Li H, Chen J, Di Zhang, Cao S, Ye J. Combination of non-thermal plasma and biotrickling filter for chlorobenzene removal. J. Chem. Technol. Biotechnol. 2016; 91:3079–87. doi:10.1002/jctb.4984.

[69] Moussavi G, Mohseni M. Using UV pretreatment to enhance biofiltration of mixtures of aromatic VOCs. J Hazard Mater. 2007; 144:59–66. doi:10.1016/j.jhazmat.2006.09.086.

[70] Xi J, Saingam P, Gu F, Hu H-Y, Zhao X. Effect of continuous ozone injection on performance and biomass accumulation of biofilters treating gaseous toluene. Appl Microbiol Biotechnol. 2014; 98:9437–46. doi:10.1007/s00253-014-5888-z.

[71] Wang C, Xi J-Y, Hu H-Y. Reduction of toxic products and bioaerosol emission of a combined ultraviolet-biofilter process for chlorobenzene treatment. J Air Waste Manag Assoc. 2009; 59:405–10. doi:10.3155/1047-3289.59.4.405.

[72] Wang C, Xi J-Y, Hu H-Y, Yao Y. Advantages of combined UV photodegradation and bio-filtration processes to treat gaseous chlorobenzene. J Hazard Mater. 2009; 171:1120–5. doi:10.1016/j.jhazmat.2009.06.129.

[73] Jianming Y, Wei L, Zhuowei C, Yifeng J, Wenji C, Jianmeng C. Dichloromethane removal and microbial variations in a combination of UV pretreatment and biotrickling filtration. J Hazard Mater. 2014; 268:14–22. doi:10.1016/j.jhazmat.2013.12.068.

[74] Hinojosa-Reyes M, Rodríguez-González V, Arriaga S. Enhancing ethylbenzene vapors degradation in a hybrid system based on photocatalytic oxidation UV/TiO$_2$-In and a biofiltration process. J Hazard Mater. 2012; 209–210:365–71. doi:10.1016/j.jhazmat.2012.01.035.

[75] Saucedo-Lucero JO, Arriaga S. Photocatalytic oxidation process used as a pretreatment to improve hexane vapors biofiltration. J. Chem. Technol. Biotechnol. 2015; 90:907–14. doi:10.1002/jctb.4396.

[76] Den W, Ravindran V, Pirbazari M. Photooxidation and biotrickling filtration for controlling industrial emissions of trichloroethylene and perchloroethylene. Chemical Engineering Science. 2006; 61:7909–23. doi:10.1016/j.ces.2006.09.015.

[77] Koh L-H, Kuhn DCS, Mohseni M, Allen DG. Utilizing ultraviolet photooxidation as a pre-treatment of volatile organic compounds upstream of a biological gas cleaning operation. J. Chem. Technol. Biotechnol. 2004; 79:619–25. doi:10.1002/jctb.1030.

[78] Mohseni M, Prieto L. Biofiltration of hydrophobic VOCs pretreated with UV photolysis and photocatalysis. IJETM. 2008; 9:47. doi:10.1504/IJETM.2008.017859.

[79] Álvarez-Hornos FJ, Martínez-Soria V, Marzal P, Izquierdo M, Gabaldón C. Performance and feasibility of biotrickling filtration in the control of styrene industrial air emissions. International Biodeterioration & Biodegradation. 2017; 119:329–35. doi:10.1016/j.ibiod.2016.10.016.

[80] Palau J, Penya-Roja JM, Gabaldón C, Álvarez-Hornos FJ, Martínez-Soria V. Effect of pre-treatments based on UV photocatalysis and photo-oxidation on toluene biofiltration performance. J. Chem. Technol. Biotechnol. 2012; 87:65–72. doi:10.1002/jctb.2683.

[81] Wei Z, Sun J, Xie Z, Liang M, Chen S. Removal of gaseous toluene by the combination of photocatalytic oxidation under complex light irradiation of UV and visible light and biological process. J Hazard Mater. 2010; 177:814–21. doi:10.1016/j.jhazmat.2009.12.106.

[82] Mohseni M, Zhao JL. Coupling ultraviolet photolysis and biofiltration for enhanced degradation of aromatic air pollutants. J. Chem. Technol. Biotechnol. 2006; 81:146–51. doi:10.1002/jctb.1371.

[83] He Z, Li J, Chen J, Chen Z, Li G, Sun G, An T. Treatment of organic waste gas in a paint plant by combined technique of biotrickling filtration with photocatalytic oxidation. Chemical Engineering Journal. 2012; 200–202:645–53. doi:10.1016/j.cej.2012.06.117.

[84] Zhu R, Mao Y, Jiang L, Chen J. Performance of chlorobenzene removal in a nonthermal plasma catalysis reactor and evaluation of its byproducts. Chemical Engineering Journal. 2015; 279:463–71. doi:10.1016/j.cej.2015.05.043.

[85] Jiang L, Li S, Cheng Z, Chen J, Nie G. Treatment of 1,2-dichloroethane and n -hexane in a combined system of non-thermal plasma catalysis reactor coupled with a biotrickling filter. J. Chem. Technol. Biotechnol. 2018; 93:127–37. doi:10.1002/jctb.5331.

[86] Wei ZS, Li HQ, He JC, Ye QH, Huang QR, Luo YW. Removal of dimethyl sulfide by the combination of non-thermal plasma and biological process. Bioresour Technol. 2013; 146:451–6. doi:10.1016/j.biortech.2013.07.114.

[87] Steinberg I, Rohde C, Bockreis A, Jager J. Increase of the purification efficiency of bio-filters by the use of a complementary ionisation step. Waste Manag. 2005; 25:375–81. doi:10.1016/j.wasman.2005.02.014.

[88] Holub M, Brandenburg R, Grosch H, Weinmann S, Hansel B. Plasma supported odour removal from waste air in water treatment plants: An industrial case study. Aerosol Air Qual. Res. 2014; 14:697–707. doi:10.4209/aaqr.2013.05.0171.

[89] Jiang L, Zhu R, Mao Y, Chen J, Zhang L. Conversion characteristics and production evaluation of styrene/o-xylene mixtures removed by DBD pretreatment. Int J Environ Res Public Health. 2015; 12:1334–50. doi:10.3390/ijerph120201334.

[90] Karatum O, Deshusses MA. A comparative study of dilute VOCs treatment in a non-thermal plasma reactor. Chemical Engineering Journal. 2016; 294:308–15. doi:10.1016/j.cej.2016.03.002.

[91] Kim H, Han B, Hong W, Ryu J, Kim Y. A new combination system using biotrickling filtration and nonthermal plasma for the treatment of volatile organic compounds. Environmental Engineering Science. 2009; 26:1289–97. doi:10.1089/ees.2008.0255.

[92] Rafflenbeul R. *Nichtthermische Plasmaanlagen (NTP) zur Luftreinhaltung in der Abfallwirtschaft.* Berlin: Müll und Abfall; 1998. doi:10.37307/j.1863-9763.1998.01.06.

[93] Schiavon M, Scapinello M, Tosi P, Ragazzi M, Torretta V, Rada EC. Potential of non-thermal plasmas for helping the biodegradation of volatile organic compounds (VOCs) released by waste management plants. Journal of Cleaner Production. 2015; 104:211–9. doi:10.1016/j.jclepro.2015.05.034.

[94] Schiavon M, Schiorlin M, Torretta V, Brandenburg R, Ragazzi M. Non-thermal plasma assisting the biofiltration of volatile organic compounds. Journal of Cleaner Production. 2017; 148:498–508. doi:10.1016/j.jclepro.2017.02.008.

[95] Quan Y, Wu H, Guo C, Han Y, Yin C. Enhancement of TCE removal by a static magnetic field in a fungal biotrickling filter. Bioresour Technol. 2018; 259:365–72. doi:10.1016/j.biortech.2018.03.031.

[96] Quan Y, Wu H, Yin Z, Fang Y, Yin C. Effect of static magnetic field on trichloroethylene removal in a biotrickling filter. Bioresour Technol. 2017; 239:7–16. doi:10.1016/j.biortech.2017.04.121.

[97] Oda T, Yamaji K, Takahashi T. Decomposition of dilute trichloroethylene by nonthermal plasma processing—Gas flow rate, catalyst, and ozone effect. IEEE Trans. Ind. Applicat. 2004; 40:430–6. doi:10.1109/TIA.2004.824440.

[98] Malhautier L, Soupramanien A, Bayle S, Rocher J, Fanlo J-L. Potentialities of coupling biological processes (biotrickler/biofilter) for the degradation of a mixture of sulphur compounds. Appl Microbiol Biotechnol. 2015; 99:89–96. doi:10.1007/s00253-014-5842-0.

[99] Greinacher P. *RLT-Anlagen: Definition, Aufbau, Wartung & Kosten*; 2019. www.klimattechniker.net/magazin/rlt-anlagen-20193668.

[100] DIN EN 779. *Particulate Air Filters for General Ventilation—Determination of the Filtration Performance.* 2012nd ed. Berlin: Beuth Verlag; 2021.

[101] Trox GmbH. *X-Cube Raumlufttechnische Geräte von TROX—Planungshandbuch*; 2018. www.trox.de/downloads/bb58453e061a1c4b/DM_2018_04_air_handling_units_DE_de.pdf.

[102] Suárez S, Carballa M, Omil F, Lema JM. How are pharmaceutical and personal care products (PPCPs) removed from urban wastewaters? Rev Environ Sci Biotechnol. 2008; 7:125–38. doi:10.1007/s11157-008-9130-2.

[103] Webster TS, Deshusses MA, Devinny JS. *Biofiltration for Air Pollution Control.* London: CRC Press; 1999.

14 De-NOx SCR
Catalysts and Process Designs in the Automotive Industry

Gerardo Coppola, Valerio Pugliese, and Sudip Chakraborty

CONTENTS

14.1 Introduction..267
14.2 De-NOx Reactions and Kinetics..268
14.3 Typical Catalysts Used in SCR Systems..272
14.4 Chemistry and Design of Automotive SCR Systems.....................................278
14.5 SCR Systems Drawbacks and Failures..281
14.6 System Reliability Predictions...282
14.7 Conclusions...283
References...283

14.1 INTRODUCTION

Catalysis is of paramount importance in the process industry and represents a key driver for sustainability, societal challenges [1, 2], and economics [3]. Accordingly, 90% of industrial processes include at least one catalytic step, representing 80% of the added value in the chemical industry [1, 3]. The catalysis market in 2020 reached a value of 33.9 billion USD $ in 2019, with an expected compounded annual growth rate of 4.4% between 2020 and 2027, while environmental catalysis represents about 25% of the market value [4]. The COVID-19 pandemic has negatively affected manufacturing, oil and gas, automotive, chemical production, petrol production, and polymer catalysis, which is highly dependent on catalysts. The Current situation has lowered the global demand for catalysts from 2020 to 2022. Moreover, due to the risk of infections in the workers and personnel, most chemical companies where catalysts are extensively employed (hydroprocessing and chemical synthesis) have shut down or downsized the volume of their processes [5].

The catalytic technologies for gaseous waste stream control represent a strategic field in the subfield of environmental catalysis. The latter, since the 1960s, has substantially contributed to lowering the environmental impact of several

DOI: 10.1201/9781003260738-14

industrial processes and the transportation sector. In December 2021, the European Commission started the revision of the Air Quality Directive, including the new Air Quality Standard, based on the guidelines of the World Health Organization published in the same year [6]. Research for catalytic materials for gas waste stream treatment started in the 1960s, with the automotive industry's push leading to the commercialization of the first catalytic car converters in the 1970s [5].

The NO_x class of airborne pollutants mainly comprises mixtures of NO, NO_2, and N_2O. Such species are responsible for the formation of smog and acid rains, ozone depletion, global warming, and eutrophication phenomena. Most of the NO_x is formed during the combustion of fuels. Their formation during combustion processes is, in order of decreasing importance, related to the direct reaction of N_2 and O_2 at high temperatures (higher than 800°C), nitrogen-containing fuels, and the combustion of HCN. Besides, NO_x can also be formed by the reaction of N_2 and hydrocarbon radicals from incomplete fuel combustion. The formation of NO_x is especially problematic in lean-burn engines, e.g., diesel engines, jeopardizing their higher combustion efficiency than stoichiometrically operated ones. Such an effect is exacerbated during the cold-start and low-load engine phases due to the low temperatures of the catalysts for their treatment. The concentration range of NO_x from typical industrial plants, prior to after-treatment processes, ranges from 150 to 3000 ppmv depending on the kind of industrial plant and operating conditions. Typical compositions and concentrations were summarized in ref. [6], depending on the type of process. Generally, in flue gas from combustion processes NO is the most abundant species, followed by NO_2 and N_2O, and others. Large amounts of NO_2 are present in the waste gas streams of nitric acid production plants [7].

Due to the strict policies for reducing NO_x emissions, many strategies were developed to meet the national emission standards. For example, in the case of boiler combustion processes, the emission of NO_x can be reduced by a) fuel denitrification processes, b) decreasing the temperature of the combustion, and c) by after-treatment of the flue gases [8]. Typically, post-combustion methods for NO_x abatement present higher NO_x removal efficiency (>80%), as compared to pre-combustion methods and combustion methods (<50%) [9, 10]. Moreover, depending on the concentration of NO_x in the waste gas stream, the volumetric flow rate, temperatures, and composition, several methods can be selected and optimized for each application. In this chapter, we will focus only on catalytic after-treatment processes.

14.2   DE-NOX REACTIONS AND KINETICS

Nitrogen oxides (NOx) are essentially composed of nitric dioxide (NO_2), nitrous oxide (N_2O), and nitric oxide (NO). The majority, 90% in weight, of nitrogen oxide emissions produced during a generic "burning" (combustion reactions) is represented by NO, while NO_2 is a by-product between 0.5% and 10% by weight [11].

The reduction of NO_x in $DeNO_x$ processes allows the removal of potentially harmful pollutants to H_2O and N_2 by using a reductant, e.g., NH_3, urea (R. 1–2), CH_4, CO, or H_2.

Using NH_3 as a reductant, in the absence of a catalyst, the following reactions can happen:

$$\left(NH_2\right)_2 CO + H_2O \rightarrow CO_2 + 2NH_3 \qquad \text{(R 1)}$$

$$4NO + 4NH_3 + O_2 \rightarrow 6H_2O + 4N_2 \qquad \text{(R 2)}$$

In the latter case, the process is known as selective non-catalytic reduction (SNCR), a thermal denitrification process developed in the 1970s using ammonia or urea as a reducing agent at 900–1100°C [11]. SNCR process and its performance are strongly affected by Nitrogen oxide concentrations, additives, nitrogen starvation response (NSR), and precise parameters such as oxygen (O_2) contents and temperatures [12]. The term "selective" was chosen to remark the preferential reaction of NH_3 with NO_x rather than O_2, a feature not previously observed using CO or CH_4 as reductants [13]. Outside the above-reported temperature range, the performance of the process decreases substantially, leading to unselective ammonia oxidation and increased formation of NO_2 at T >1100 °C. Conversely, below 900°C, ammonia slip increases due to incomplete ammonia conversion. Typical side reactions are reported in reactions 3–5.

$$4NH_3 + 5O_2 \rightarrow 6H_2O + 4NO \qquad \text{(R 3)}$$

$$4NH_3 + 3O_2 \rightarrow 6H_2O + 2N_2 \qquad \text{(R 4)}$$

$$2NH_3 + 5O_2 \rightarrow 6H_2O + 2NO_2 \qquad \text{(R 5)}$$

Such a process presents the advantage of low reactor volumes and no dead times for catalysts replacements. However, it presents a maximum NO_x removal efficiency of 50–60% and large ammonia losses (ammonia slip). Therefore, it is rarely used for removing NOx as a one-standing solution in industrial practice.

In the past, in the presence of stoichiometric excess of oxygen and large amounts of NO_2, e.g., in the case of gas waste streams from HNO_3 production facilities, NO_x was reduced by CH_4 at 350–800 °C, using low-loaded noble metal catalysts, a process known as non-selective catalytic reduction of NOx (**NSCR**, R1 – R2) [14, 15]. This technology is derived from the automotive sector and employs a three-way catalysis concept [16]. Also, CO and H_2 were used, e.g., in the case of fuel-rich recirculating engines [15] (reactions 6–9). 0.2–0.5% Rh promoted Pt/Al_2O_3 or Pd/Al_2O_3 catalysts were typically wash coated on monolithic honeycombs [14]. However, due to high fuel costs and catalyst deactivation issues, the NSCR process was replaced by the selective catalytic reduction process (SCR).

$$CH_4 + 2O_2 \rightarrow CO_2 + 2H_2O \qquad \text{(R 6)}$$

$$2NO_2 + CH_4 \rightarrow 2H_2O + CO_2 + N_2 \qquad \text{(R 7)}$$

$$2NO + CO \rightarrow CO_2 + N_2 \qquad \text{(R 8)}$$

$$2NO + 2H_2 \rightarrow 2H_2O + N_2 \qquad \text{(R 9)}$$

In the presence of a catalyst, the following phenomena occur [17, 18]:

(A) NO adsorption, which is a green reaction with the production of N_2 and H_2O;
(B) NO redox process;
(C) Dual function reaction process.

The SCR process, patented for the first time by Engelhard in 1957 [9], and its short version are currently the most used after treatment technologies in today's industry for NOx treatment [19]. As in the corresponding SNCR version, NH_3 or urea are reacted selectively with NOx to N_2 and H_2O over SO_2 or oxygen excess in the SCR process. However, in this case, the presence of redox catalyst allows operation at lower temperatures than the SNCR process, with increased NOx removal efficiency (>90%), a drastically reduced ammonia slip, and the conversion of SO_2 to SO_3 is minimized [14, 15]. The stoichiometric reactions in the presence of conventional waste gas streams from combustion processes are the same as for the SNCR process (reactions 7 and 8). NH_3 or urea are typically dosed to get a stoichiometric composition (NH_3/NOx = 1) [20] to get the highest NOx conversion efficiency and lowest ammonia slip. As in the NSCR, the operating temperature should be optimized depending on the used catalyst (Figure 14.1) [16, 19, 21]. Specifically, the conversion rate of ammonia and NOx increases with increasing the operating temperature up to a catalyst's dependent temperature (optimal point), minimizing ammonia slip. Then, further increasing the temperature leads to a decrease in NOx conversion and increased NH_3 consumption

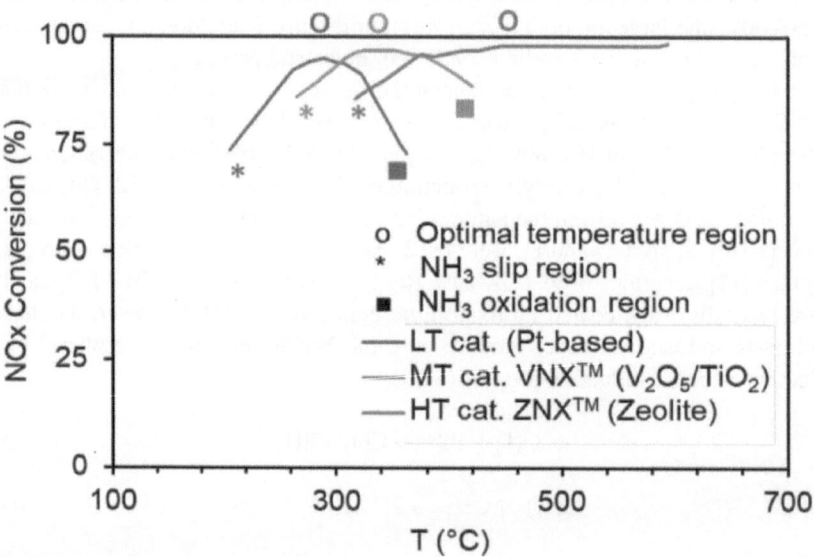

FIGURE 14.1   Performance of typical commercial catalysts form Engelhard employed for the SCR process, concerning the temperature, depending on the catalyst formulation.

(*Source:* Adapted from [20, 23]).

due to oxidation. At high temperatures, the oxidation of SO_2 becomes important (R 10). Moreover, in excess oxygen, NO can be further oxidized by O_2 and quickly converted to N_2 and H_2O (Fast SCR) [11].

The presence of SO_3 or HCl, depending on their concentration and temperatures, leads to side ammonia consumption further, catalysts' pore plugging, and fouling of downstream processes, e.g., by the formation of NH_4HSO_4, $(NH_4)_2SO_4$ and NH_4Cl (R 11–13). In the presence of a catalyst, the conversion rate of NO_x to N_2 is mass-transfer controlled by the diffusion of the reactants in the gas phase and within the pores of the catalyst. In contrast, the overoxidation of NH_3 and SO_2 are kinetically controlled and are strictly dependent on the catalyst formulation [22]. Therefore, the temperature and catalysts' selectivity are among the most important parameters to optimize.

Generally, catalysts must ensure high mass transfer rates for NO_x on the catalyst surface, low SO_2 and NH_3 oxidation activity, high resistance to SO_2 and poisoning, high mechanical stability, and long service life.

$$2SO_2 + O_2 \rightarrow 2SO_3 \qquad \text{(R 10)}$$

$$2NH_3 + SO_3 + H_2O \rightarrow (NH_4)_2 SO_4 \qquad \text{(R 11)}$$

$$2NH_3 + 2SO_3 + 2H_2O \rightarrow 2NH_4HSO_4 \qquad \text{(R 12)}$$

$$NH_3 + HCl \rightarrow NH_4Cl \qquad \text{(R 13)}$$

All the catalytic materials active in the partial oxidations can potentially be used for the NH_3-SCR reaction [23, 24].

Instead, using H_2 as a reducing agent, the following reactions can occur:

$$2NO + 4H_2 + O_2 \rightarrow N_2 + 4H_2O \; (A) \qquad \text{(R 14)}$$

$$2NO + 3H_2 + O_2 \rightarrow N_2O + 3H_2O \; (B) \qquad \text{(R 15)}$$

$$2NO + H_2 \rightarrow N_2O + H_2O \; (C) \qquad \text{(R 16)}$$

$$NO + 5/2H_2 \rightarrow NH_3 + H \qquad \text{(R 17)}$$

$$NO + 1/2O_2 \rightarrow NO_2 \qquad \text{(R 18)}$$

$$H_2 + 1/2O_2 \rightarrow H_2O \qquad \text{(R 19)}$$

Cu^{2+} ions are the most suitable catalyst prospects for carrying on the NO removal process via H_2 reduction on Cu/SiO_2 [25].

Cu^{2+} creates active sites near the reducing component, while NOx is subjected to decomposition and adsorption, as shown in Figure 14.2. The absorption/decomposition process originates NO and copper(metallic) on the catalyst surface, which is present in the form of CuO [11, 26].

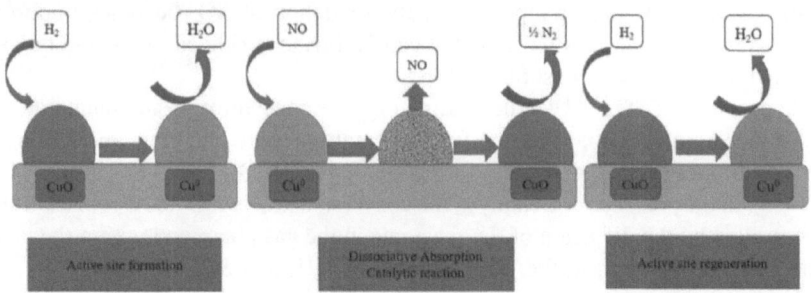

FIGURE 14.2   Redox fatly cycle for the selective reduction of NO with H_2 on Cu/SiO_2 [11].

14.3   TYPICAL CATALYSTS USED IN SCR SYSTEMS

Typical commercial, industrial catalysts include V_2O_5, MoO_3 (and/or WO_3), supported on TiO_2 [23, 27]. These kinds of catalysts are generally active between 300 and 400 °C.

To improve the mass-transfer rates of reactants on the catalysts' surface, it is necessary to operate at high gas hourly space velocities (GHSV). At such high GHSV, pressure drops are kept low using structured catalysts. Therefore, SCR catalysts are generally prepared as monolithic honeycombs, prepared by extrusion of the catalytic components, or by coating the catalyst on pre-shaped ceramic or metallic monoliths [23]. Besides improving mass transfer rates of reactants on the catalytic surface and allowing them to operate at reduced pressure drops, such a design ensures high mechanical stability and avoids the deposition of soot on the catalyst (Figure 14.3).

V-based catalysts present high tolerance to SO_2 deactivation in the underlined temperature region. However, they can get deactivated by high-temperature thermal aging, pore-plugging (alkali metal and sulfates), and poisoning (e.g., by As) [27, 28]. To avoid the oxidation of SO_2 to SO_3, the V_2O_5 loading is kept low (0.1–4 %) [23, 29, 30]. Moreover, although the operating temperature window is enlarged at the increase of V_2O_5 loadings, e.g., at V_2O_5 loadings >2 wt%, the thermal stability of the catalyst is strongly decreased [31]. At such low loadings used in typical catalysts, V_2O_5 on the catalyst's surface is generally present as a highly dispersed monomeric and polymeric species in an amorphous monolayer [31]. Such a high dispersion is related to the TiO_2 (anatase) ability to disperse and stabilize the V_2O_5 phase. Moreover, TiO_2, due to its weak chemisorption ability for SO_2, helps avoid its oxidation [23, 32]. To further reduce the SO_2 oxidation and stabilize the catalyst, MoO_3 and WO_3 are used as chemical and structural promoters [23, 29]. Specifically, it has been found that MoO_3 and WO_3 promote the activity of V_2O_5 by improving its redox properties and increasing the density of Bronsted acidic sites, linearly correlated with the former, while, conversely, it is depressed in the presence of basic materials, e.g., ZnO [33]. Therefore, the latter components promote the reaction by increasing the selectivity to N2 and enlarging the temperature window in which the catalyst is active.

The importance of the intimate contact between the V_2O_5 phase and TiO_2 and the surface redox properties were highlighted by the superior low-temperature

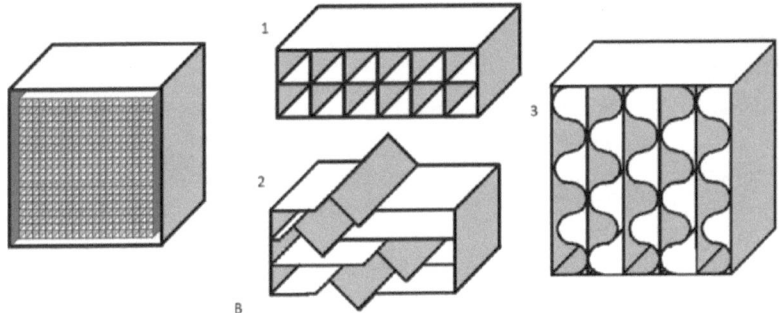

FIGURE 14.3 A) Catalyst monolith, B) Three of the most used monolith arrangement designs.

performance of $V_2O_5(WO_3)/TiO_2$ catalysts prepared by a solvothermal method (precursor $VO(acac)_2$), as compared with a similar one prepared by incipient wetness (precursor NH_4VO_3), related by the authors to the improved redox (higher V^{4+}/V^{5+}) and acidic properties in the first catalyst (Table 14.1 #9–10) [34].

To increase the stability at high temperatures (600°C) of V-based catalysts, the addition of SiO_2 to the catalyst's formulation has been reported. To preserve the catalysts' activity by protecting from the formation of $V_xT_{1-x}O_2$ solid solution at high temperature (typical NO_x conversion efficiencies of this kind of catalysts are reported in Table 14.1, #1–2) [31]. However, improved catalytic activity on V_2O_5/TiO_2 catalysts upon doping with Si was also found in the past and attributed to improved textural and redox properties, together with acidity and basicity of the catalysts [35].

TABLE 14.1
Selected SCR Catalysts and Their Performance

#	Catalyst	Inlet NOx mixture	Temp [°C]	GHSV	Removal efficiency	Notes	Ref
1	V_2O_5-WO_3/TiO_2 (loading 2% V_2O_5)	500 ppm NO, 500 ppm NH_3, 5% O_2, N_2 balance	350	50 000	98		[31]
2	V_2O_5-WO_3-SiO_2/ TiO_2 (loading 2% V_2O_5)	500 ppm NO, 500 ppm NH_3, 5% O_2, N_2 balance	350	50 000	98		[31]
3	V_2O_5-WO_3/TiO_2 (loading 1% V_2O_5)	500 ppm NO, 500 ppm NH_3, 3% O_2, N_2 balance	200	28 000	50		[30]
4	V_2O_5-WO_3/TiO_2 (loading 1% V_2O_5, CeO_2 10% wt.)	500 ppm NO, 500 ppm NH_3, 3% O_2, N_2 balance	200	28 000	90		[30]
5	V_2O_5-WO_3/TiO_2	500 ppm NO, 500 ppm NH_3, 5% O_2, 5% H_2O	200–400	60000 mL· g-1·h-1.	17–81	Fresh	[40]

(Continued)

TABLE 14.1
Continued

#	Catalyst	Inlet NOx mixture	Temp [°C]	GHSV	Removal efficiency	Notes	Ref
6	V_2O_5-WO_3/TiO_2	500 ppm NO, 500 ppm NH_3, 5% O_2, 5% H_2O	200–400	60000 mL·g–1·h–1.	0	Poisoned by K	[40]
7	V_2O_5(WO_3-CeO_2-ZrO_2)/TiO_2		200–400	60000 mL·g–1·h–1.	87–92	Fresh	[40]
8	V_2O_5(WO_3-CeO_2-ZrO_2)/TiO_2		200–400	60000 mL·g–1·h–1.	41–54	Poisoned by K	[40]
9	V_2O_5-WO_3/TiO_2		200–400		54–82	Prepared by incipient wetness	[34]
10	V_2O_5-WO_3/TiO_2		200–400		80–80	Prepared by solvothermal method	[34]
11	H-ZSM-5		300–350	42 500 mL/h/ gcat	80–100		[43]
12	H-SSZ-16		300–350	42 500 mL/h/ gcat	70		[43]
13	H-SSZ-13		300–400	42 500 mL/h/ gcat	90		[43]
14	Cu/ZSM-5	500 ppm NO, 500 ppm NH_3 10 % O_2, Balance N_2	200–450/500	42 500 mL/h/ gcat	95–100/80	Fresh	[43]
15	Cu/SSZ-16	500 ppm NO, 500 ppm NH_3 10 % O_2, Balance N_2	200–500	42 500 mL/h/ gcat	95–100	Fresh	[43]
16	Cu/SSZ-13	500 ppm NO, 500 ppm NH_3 10 % O_2, Balance N_2	200–500	42 500 mL/h/ gcat	95–100	Fresh	[43]
17	Cu/ZSM-5	500 ppm NO, 500 ppm NH_3 10 % O_2, Balance N_2	200–400/500	42 500 mL/h/ gcat	85–95/90	Steamed, 7 h	[43]
18	Cu/SSZ-16	500 ppm NO, 500 ppm NH_3 10 % O_2, Balance N_2	200–400/500	42 500 mL/h/ gcat	95–100/95	Steamed, 7 h	[43]
19	Cu/SSZ-13	500 ppm NO, 500 ppm NH_3 10 % O_2, Balance N_2	200–350/500	42 500 mL/h/ gcat	98–98 / 80	Steamed, 7 h	[43]

Due to the high toxicity of V_2O_5, efforts have been devoted to decreasing the amount of the active phase while preserving the catalytic activity, especially at low temperatures. To meet this challenge, recently, studying the promotional effect of Fe, Mn, and Ce for V_2O_5-WO_3/TiO_2 catalysts with 1 wt. % V_2O_5 loading, it was demonstrated that CeO_2, which, although not the most active one, is an active catalyst by itself toward the NH_3-SCR process [36–38], increased the NO_x conversion efficiency at 200°C by 40% as compared to the respective unpromoted catalyst (Table 14.1, #3–4) [30]. The authors attributed such a promotional effect to the synergic behavior of Ce and V, and the enhanced basicity, acidity, and oxygen storage capacity of the catalyst, enhancing the adsorption of NO_x and NH_3 through the Fast-SCR process.

Recently, CeO_2, besides improving the activity of the catalysts, has been found to improve the deactivation stability against Na_2O doping on V_2O_5(WO_3-CeO_2)/TiO_2 catalysts [39]. Similarly, CeO_2 and ZrO_2 as promoters, introduced during the preparation of the TiO_2 support, improve the activity of V_2O_5-WO_3/TiO_2-CeO_2-ZrO_2 catalysts and their stability toward K_2O poisoning [40]. Specifically, while the V_2O_5-WO_3/TiO_2 catalysts were completely deactivated upon K poisoning, V_2O_5(WO_3-CeO_2-ZrO_2)/TiO_2 catalysts were more active than the unpromoted catalyst at low temperature (Table 14.1 #5–8).

Copper-based catalysts are among the most studied ones in the NH_3-SCR reaction. In the past, CuO supported on TiO_2, Al2O3, carbon, and $CuSO_4$, due to the sensitivity of copper to SO_3, have been reported. However, compared to V-based catalysts, their activity was quite low [41]. Better performances were instead achieved by Cu/zeolites. Although H-zeolites present a significant NO_x removal efficiency, this is enhanced by Cu (Table 14.1, # 1–16 vs. #14–16). Several Cu-based zeolites were reported for the NH3-SCR reaction. Although interesting activities were obtained with the Cu/ZSM5, because of its low hydrothermal stability, in the presence of steam at T > 700°C, it deactivates due to dealumination, leading to a loss in Bronsted acidic sites and loss of Cu surface area, e.g., due to formation of copper aluminates [42]. Small-pore Cu/zeolites, like SSZ-13, SSZ-16, and SAPO-34 present better performance than medium-pores Cu/ZSM-5 with high removal efficiency operating at 150–500°C and improved hydrothermal stability (Table 14.1, #14–16 vs. 17–19) [43]. The latter results, as compared to V-type catalysts, enlarge the operational temperature window of the NH_3-SCR reaction (Table 14.1, #1–10 vs. #14–16). These developments also led to the commercialization of these catalysts for the automotive market and increased the scientific interest of the community [44].

Although the reaction has been deeply studied, the mechanism is still unclear. However, recently, a comprehensive spectroscopic study, employing in-situ and operando IR and UV/vis, in transient conditions, together with in-situ V K-edge XANES measurements, on a) V_2O_5/TiO_2, b) V_2O_5(WO_3), and, Cu loaded on c) CHA and d) AFX zeolites, proposed a unified view, bridging mechanistic redox cycles on Cu and V, according to the following steps [45]: 1) adsorption of NH_3 on Lewis oxidized sites, and 1b) on Bronsted sites. Then, 2) NH_3 on Lewis sites reacts with NO and H^+ on the surface, e.g., via NH_2NO, generating reduced sites. At the same time, 2b) NH_3 on Bronsted sites can migrate on free Lewis oxidized sites and react as in step (2). In step (3), O_2 restores oxidized sites.

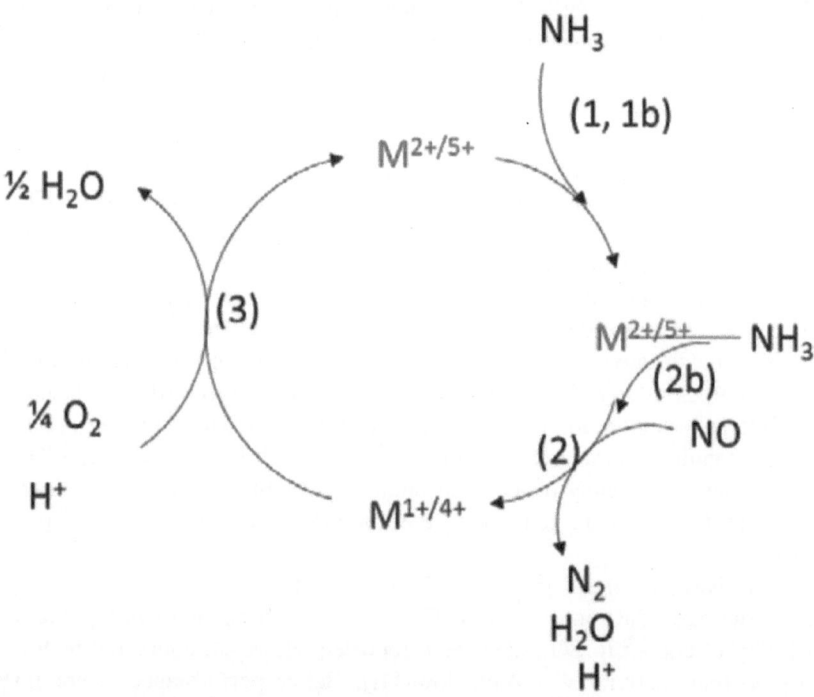

FIGURE 14.4 NH3-SCR mechanism on (M) Cu and V sites, adapted from [45].

Mn-based catalysts for the NH3-SCR process have been recently reviewed [46]. These catalysts suffer from SO_2 and H_2O inhibition and poisoning, this being the main reason for the growing research interest in Mn-based multiple metal oxides. For instance, MnO_x–CeO_2 catalysts have exceptional SCR performances thanks to the combined effect of Mn and Ce, enhancing the quantity and the acidity of acid sites and the capability to store and release oxygen [47]. A study by Yang et al. showed a 95% NO conversion at 393K using a Ce–Mn–Ox catalyst prepared by citric acid method with 100ppm of SO_2 and 2.5% of H_2O at a GHSV of $42 \cdot 10^3$ h^{-1} [48]. Liu et al. (2013) prepared an Mn–Ce–O$_x$ catalyst by a surfactant template reaching a NOx conversion of more than 90% at 423–473K with a 5 %vol of H_2O and 50 ppm of SO_2 at $64 \cdot 10^3$ h^{-1} [49]. Nevertheless, the consequences on the catalyst activity of using high concentrations in H_2O and SO_2 demand further studies. Chang et al. found an SCR activity decrease on an Mn(0.4)CeO$_x$ catalyst while using a co-precipitation method; in particular, this led to a 68% decrease at 383K with the addition of 100 ppm of SO_2 at $35 \cdot 10^3$ h^{-1} [50].

Mn-based catalyst's main feature is the low-temperature utilization in NO$_x$ SCR with an excess of oxygen and SO_2 with a considerably wide temperature range, without any major efficiency collapsing. In the most recent studies, many Mn-based catalysts have been studied to increase the resistance towards H_2O and SO_2, coming up with such formulations as modified MnO_x-CeO_2, (multi) metal oxides with special crystal structures and shapes (hollow nanofibers, nanocages, MOFs, etc.), TiO_2 and carbon supports. The above-mentioned catalysts exhibit good resistances because of the pre-sulfating of dopant compounds, altering H_2O/SO_2 resistance.

Although considerable advances have been attained, catalysts used in low-temperature NO_x elimination still lack long-term stability. It is still worth investigating the H_2O/SO_2 inhibition by DRIFTS and DFT analysis to route new catalysts designed with different crystal and shape assemblies [46].

Ce-added catalysts have reached a peak of interest in the SCR catalysts world. Liu et al. (2022) prepared some modified catalysts using an enhanced impregnation process starting from a commercial catalyst as a supporter—a V_2O_5-WO_3-TiO_2 catalyst – ground in 10 g powder groups. For each group, different amounts of $Ce(NO_3)_3 \cdot 6H_2O$ (0.31 g to 3.1 g) and $Cu(NO_3)_2 \cdot 3H_2O$ (0.38 g) were dissolved in an aqueous solution as modifiers. Afterward, the catalyst and the solutions were mixed, dried (378K overnight), and calcined at 773K. After calcination, the catalysts were indicated as Ce_x/SCR, Cu/SCR, and Cu-Ce_x/SCR (with values for x varying from 1 to 10 wt.%) [51].

The performance results showed how the NH_3 conversion considerably improved with temperature, while the efficiency of commercial catalysts – without the presence of NO—was lower than 15% up to 400°C. Against the Cu catalyst, Ce loaded in Cu catalysts upgraded their performances at all temperatures, with conversion rates increasing with Ce concentration. NH_3 conversion of Cu-Ce_x catalysts was low at low temperatures, but as soon as reaching 300°C, they exhibited very steep increases in the range of 78 to 92%. For temperatures higher than 400 °C, the conversions only grew a little. In conclusion, compared to Cu/SCR catalysts, Ce addition enlarged the temperature range where the catalysts are the most active.

In Figure 14.5, the NH_3 oxidation and NO reduction mechanisms on Cu-Ce/SCR catalysts.

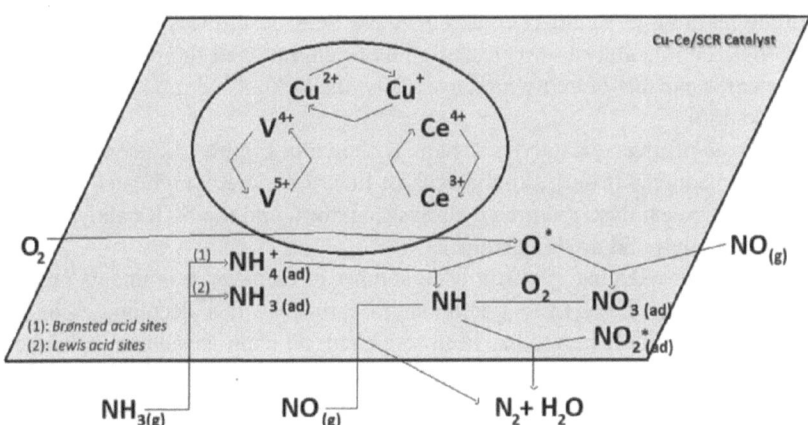

FIGURE 14.5 NH_3 oxidation and NO reduction mechanisms on a Cu-Ce/SCR catalyst, adapted from [51].

As a recapitulation, a collection of the main SCR catalysts and their performances in terms of removal efficiencies is presented in the following table.

14.4 CHEMISTRY AND DESIGN OF AUTOMOTIVE SCR SYSTEMS

The most recent severe emissions legislation for diesel-fueled vehicles made researchers face the need for exhaust post-treatment development—the most promising method when dealing with NO_x reduction is the Urea-SCR.

SCR catalysts can be applied following two main designs:

1) A single (full) catalyst with a channel wall consisting of the actual active catalyst.
2) A metallic/cordierite substrate coated with the active catalyst.

Using the substrates results in lower amounts of active catalysts with consequent inferior low-temperature performances regarding the design of the full catalyst. A $NO:NO_2$ ratio at about 1 on a metal or zeolitic catalyst can improve low-temperature performances [52, 53].

Walker et al. verified a series constituted of a pre-oxidation catalyst – an SCR-catalyst – and an NH_3 slip oxidation catalyst combined with a soot filter. This system exhibited relatively acceptable low-temperature performances, which was improved by the pre-catalyst [54].

Gekas et al. studied a comparison between three different systems: a pure SCR system, the addition of a slip oxidation catalyst, and the addition of a pre-oxidation catalyst. The pre-oxidation catalyst enhanced low-temperature performances but with very limited effects, reaching comparable efficiencies by swapping the pre-oxidation catalyst with an SCR catalyst of the same volume. The same study also exposed better catalyst performances with a cell density increase of 170cpsi, reducing the catalyst volume by 1/3 with no change in efficiency [55].

In the research of Lambert et al. a different SCR design was studied: a group of metal and zeolite catalysts was tested on a car engine, and the results showed that the system is capable of cutting, on averagely, the 83% of the total NO_x (US federal test cycle) [56].

A scheme of diverse catalytic designs is shown in Figure 14.6, consisting of different arrangements in series in the exhaust line, the more complicated one having: a pre-oxidation catalyst, the urea hydrolysis catalyst, and the SCR catalyst, the NH_3 oxidation catalyst and urea injection.

Dealing with reducing agents, a wide number of them, even ammonia precursors, have been tested. NH_4NH_2COO, ammonium carbamate that decomposes into urea, $(NH_2)_2CO$, and H_2O if heated has been suggested, its main advantage is the fact that it is a solid with a consequent onboard reduction of volume needed, compared to the standard 32.5% urea in water solution; the solid-state also brings out the main drawback which is the complexity of the injection design. Stieger and Weisweiler suggested to dose NH_4NH_2COO powder in an oil bath to decompose to urea first and to NH_3 later, while the ammonia would then be let in the exhaust upstream of the SCR multiple times [58].

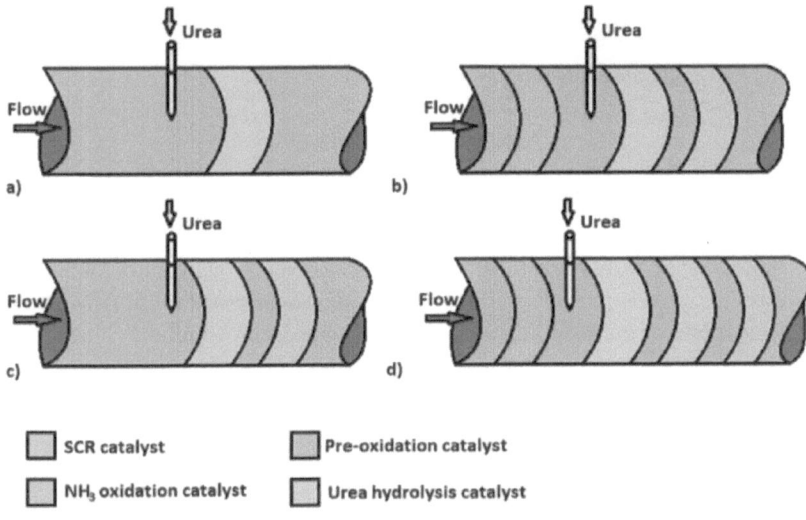

FIGURE 14.6 Scheme of 4 catalytic designs, adapted from [57].

Currently, the favored option is the above-mentioned 32.5% urea water solution, being even prescribed by DIN norms. This particular concentration was selected because it represents the concentration with the minimum crystallization point (−11°C). This also reflects the standards of Europe, when, according to the European Automobile Manufacturers' Association (ACEA), an appropriate infrastructure for the reducing agent and the urea filling nozzle is established [59]. The urea-water solution for the European market has been called *AdBlue*, while in the US the same solution is sold as *Diesel Exhaust Fluid* (DEF).

In the SCR systems used nowadays, a series of reactions can be traced throughout the whole system, as listed below [60].

1) Water evaporation:

$$CO(NH_2)_{2(aq)} \rightarrow 0.126CO(NH_2)_{2(m)} + 0.874H_{2(g)} \qquad (R\ 20)$$

2) Urea thermolysis to isocyanic acid and ammonia:

$$CO(NH_2)_2 \rightarrow NH_3 + HCNO \qquad (R\ 21)$$

3) HCNO hydrolysis with water producing NH_3 and CO_2:

$$HCNO + H_2O \rightarrow NH_3 + CO_2 \qquad (R\ 22)$$

4) Besides that, due to the relatively high stability of the isocyanic acid in the humid gaseous phase, hydrolysis would slowly in the gas flow beneath 400 °C [61]:

$$CO(NH_2)_2 \rightarrow 2NH_2 + CO \qquad\qquad (R\ 23)$$

5) The NH_2 radical reacts with NO:

$$NH_2 + NO \rightarrow N_2 + H_2O \qquad\qquad (R\ 24)$$

6) The ammonia produced in the previous steps gets in contact with the catalyst. The consequent reaction takes three paths: the *standard SCR reaction* (6a); a *fast SCR reaction* (6b), which includes NO in the reaction, which is a species already present in the exhaust; a *slow SCR reaction* (6c) that takes place if NO_2 is consistently present and at higher temperatures:

$$4NO + 4NH_3 + O_2 \rightarrow 4N_2 + 6H_2O \qquad\qquad (R\ 24a)$$

$$NO + NO_2 + 2NH_3 \rightarrow 2N_2 + 3H_2O \qquad\qquad (R\ 24b)$$

$$6NO_2 + 8NH_3 \rightarrow 7N_2 + 12H_2O \qquad\qquad (R\ 24c)$$

7) Engine tests suggested that even the following two reactions take place [62]:

$$2(NH_2)_2 CO + 6NO \rightarrow 5N_2 + 4H_2O + 2CO_2 \qquad\qquad (R\ 25a)$$

$$4HCNO + 6NO \rightarrow 5N_2 + 2H_2O + 4CO_2 \qquad\qquad (R\ 25b)$$

8) More than one reaction giving N_2 as a product were observed [63]:

$$6NO + 4NH_3 \rightarrow 5N_2 + 6H_2O \qquad\qquad (R\ 26a)$$

$$2NO_2 + 4NH_3 + O_2 \rightarrow 3N_2 + 6H_2O \qquad\qquad (R\ 26b)$$

$$4NH_3 + 3O_2 \rightarrow 2N_2 + 6H_2O \qquad\qquad (R\ 26c)$$

9) The unwanted reactions which consume NH_3 causing secondary emissions that can take place are listed below [64]:

$$4NH_3 + 3O_2 \rightarrow 4N + 6H_2O \qquad\qquad (R\ 27a)$$

$$2NH_3 + 2O_2 \rightarrow N_2O + 3H_2O \qquad\qquad (R\ 27b)$$

$$8N_2O + 6NH_3 \rightarrow 7N_2O + 9H_2O \qquad\qquad (R\ 27c)$$

$$4N_2O + 4NH_3 + O_2 \rightarrow 4N_2O + 6H_2O \qquad\qquad (R\ 27d)$$

10) Some of the listed main reactions suffer from H_2O inhibition, which can react with NH_3 and NO_2 to form ammonium nitrate, which can form solid deposits on the catalyst and a transitory deactivation [65]:

$$2NH_3 + 2NO_2 + 2H_2O \rightarrow NH_4NO_3 + NH_4NO_2 \qquad \text{(R 28)}$$

11) Other undesired byproducts can be generated: NH_3 can react with SO_3 from the oxidation catalyst (modern diesel vehicles have an oxidation catalyst acting before the SCR, exposing to sulphur oxides), which can lead to catalyst fouling:

$$NH_3 + SO_3 + H_2O \rightarrow NH_4HSO_4 \qquad \text{(R 29a)}$$

$$2NH_3 + SO_3 + H_2O \rightarrow \left(NH_4\right)_2 SO_4 \qquad \text{(R 29b)}$$

14.5   SCR SYSTEMS DRAWBACKS AND FAILURES

NO is the most predominant form of NO_x compound present in the automotive sector and engines. NO, in contact with the atmosphere, oxidizes on Pt which is adsorbed on an oxide surface to generate NO_2, which can later react with numerous hydrocarbons (HC) to generate NO_x ozone. Selective catalytic reduction (SCR) is the primary NO_x removal method for lean diesel conditions [66].

Originally, SCR represented a comprehensive technology used for different technical/engineering purposes and operations such as steam and diesel engines, gas turbines, heavy/light-duty diesel, and energy plants. It is known that this method has evolved and has found its main application in mobile diesel engines in recent decades. These applications involve specific challenges due to rigid emission limits, new internal combustion engine technologies, and alternative fuels [67, 68].

NOx levels in diesel engines are regulated and monitored through three-way catalysts (TWCs). However, their incapability to operate in different stoichiometric conditions is a limiting factor for many applications such as "lean-burn" engines, which perform at high air/fuel proportions [69].

This drawback blocked their market uptake for the inability of the three-way catalyst to reduce NOx emissions at high air/fuel ratios.

In this regard, SCR emerged as the leading remediation method for lean diesel conditions. Urea, chosen as a preferred reducing agent for its safety and low toxicology [70], decomposes into NH_3 and CO_2 and anticipates the whole SCR process. Subsequently, NH_3 is blended with the exhaust gas and simultaneously carries on a selective reduction of NOx. The Selective reduction made by NH_3 is an alternative to the oxidation of excess oxygen, performed by innate discharge reducers, HC and CO [71].

Several studies concentrated on lean-NOx, which refers to the selective catalytic reduction of NOx by hydrocarbons emission control. In this regard, many signs of progress have been achieved thanks to the recent use of selective catalytic reduction (NH3-SCR) and NOx storage and reduction (NSR) catalysts [72, 73]. These catalytic applications represent suitable tools based on nanomaterials technology destined for diesel motors.

One of the significant drawbacks of using SCR technology in vehicles is monitoring the temperature threshold (about 200°C) of the injection/mixing of urea solution in the hot exhaust gas, which is higher than the temperature for some modes of engine operation exhaust gases. Typically, this process occurs during the cold start process, in which additional fuel must be consumed to increase the temperature of the exhaust gas [70]. The immediate consequence leads to higher particle emissions after the engine has reached its average operating temperature [74]. Previous studies demonstrated that the preheating and evaporation of urea solution before injection into the motor's exhaust gas is a significant problem that affects the efficiency of the process [70].

Catalytic converters emerged as the most effective exhaust gas removal method employed to transform dangerous NO_x, CO, and HC discharge into N_2, O_2, and H_2O in long-lasting engine operating conditions. Catalysts must reach light-off temperatures of 200–370°C before achieving substantial contaminant conversion. In addition, the heating of the catalyst should occur faster when the engine is working under heavier loads than when it is placed nearer to the engine [74].

Nevertheless, it is problematic to implement an SCR system onboard a car, as it needs infrastructure deployment with NO_x/NH_3 catalysts, a urea tank, and detectors. To reduce the size of the post-treatment system, SCR should be covered in a Diesel Particulate Filter (DPF), which reduces the particulate matter emissions from the engine exhaust gases [75, 76].

Another critical aspect that negatively affects the performance of SCR systems is the nature and the distribution of the reducing agent distribution across the surface of the catalytic exhaust emission control device [70, 74, 76, 77]. The distribution and uniformity of ammonia entering the SCR reactor have become even more crucial since SCR technology is widely involved in the light-duty cars sector, requiring a reliable urea uniformity evaluation method [76].

Furthermore, studies were conducted to estimate the blending performance and increase urea uniformity by enhancing the blender/mixer configuration and upgrading the urea solution injection procedure [19, 78, 79, 20].

Recently, an automated analysis system for the SCR of passenger cars was designed to observe how the concentration of NO_x and NH_3 evolve along the downstream flank of the converter [76].

Rare-earth metal oxides have started to arise as SCR catalysts in the last few years. Amongst the others niobium-modified ceria–titania system shows low-temperature activity and good nitrogen selectivity [80]. The Niobium employment still presents opposite effects as it has decreased the surface area but, due to robust titania-ceria interactions, has shown a substantial increase in superficial acidity. MnO catalysts represent another fascinating application due to their reasonable costs and unusual SCR activity. The bottleneck of this class of catalysts is their low nitrogen selectivities and inadequate sulfur and water resistance. However, much research is still needed to reach a level where the commercial application of MnO-based catalysts [81].

14.6   SYSTEM RELIABILITY PREDICTIONS

SCR is an intricate system that involves several chemical reactions that are not easy to take into control. More than one study focused on SCR control improvement, NO_x conversion enhancement, and NH_3 slip inhibition. To solve an intricate nonlinear system like this, computer-based high-performance algorithms can help with their resolution. Some potent control devices, such as integrated control, have been applied in post-treatment systems. The main difficulties when dealing with diesel particulate filters are to give a precise valuation of soot and a consistent strategy to control the regeneration phase. High-efficiency systems are required to fulfill the stringent emissions protocols, thus requiring very reliable NO_x sensors. In this context, a system for reliability prediction can be useful by using the information on the life-cycle and the failure mechanisms of a system to accomplish reliability modeling. In this context, fault diagnostics procedures can integrate database builds, computing technologies, artificial intelligence, and control theory.

One of the greatest substantial difficulties of system reliability is the strong interdependency of the multi-component systems. Actually, it has often been verified that, in real applications, assuming independence in the systems frequently leads to miscalculations when evaluating the system lifetime. Reliability can be enhanced by fault tolerance, the main method being redundancy. Even though applying redundancy always translates into higher cost and complexity. The use of alternative sensing devices—other than sensors' direct feedback—could enhance reliability [82].

The most significant concern in system reliability is to deliver proper assurance for intent accomplishment. The design of a post-treatment system has to comply with national legislations in terms of emissions while being labeled as a reliable product. This concept will easily evolve into a demand for post-treatment systems that can even provide self-certification. The post-treatment systems will be expected to self-certify beside mutable targets received by connecting with the environment. In this context, big data and machine learning will serve as critical tools for reliability [83].

14.7 CONCLUSIONS

The SCR technology is well known to be very effective in reducing NOx of diesel-driven engines in the automotive sector, with NOx reduction often ranging in the 0.80–0.9 of the total NOx content. Results through-out years of research show how SCR technology is able to keep up with air quality regulations worldwide, as per the Air Quality Standard stated by the European Commission. Urea has been selected as the reducing agent in Europe, so the design and control of its injection system is one of the most critical topics in achieving the best SCR performances, so the ongoing researches still try to implement systems with reduced sizes and to achieve good performances at lower temperatures in respect to the current available catalysts.

REFERENCES

[1] K. Umar, A. A. Yaqoob, M. N. M. Ibrahim, T. Parveen, and M. T. Safian, "Chapter Thirteen—Environmental applications of smart polymer composites," in *Smart Polymer Nanocomposites*, S. A. Bhawani, A. Khan, and M. Jawaid, Eds. Woodhead Publishing, 2021, pp. 295–312.

[2] G. Centi, and S. Perathoner, "Catalysis, a driver for sustainability and societal challenges," *Catal. Today*, vol. 138, no. 1–2, pp. 69–76, 2008, doi: 10.1016/j.cattod.2008.04.037.

[3] J. Hagen, "Economic Importance of Catalysts," *Ind. Catal.*, pp. 425–428, 2006, doi: 10.1002/3527607684.ch15.

[4] "Catalyst Market Size & Share | Industry Report, 2020–2027." CA, US: Grand View Research. Report ID: 978-1-68038-228-0.

[5] "Catalyst Market Statistics, Trends | Industry Analysis 2030." OR, US: Allied Market Research. Report ID: A01408.

[6] A. Robotto, S. Barbero, R. Cremonini, E. Brizio, and E. Brizio, "Improving air quality and health in Northern Italy : limits and perspectives," pp. 1–38, 2022.

[7] C. H. Bartholomew and R. Farrauto, "Chapter 11—Environmental catalysis: Stationary sources," in *Fundamentals of Industrial Catalytic Processes*, 2nd Edn. 2005. NJ: John Wiley & Sons, Inc, 2010.

[8] M. Si et al., "Review on the NO removal from flue gas by oxidation methods," *J. Environ. Sci. (China)*, vol. 101, no. x, pp. 49–71, 2021, doi: 10.1016/j.jes.2020.08.004.

[9] F. Gholami, M. Tomas, Z. Gholami, and M. Vakili, "Technologies for the nitrogen oxides reduction from flue gas: A review," *Sci. Total Environ.*, vol. 714, p. 136712, 2020.

[10] P. M. Park, Y. Park, and J. Dong, "Reaction characteristics of NOx and N_2O in Selective non-catalytic reduction using various reducing agents and additives," *Atmosphere (Basel)*, vol, 12, p. 1175, 2021.

[11] Y. Guan, Y. Liu, Q. Lv, B. Wang, and D. Che, "Review on the selective catalytic reduction of NOx with H_2 by using novel catalysts," *J. Environ. Chem. Eng.*, vol. 9, no. 6, p. 106770, 2021, doi: 10.1016/j.jece.2021.106770.

[12] Z. Guan and D. Chen, "NO × Removal in the Selective Non-catalytic Reduction (SNCR) Process and Combined NO × and PCDD/Fs Control," no. 50874134, pp. 7–10.

[13] P. Forzatti, "Present status and perspectives in de-NOx SCR catalysis," *Appl. Catal. A Gen.*, vol. 222, no. 1–2, pp. 221–236, 2001, doi: 10.1016/S0926-860X(01)00832-8.

[14] "Environmental Catalysis: Stationary Sources," in *Fundamentals of Industrial Catalytic Processes*, John Wiley & Sons, Ltd, 2005, pp. 753–819.

[15] R. L. Berglund, "Emission Control, Industrial," in *Kirk-Othmer Encyclopedia of Chemical Technology*, vol. 10, John Wiley & Sons, Inc., 2004.

[16] R. M. Heck, "Catalytic abatement of nitrogen oxides-stationary applications," *Catal. Today*, vol. 53, no. 4, pp. 519–523, 1999, doi: 10.1016/S0920-5861(99)00139-X.

[17] Z. Liu, J. Li, and S. I. Woo, "Recent advances in the selective catalytic reduction of NOx by hydrogen in the presence of oxygen," *Energy Environ. Sci.*, vol. 5, no. 10, pp. 8799–8814, 2012, doi: 10.1039/c2ee22190j.

[18] J. Shibata, M. Hashimoto, K. I. Shimizu, H. Yoshida, T. Hattori, and A. Satsuma, "Factors controlling activity and selectivity for SCR of NO by hydrogen over supported platinum catalysts," *J. Phys. Chem. B*, vol. 108, no. 47, pp. 18327–18335, 2004, doi: 10.1021/jp046705v.

[19] I. V. Yentekakis, A. G. Georgiadis, C. Drosou, N. D. Charisiou, and M. A. Goula, "Selective Catalytic reduction of NOx over Perovskite-based catalysts using CxHy(Oz), H_2 and CO as reducing agents—A review of the latest developments," *Nanomaterials*, vol. 12, no. 7, p. 1042, 2022, doi: 10.3390/nano12071042.

[20] C. E. Romero, and X. Wang, "Chapter Three—Key technologies for ultra-low emissions from coal-fired power plants," in *Advances in Ultra-Low Emission Control Technologies for Coal-Fired Power Plants*, Y. Zhang, T. Wang, W.-P. Pan, and C. E. Romero, Eds. Woodhead Publishing, 2019, pp. 39–79.

[21] R. M. Heck, J. M. Chen, B. K. Speronello, and L. Morris, "Family of versatile catalyst technologies for NOχ removal in power plant applications," in *Environmental Catalysis*, vol. 552, American Chemical Society, 1994, pp. 17–215.

[22] W. Kamela and S. W. Kruczyński, "Examination of the ammonia dose influence on nitric oxides transformations into combined oxide—platinum SCR catalyst," *J. Kones*, vol. 19, pp. 253–258, 2015.

[23] P. Gabrielsson and H. G. Pedersen, "Flue Gases from Stationary Sources," in *Handbook of Heterogeneous Catalysis*, no. x, Wiley-VCH Verlag GmbH & Co. KGaA, 2008.

[24] S. R. Christensen, B. B. Hansen, K. Johansen, K. H. Pedersen, J. R. Thøgersen, and A. D. Jensen, "SO$_2$ Oxidation Across Marine V$_2$O$_5$-WO$_3$-TiO$_2$ SCR Catalysts: A Study at Elevated Pressure for Preturbine SCR Configuration," *Emiss. Control Sci. Technol.*, vol. 4, no. 4, pp. 289–299, 2018, doi: 10.1007/s40825-018-0092-8.

[25] L. Sun et al., "Fabrication of Cu-CHA composites with enhanced NH$_3$-SCR catalytic performances and hydrothermal stabilities," *Microporous Mesoporous Mater.*, vol. 309, no. 3, p. 110585, 2020, doi: 10.1016/j.micromeso.2020.110585.

[26] E. Gioria, F. A. Marchesini, and A. Giorello, "Green synthesis of a Cu/SiO$_2$ catalyst for efficient H$_2$-SCR of NO," *Appl. Sci.*, vol. 9, no. 19, p. 4075, 2019.

[27] P. Granger, H. W. Siaka, and S. B. Umbarkar, "What news in the surface chemistry of bulk and supported vanadia based SCR-catalysts: Improvements in their resistance to poisoning and thermal sintering," *Chem. Rec.*, vol. 19, no. 9, pp. 1813–1828, Sep. 2019, doi: 10.1002/tcr.201800092.

[28] W. Zhang, S. Qi, G. Pantaleo, and L. F. Liotta, "WO3–V2O5 Active Oxides for NOx SCR by NH$_3$: Preparation methods, catalysts' composition, and deactivation mechanism—a review," *Catalysts*, vol. 9, no. 6, p. 527, 2019, doi: 10.3390/catal9060527.

[29] I. Nova, L. Lietti, E. Tronconi, and P. Forzatti, "Dynamics of SCR reaction over a TiO$_2$-supported vanadia-tungsta commercial catalyst," *Catal. Today*, vol. 60, no. 1, pp. 73–82, 2000, doi: 10.1016/S0920-5861(00)00319-9.

[30] Z. Song et al., "Promotional effect of acidic oxide on catalytic activity and N$_2$ selectivity over CeO$_2$ for selective catalytic reduction of NOx by NH$_3$," *Appl. Organomet. Chem.*, vol. 33, no. 6, p. e4919, 2019, doi: 10.1002/aoc.4919.

[31] Z. Yan et al., "The way to enhance the thermal stability of V2O5-based catalysts for NH3-SCR," *Catal. Today*, vol. 355, pp. 408–414, 2020, doi: 10.1016/j.cattod.2019.07.037.

[32] Y. Liu, J. Zhao, and J.-M. Lee, "Conventional and New Materials for Selective Catalytic Reduction (SCR) of NOx," *ChemCatChem*, vol. 10, no. 7, pp. 1499–1511, 2018, doi: 10.1002/cctc.201701414.

[33] M. D. Amiridis, R. V. Duevel, and I. E. Wachs, "The effect of metal oxide additives on the activity of V$_2$O$_5$/TiO$_2$ catalysts for the selective catalytic reduction of nitric oxide by ammonia," *Appl. Catal. B Environ.*, vol. 20, no. 2, pp. 111–122, 1999, doi: 10.1016/S0926-3373(98)00101-5.

[34] L. Gan, F. Guo, J. Yu, and G. Xu, "Improved low-temperature activity of V$_2$O$_5$-WO$_3$/TiO$_2$ for denitration using different vanadium precursors," *Catalysts*, vol. 6, no. 2, p. 25, 2016, doi: 10.3390/catal6020025.

[35] Y. Pan, W. Zhao, Q. Zhong, W. Cai, and H. Li, "Promotional effect of Si-doped V$_2$O$_5$/TiO$_2$ for selective catalytic reduction of NO × by NH$_3$," *J. Environ. Sci. (China)*, vol. 25, no. 8, pp. 1703–1711, 2013, doi: 10.1016/S1001-0742(12)60181-8.

[36] J. Zhou et al., "Cerium oxide-based catalysts for low-temperature selective catalytic reduction of NOx with NH$_3$: A review," *Energy and Fuels*, vol. 35, no. 4, pp. 2981–2998, 2021, doi: 10.1021/acs.energyfuels.0c04231.

[37] Y. Ke et al., "Surface acidity enhancement of CeO$_2$ catalysts: Via modification with a heteropoly acid for the selective catalytic reduction of NO with ammonia," *Catal. Sci. Technol.*, vol. 9, no. 20, pp. 5774–5785, 2019, doi: 10.1039/c9cy01346f.

[38] Y. Zeng et al., "Recent Progress of CeO$_2$–TiO$_2$ Based Catalysts for Selective Catalytic Reduction of NOx by NH$_3$," *ChemCatChem*, vol. 13, no. 2, pp. 491–505, 2021, doi: 10.1002/cctc.202001307.

[39] G. Hu et al., "Effect of Ce doping on the resistance of Na over V$_2$O$_5$-WO$_3$/TiO$_2$ SCR catalysts," *Mater. Res. Bull.*, vol. 104, pp. 112–118, 2018, doi: https://linkinghub.elsevier.com/retrieve/pii/S0025540817332361.

[40] J. Cao *et al.*, "Improving the denitration performance and K-poisoning resistance of the V$_2$O$_5$-WO$_3$/TiO$_2$ catalyst by Ce4+ and Zr4+ co-doping," *Chinese J. Catal.*, vol. 40, no. 1, pp. 95–104, 2019, doi: 10.1016/S1872-2067(18)63184-5.

[41] V. I. Pârvulescu, P. Grange, and B. Delmon, "Catalytic removal of NO," *Catal. Today*, vol. 46, no. 4, pp. 233–316, 1998, doi: 10.1016/S0920-5861(98)00399-X.

[42] R. A. Grinsted, H. W. Jen, C. N. Montreuil, M. J. Rokosz, and M. Shelef, "The relation between deactivation of CuZSM-5 in the selective reduction of NO and dealumination of the zeolite," *Zeolites*, vol. 13, no. 8, pp. 602–606, 1993, doi: 10.1016/0144-2449(93)90130-U.

[43] D. W. Fickel, E. D'Addio, J. A. Lauterbach, and R. F. Lobo, "The ammonia selective catalytic reduction activity of copper-exchanged small-pore zeolites," *Appl. Catal. B Environ.*, vol. 102, no. 3–4, pp. 441–448, 2011, doi: 10.1016/j.apcatb.2010.12.022.

[44] C. Paolucci, J. R. Di Iorio, F. H. Ribeiro, R. Gounder, and W. F. Schneider, *Catalysis Science of NOx Selective Catalytic Reduction with Ammonia Over Cu-SSZ-13 and Cu-SAPO-34*, 1st ed., vol. 59, no. x. Elsevier Inc., 2016.

[45] H. Kubota *et al.*, "Analogous Mechanistic Features of NH3-SCR over Vanadium Oxide and Copper Zeolite Catalysts," *ACS Catal.*, vol. 11, no. 17, pp. 11180–11192, 2021, doi: 10.1021/acscatal.1c02860.

[46] F. Gao et al., A Review on selective catalytic reduction of NOx by NH$_3$ over Mn–based Catalysts at low temperatures. *Catalysts, Mechanisms, Kinetics and DFT Calculations*, vol. 7, no. 7, 2017.

[47] X. Yao *et al.*, "Influence of preparation methods on the physicochemical properties and catalytic performance of MnOx-CeO$_2$ catalysts for NH3-SCR at low temperature," *Chinese J. Catal.*, vol. 38, no. 1, pp. 146–159, 2017, doi: 10.1016/S1872-2067(16)62572-X.

[48] G. Qi and R. T. Yang, "Performance and kinetics study for low-temperature SCR of NO with NH$_3$ over MnOx–CeO$_2$ catalyst," *J. Catal.*, vol. 217, no. 2, pp. 434–441, 2003, doi: 10.1016/S0021-9517(03)00081-2.

[49] Z. Liu, Y. Yi, S. Zhang, T. Zhu, J. Zhu, and J. Wang, "Selective catalytic reduction of NOx with NH$_3$ over Mn-Ce mixed oxide catalyst at low temperatures," *Catal. Today*, vol. 216, pp. 76–81, 2013, doi: 10.1016/J.CATTOD.2013.06.009.

[50] H. Chang *et al.*, "Effect of Sn on MnOx–CeO$_2$ catalyst for SCR of NOx by ammonia: Enhancement of activity and remarkable resistance to SO$_2$," *Catal. Commun.*, vol. 27, pp. 54–57, 2012, doi: 10.1016/J.CATCOM.2012.06.022.

[51] W. Liu *et al.*, "Promotional effect of Ce in NH$_3$-SCO and NH$_3$-SCR reactions over Cu-Ce/SCR catalysts," *J. Ind. Eng. Chem.*, vol. 107, pp. 197–206, 2022, doi: 10.1016/J.JIEC.2021.11.045.

[52] J. G. M. Brandin, L. A. H. Andersson, and C. U. I. Odenbrand, "Catalytic reduction of nitrogen oxides on mordenite some aspect on the mechanism," *Catal. Today*, vol. 4, no. 2, pp. 187–203, 1989, doi: 10.1016/0920-5861(89)85050-3.

[53] L. H. Andersson, *Selektiv katalytisk reduktion av kva¨veoxider*. University of Lund, LUTKDH, 1989.

[54] A. P. Walker et al., "The development and performance of the compact SCR-trap system: A 4-way diesel emission control system," *SAE Tech. Pap.*, 2003, doi: 10.4271/2003-01-0778.

[55] "Urea-SCR Catalyst System Selection for Fuel and PM Optimized Engines and a Demonstration of a Novel Urea Injection System on JSTOR." www.jstor.org/stable/44743093?seq=1 (accessed May 27, 2022).

[56] C. Lambert, J. Vanderslice, R. Hammerle, and R. Belaire, "Application of urea SCR to light-duty diesel vehicles," *SAE Tech. Pap.*, Sep. 2001, doi: 10.4271/2001-01-3623.

[57] P. L. T. Gabrielsson, "Urea-SCR in automotive applications," *Top. Catal. 2004 281*, vol. 28, no. 1, pp. 177–184, 2004, doi: 10.1023/B:TOCA.0000024348.34477.4C.

[58] "Ammoniak-Generator für die NOx-Minderung in Diesel-Abgasen: Modellierung der Ammoniumcarbamat-Thermolyse—Stieger—2001 — Chemie Ingenieur Technik—Wiley Online Library." https://onlinelibrary.wiley.com/doi/abs/10.1002/1522-2640(200101)73: 1/2%3C123::AID-CITE123%3E3.0.CO;2-F?casa_token=ZmyvKOHY5WcAAAAA:W-jCeoKxEeg7V1esyNtXzQMVis-p4dFygL7xy2nDa8hS-L7DaZ2K5rZGn65_E5N06VkeIT6m_HmLWwg (accessed May 20, 2022).

[59] "ACEA Statement on the Adoption of SCR Technology to Reduce Emissions Levels of Heavy-Duty Vehicles,"

[60] D. Maizak, T. Wilberforce, and A. G. Olabi, "DeNOx removal techniques for automotive applications – A review," *Environ. Adv.*, vol. 2, p. 100021, 2020, doi: 10.1016/j.envadv.2020.100021.

[61] G. Piazzesi, O. Kröcher, M. Elsener, and A. Wokaun, "Adsorption and hydrolysis of isocyanic acid on TiO_2," *Appl. Catal. B Environ.*, vol. 65, no. 1–2, pp. 55–61, 2006, doi: 10.1016/J.APCATB.2005.12.018.

[62] H. L. Fang and H. F. M. DaCosta, "Urea thermolysis and NOx reduction with and without SCR catalysts," *Appl. Catal. B Environ.*, vol. 46, no. 1, pp. 17–34, 2003, doi: 10.1016/S0926-3373(03)00177-2.

[63] M. Koebel, M. Elsener, and M. Kleemann, "Urea-SCR: A promising technique to reduce NOx emissions from automotive diesel engines," *Catal. Today*, vol. 59, no. 3–4, pp. 335–345, 2000, doi: 10.1016/S0920-5861(00)00299-6.

[64] G. Madia, M. Koebel, M. Elsener, and A. Wokaun, "Side reactions in the selective catalytic reduction of NOx with various NO_2 fractions," *Ind. Eng. Chem. Res.*, vol. 41, no. 16, pp. 4008–4015, Aug. 2002, doi: 10.1021/IE020054C.

[65] A. Kumar, M. A. Smith, K. Kamasamudram, N. W. Currier, and A. Yezerets, "Chemical deSOx: An effective way to recover Cu-zeolite SCR catalysts from sulfur poisoning," *Catal. Today*, vol. 267, pp. 10–16, Jun. 2016, doi: 10.1016/J.CATTOD.2016.01.033.

[66] Z. Ma et al., "Characteristics of NOx emission from Chinese coal-fired power plants equipped with new technologies," *Atmos. Environ.*, vol. 131, no. x, pp. 164–170, 2016, doi: 10.1016/j.atmosenv.2016.02.006.

[67] R. Villamaina, I. Nova, E. Tronconi, T. Maunula, and M. Keenan, "Effect of the NH4NO3 addition on the Low-T NH3-SCR performances of individual and combined Fe- and Cu-Zeolite Catalysts," *Emiss. Control Sci. Technol.*, vol. 5, 2019, doi: 10.1007/s40825-019-00140-3.

[68] O. Kröcher, "Selective catalytic reduction of NOx," *Catalysts*, vol. 8, no. 10, pp. 10–13, 2018, doi: 10.3390/catal8100459.

[69] A. M. Beale, F. Gao, I. Lezcano-Gonzalez, C. H. F. Peden, and J. Szanyi, "Recent advances in automotive catalysis for NOx emission control by small-pore microporous materials," *Chem. Soc. Rev.*, vol. 44, no. 20, pp. 7371–7405, 2015, doi: 10.1039/c5cs00108k.

[70] R. Sala, P. Bielaczyc, and M. Brzezanski, "Concept of vaporized urea dosing in selective catalytic reduction," *Catalysts*, vol. 7, no. 10, 2017, doi: 10.3390/catal7100307.

[71] T. V. Johnson, "Review of selective catalytic reduction (SCR) and related technologies for mobile applications," in *Urea-SCR Technology for deNOx After Treatment of Diesel Exhausts*, Springer, New York, NY, pp. 3–31, 2014, doi: 10.1007/978-1-4899-8071-7.

[72] B. Moden, J. M. Donohue, W. E. Cormier, and H.-X. Li, "Effect of Cu-loading and structure on the activity of Cu-exchanged zeolites for NH3-SCR," in *Zeolites and related Materials: Trends, Targets and Challenges*, vol. 174, A. Gédéon, P. Massiani, and F. Babonneau, Eds. Elsevier, 2008, pp. 1219–1222.

[73] A. G. Greenaway et al., "Detection of key transient Cu intermediates in SSZ-13 during NH3-SCR deNOx by modulation excitation IR spectroscopy," *Chem. Sci.*, vol. 11, no. 2, pp. 447–455, 2020, doi: 10.1039/c9sc04905c.

[74] M. S. Reiter and K. M. Kockelman, "The problem of cold starts: A closer look at mobile source emissions levels," *Transp. Res. Part D Transp. Environ.*, vol. 43, no. x, pp. 123–132, 2016, doi: 10.1016/j.trd.2015.12.012.

[75] T. Johnson, "Vehicular emissions in review," *SAE Int. J. Engines*, vol. 9, no. 2, pp. 1258–1275, Apr. 2016, doi: https://doi.org/10.4271/2016-01-0919.

[76] R. Sala, J. Dzida, and J. Krasowski, "Ammonia concentration distribution measurements on selective catalytic reduction catalysts," *Catalysts*, vol. 8, no. 6, 2018, doi: 10.3390/catal8060231.

[77] M. K. Robinson, K. M. Sentoff, and B. A. Holmén, "Particle number and size distribution of emissions during light-duty vehicle cold start data from the total onboard tailpipe emissions measurement system," *Transp. Res. Rec. J. Transp. Res.*, vol. 2158, pp. 86–94, 2010, doi: 10.3141/2158-11.

[78] Y.-G. Zhao, L. Hua, J. Hu, T. Tang, S. Shuai, and J.-X. Wang, "Experimental study of urea solution spray characteristics in SCR system of diesel engine," vol. 33, pp. 22–27, 2012.

[79] X. Zhang and M. Romzek, "3-D Numerical study of flow mixing in front of SCR for different injection systems," SAE World Congress & Exhibition, SAE International, United States, 2007.

[80] J. Mosrati *et al.*, "Nb-modified Ce/Ti oxide catalyst for the selective catalytic reduction of NO with NH_3 at low temperature," *Catalysts*, vol. 8, no. 5, 2018, doi: 10.3390/catal8050175.

[81] K. Zhang *et al.*, "Enhanced low temperature NO reduction performance via MnOx-Fe2O3/vermiculite monolithic honeycomb catalysts," *Catalysts*, vol. 8, no. 3, 2018, doi: 10.3390/catal8030100.

[82] M. Soleimani, F. Campean, and D. Neagu, "Reliability Challenges for Automotive Aftertreatment Systems: A State-of-the-art Perspective," *Procedia Manuf.*, vol. 16, pp. 75–82, 2018, doi: 10.1016/J.PROMFG.2018.10.174.

[83] T. J. Cui and S. sha Li, "Study on the relationship between system reliability and influencing factors under big data and multi-factors," *Cluster Comput.*, vol. 22, no. 4, pp. 10275–10297, 2019, doi: 10.1007/S10586-017-1278-5/FIGURES/11.

15 Advanced Technology for Cleanup of Syngas Produced from Pyrolysis/Gasification Processes

Saleh Al Arni

CONTENTS

15.1 Introduction...289
15.2 Types of Waste Gases 290
15.3 Waste Gas Abatement Techniques..292
    15.3.1 Treatment Technologies..292
    15.3.2 Application of Gas Cleaning 293
    15.3.3 Methods of Syngas Cleaning..295
15.4 Equipment of Syngas Cleaning 298
15.5 Conclusion 302
Notes 303
References..303

15.1 INTRODUCTION

Treatment of waste materials by thermochemical conversion processes, especially pyrolysis and gasification, can produce gases, liquid, char, and ashes [1–3]. The gases are a source of emission of pollutants. Also, there are different industrial sources of pollutant emissions such as industrial incinerators (stacks) [4, 5].

The typical volume of flue gas[1] produced by Municipal Solid Waste (MSW) incinerators is between 4000 and 6000 m^3 per ton of waste [6–8]. The major types of pollutants emitted by combustion plants take the form of particulates and gases; these contain carbon oxides (CO_x), nitrogen oxides (NO_x), water vapor (H_2O), sulfur oxides (SO_x), hydrogen chloride (HCl), hydrogen fluoride (HF) and other hydrocarbons such as polycyclic aromatic hydrocarbons, dioxins and furans, and heavy metals such as mercury, cadmium and lead [9]. Emission of pollutants to the atmosphere is established and regulated by environmental protection legislations that set emission limits [9]. These limits are very similar in most countries of the world, and they are independent of the operation conditions of the plant and the efficiency of treatment [9].

DOI: 10.1201/9781003260738-15

Furthermore, industrial control of the discharges of atmospheric pollutants could be arranged before and after the production process by minimizing emissions of pollutants [4–9]. This makes an integral part of the design of the plant and considered, after emissions, as a treatment and efficient removal technology [4–9].

There are numerous publications in the literature that deal with the issues of air pollutants emission, treatments and controls, and describe the technologies used in cleaning up syngas [10–13]. This chapter, therefore, comes to provide a review of these technologies that are applied into gas streams during and after production, and before using syngas or discharge it to the air. Gaseous waste from fuel cycle facilities, reactor vitrification, incineration, pyrolysis, and gasification can be cleaned by filtration, adsorption and absorption. A syngas cleanup system depends on the type and quantity of waste in the gas, effluent and the regulation emission limit into the atmosphere and the required treatment efficiency.

15.2   TYPES OF WASTE GASES

Gaseous waste contains several types of gases such as CO_x, NO_x, SO_x, water vapor, polycyclic aromatic hydrocarbons, furans, dioxins, unburned particles and other pollutants, such as dust or particulates.

In the following, a brief description of various types of gaseous contaminants will be presented.

Dusts: dusts are particulate materials found in the gas effluents. They could be organic or inorganic (metallic or non-metallic as silica) and aerosols/droplets. The particulates in the flue gases usually have a size in the range between 1 and 75 μm. The particulate control of flue gases depends on the particle load in the gas stream, the flowrate of gas, the average particle size, the particle-size distribution, the flue gas temperature and the required outlet gas concentration. Based on these parameters, the selection of gas cleaning equipment can be determined.

The combustion process produces a very dusty waste with heterogeneous and toxic nature, usually composed of ash, and requires efficient and high levels of pollutants removal. Also associated with dust particulates, there are more toxic pollutants such as dioxins and furans and heavy metals.

Soot: soot is a black powder or flake substance consisting of a large amorphous carbon that is produced by the incomplete burning of organic matter in conditions of high temperature and low oxygen content. Soot contains molecules such as acetylene radicals and polycyclic aromatic hydrocarbons. The reduction of soot formation can use secondary combustion air to complete burn of carbon that is not burned.

Smuts: smut is a small flake of soot or smudge that is a product of a combustion containing corrosive acids such as hydrochloric, sulphuric or hydrofluoric acid.

Acidic and corrosive pollutants: municipal waste incinerators produce acidic and toxic or corrosive gases such as hydrogen chloride, hydrogen fluoride and sulphur dioxide (SO_2). Chlorine in flow gas can be derived from plastic material waste such as PVC (polyvinyl chloride) or other source materials such as rubber, leather, paper and vegetable matter that contain metal chlorides like NaCl or $CaCl_2$. On the other hand, fluorine can be derived from PTFE (polytetrafluoroethylene) waste materials, while sulphur dioxide can be derived from combustion of sulfurous compounds. The NO_x can be derived from part of the nitrogen in the air and the nitrogen in the waste.

Nitrogen oxides (NO_x): NO_x emissions in flue gases consist of two types; the first type is produced by waste incineration (known as fuel NO_x) and the second is produced by oxidation of nitrogen present in air at high temperatures (known as thermal NO_x). The thermal NO_x represents about 25% of the total nitrogen oxides [14].

In large combustion plants, these are typically composed of about 90% NO and 10% NO_2 [7]. Furthermore, wastewater treatment and agricultural industries are additional sources for NO_x emissions. Nitrogen oxides affect health and environment by the formation of photochemical smog and they contribute to acid rain. These oxides (NO_x) can be reduced by recirculation of the flue gases into the combustion chamber or by ammonia addition or by bioprocesses [15].

Volatile organic compounds (VOCs): they are organic chemical compounds that are easy to be volatile at low temperatures because of their physicochemical properties and they have been classified as hazardous materials and have healthy effects. They need to be treated from waste gases.

Dioxins and Furans: Dioxin ($C_4H_4O_2$) and furan (C_4H_4O) are organic compounds that combine with chlorine atoms to formulate a family of organic compounds called polychlorinated dibenzodioxins (PCDD) and polychlorinated dibenzofurans (PCDF), such as Dioxin molecule, Furan molecule, and their congeners with chlorine atoms such as 2,3,7,8-tetrachlorodibenzo-p-dioxin, and 2,3,7,8-Tetrachlorodibenzofuran. All these compounds are toxic and have health effects to humans, animals and the environment [7, 16, 17]. The European Community (EC) Waste Incineration Directive, 2000, determined the emission-limit value for old municipal solid waste incinerator for these compounds PCDD and PCDF at 0.1 ngTEQ /$Nm^{3.2}$ while for the modern plants the emissions-limit value is between 0.0002 and 0.08 ngTEQ/Nm^3 [18].

The main sources of dioxins are the waste and the combustion processes. The CO concentration is an indicator on the formation of dioxins in a combustion operation. The relationship between the temperature and CO concentration is given by the following empirical equation [17].

$$PCDDs + PCDFs = a - bT + cCO \qquad (Eq. 15.1)$$

Where: a, b and c are constants

a = 2670.2 and 4754.6 for modular combustors and water-wall combustors, respectively.

b = 1.37 and 5.14 for modular combustors and water-wall combustors, respectively.

c = 100.06 and 103.41 for modular combustors and water-wall combustors respectively.

T = the temperature (in degrees Celsius) in the secondary chamber for modular combustors and the furnace temperature in water-wall combustors, respectively.

CO = the concentration of CO in percent of gases.

The production of PCDDs is proportional to the CO concentration according to the following relationship:

$$PCDDs = (CO / A)^2 \qquad \text{(Eq. 15.2)}$$

Where:

PCDDs = concentration of dioxins in the off-gases, ng/Nm^3

CO = concentration of carbon monoxide in the off gases as percent of total gas

A = a constant, function of operating system

Heavy metals: heavy metals (such as arsenic (As), cadmium (Cd), chromium (Cr), cobalt (Co), copper (Cu), mercury (Hg), manganese (Mn), nickel (Ni), lead (Pb), antimony (Sb), thallium (Tl), vanadium (V), and zinc (Zn)) are produced from incinerators as flue gas, fly ash and bottom ash during the combustion of waste process. They are released during the combustion process as ash metals particulate and or metal compounds or their derived compounds, and they are a function of their physio-chemical properties (e.g., the effect of high temperature, depending on their volatility). Heavy metals may also react with oxygen or hydrogen chloride, and/or other compounds or elements to form other compounds or as adsorption onto the fine particulates [5, 9].

Following the combustion of waste, the temperature of flowing gas decreases, which will cause the condensation of metal volatiles. These volatiles will be associated with the abatement particulates. Because heavy metals have a toxic health effect, they need to be removed from waste and waste gases; electrostatic precipitators and fabric filters can be used for abatement.

15.3   WASTE GAS ABATEMENT TECHNIQUES

The treatment techniques depend on the separation technology, the method of operation, the operating conditions and the efficiency of treatment required. The separation technology depends on the properties of the gas stream as chemical-physical characteristics, which include type of components, temperature, pressure, density, viscosity, surface tension, electrostatic forces, particle size, and several other properties.

15.3.1   TREATMENT TECHNOLOGIES

The aim of using the treatment techniques is to reduce the emission of air pollutants to the atmosphere. The abatement techniques include recovery and destruction techniques of pollutant emissions. These depend on the pollutant emissions source temperature. Treatment technologies of gaseous waste classify the pollutant emissions on the base source temperature into low and high temperature emissions:

- Sources at low-temperature such as production processes and work-up of products (handling and storage activities that cause emissions). The emissions may include solid raw materials, VOCs, inorganic volatile compounds or both, with or without dusty content. In this case, the majority of devices used in these techniques are based on using wet scrubbers.
- Sources at *high temperature* such as combustion processes and their facilities (power plants, incinerators, boilers, catalytic oxidizers and thermal). The emissions may include a mixture of particulate matter or dust, heavy metals, carbon oxides, nitrogen oxides, sulphur oxides, halogen compounds such as HCl, HF and Cl_2, dioxins and furans. For example, air pollutants of incineration include gases, solid particulate or aerosol and their associated forms including dusts, vapors, fogs, mists, fumes, and smokes. In this case, techniques used are inertial separation devices, cyclones, barrier filtration, and electrostatic precipitators' interaction.

In the case of destruction, flares can be used. Flares are simply a vent to the atmosphere where the flammable gases can be burned. The process of destruction is a thermal oxidation with efficiency of over 99%, used, usually, for continuous emission streams in oil refineries, power stations and other several industries. Sometime, before using this system, another recovery technique or preliminary treatment can be used; this includes cryogenic systems at low temperature or refrigeration and condensation techniques. Destruction process is also used in case of maintenance or upset systems. It usually uses flaring to dispose of gases safely without connecting the surplus combustible gases to abatement systems.

On the other hand, recovery techniques include gravity separators, cyclones, membrane separation, condensation, adsorption, wet scrubbers, dry, semi-dry and wet sorption, mist filters, and electrostatic precipitator (ESP).

In general, the abatement techniques used for treatment of waste gases include flaring, thermal oxidation, catalytic oxidation, gravity separators, cyclones, electrostatic precipitators, mist filters (such as 2-stage dust filters, fabric filters, catalytic filtration, absolute filters, high efficiency air filter), dry, semi-dry and wet sorption, bio-filtration, bio-scrubbing, bio-trickling, selective non catalytic reduction (SNCR), and selective catalytic reduction (SCR).

The treatment techniques, used in the cases of low-temperature sources, firstly remove solid material or mists, then remove the gaseous pollutants. If necessary, further abatement can be made that is applied in base of emission levels required.

15.3.2 Application of Gas Cleaning

The adequate treatment method depends on the types of pollutants in the waste stream. Air pollutants have different physical states and chemical compositions, and they originate from different sources. There are several methods of cleaning gaseous pollutants from gas streams; examples include gravity separation, absorption, adsorption, condensation, incineration and bio-filtration [19]. Thermal treatment in the incineration methods include open burning which is usually made without any air pollution control.

The method is used to treat MSW by burning piles of waste outdoors or burning them in a burn barrel in several urban centers to reduce the volume of waste received at the dump (open dump). The open burning could have serious effects on human health and environment. In the closed system, particulate emissions from incinerators can be collected by cyclones, electrostatic precipitators and fabric filters.

Dust can be removed from waste gas streams by adequate techniques according to the actual situation. For example, the particulates dusts can be removed from the flue gas by cyclones, filters, and/or electrostatic precipitators. In certain cases, pretreatment processes become necessary in order to protect equipment that are used in successive treatment phases. The pre-treatment could include gravity separation which could help prevent filter clogs in successive filtering processes.

Treatment technologies of VOCs depend on the types of pollutants in the waste stream and the final aim of treatment, which is recovery or destruction. A choice of a technique or another depends on the pollutant characteristics influencing the emission stream.

Scrubber can be used to treat the gaseous pollutants. For the removal or recovery of acidic pollutants such as HCl, HF, SO_2 and NO_x from waste gas streams can use scrubbers in three different techniques; these are the wet, wet-dry and dry-scrubbing. Dry or semi-dry scrubber techniques can use lime or limestone to produce gypsum (calcium sulfate, $CaSO_4$) according to following equation:

$$2SO_2 + 2CaCO_3 + O_2 => 2CaSO_4 + 2CO_2 \qquad \text{(Eq. 15.3)}$$

SO_2 treatment or Fuel Gas Desulphurisation (FGD): FGD techniques are mainly used to treat gases with low concentration of sulphur dioxide (up to 1%) which are usually produced by power plants and large combustion plants.

The emissions of dioxins and furans can be reduced by thermal post combustion, combustion conditions, reducing of the organic contents, and active coke technique. In addition, dioxins can be destroyed at high temperature above 850 °C in the presence of oxygen or by catalytic oxidation systems, or by the process of de-novo synthesis (biological processes) or by adsorbed onto solid matter (as activated carbon) by scrubber or filter. Using wet scrubbing by aqueous solvent, acidic or alkaline solution for hydrogen halides such as Cl_2, SO_2, H_2S, NH_3; or dry scrubbing to treat CS_2, COS; or adsorption for CS_2, COS, Hg; using incineration to destroy H_2S, CS_2, COS, HCN, CO; using SNCR/SCR for NOx; using biological gas treatment for NH_3, H_2S, CS_2.

Furthermore, biological treatment can be applied to remove VOC from waste gas streams. If the recovery of VOCs has a commercial value, other techniques can be used; these include condensation, membrane separation or adsorption by wet scrubbing on the solid material then regenerated. The condensation techniques are usually used as an integral part of tanks for pollution control. Adsorption by activated carbon is the effective technology for removal of organic pollutants and VOCs at low temperatures. The microporous aluminosilicate minerals (Zeolites) are also used.

Other compounds: For recovery or treatment purposes, other compounds of waste gas pollutants can be treated by the appropriate techniques.

Odor problem: it can be managed by chemical stripping, thermal destruction, and bio-filtration. The applied techniques include scrubber, fan or flare.

After the treatment, gases will be disposed of through the stack (dispersion of emissions through the chimney stack). The pollutants emitted to the atmosphere must be under the regulations and legislation of the environment. For more detailed information about gas cleanup equipment, the reader is invited to see section 15.4. To be noted that the syngas cleanup has a significant capital and operating cost.

15.3.3 METHODS OF SYNGAS CLEANING

Syngas is a one product of the gasification and or pyrolysis processes. Pyrolysis syngas can be used as fuel for the pyrolysis process itself and or as fuel gas that is rich with hydrogen and hydrocarbon compounds, depending on the operation conditions and on the natural raw material in exercise (See Chapter 3). In general, pyrolysis syngas contains CO, CO_2, H_2 and hydrocarbon compounds such as CH_4, C_2H_8, and trace compounds such as NH_3, H_2S, and HCl. The syngas prior to use in recovery systems needs a cleanup. The commercial equipment for pyrolysis processes is equipped with emission abatement devices to ensure the cleanliness of the syngas product [20].

During the phase of the pyrolysis process, the solid-state biomass decomposes into a vapor state that contains a large variety of volatile organic compounds due to the heat effect. The methods used to separate the particles from the vapor include physical separation through a cyclone and a filtration. The condensed portion of vapor state can be obtained by the reduction of temperature of the vapor. The removal of syngas contaminants can be done by physical or chemical treatment such as oil absorption, char adsorption and catalytic conversion over the char-supported catalysts. The light tar compounds such as VOCs can be removed by adsorption of bio-char, while heavy tars are removed by bio-oil absorption. In addition, catalyst methods on char-supported catalysts can be used to convert tar into the additional gas products. Examples of such catalysts include nickel, dolomite, and iron based and/or catalytic cracking. To be emphasized that a catalyst needs a regeneration after its deactivation [21].

The syngas produced from physical-chemical decomposition by heating in the absence of oxygen (pyrolysis process) is contaminated with a large variety of volatile organic compounds (volatile material) such as char particles and condensable gases (i.e.tar) that need to be cleaned from the produced syngas. Syngas cleaning of contaminants can be done by:

a) primary methods of treatment that occur during the process and take place inside the reactors (Figure 15.1). This can be done by the optimization of the operating conditions, modification of the gasifier design and addition of catalysts that are used in the reactor.

b) secondary methods of treatment occur after the pyrolysis or gasification of a process is done (Figure 15.2) and takes place by mechanical or physical processes and reforming processes.

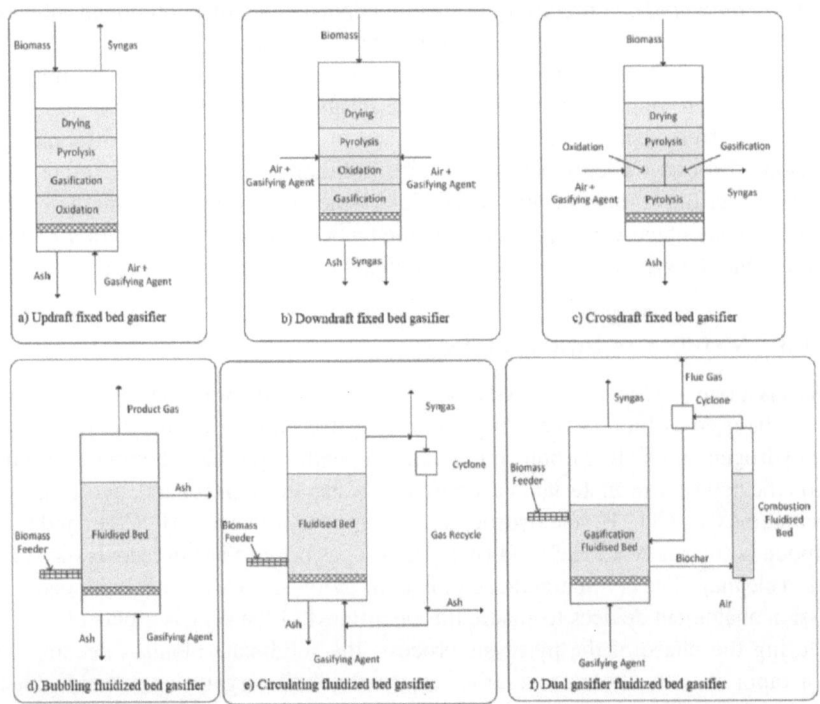

FIGURE 15.1   Schematic illustration of the fixed bed gasifiers: a) updraft, b) downdraft, c) crossdraft, and fluidized bed gasifiers: d) bubbling, e) circulating and f) dual gasifier.

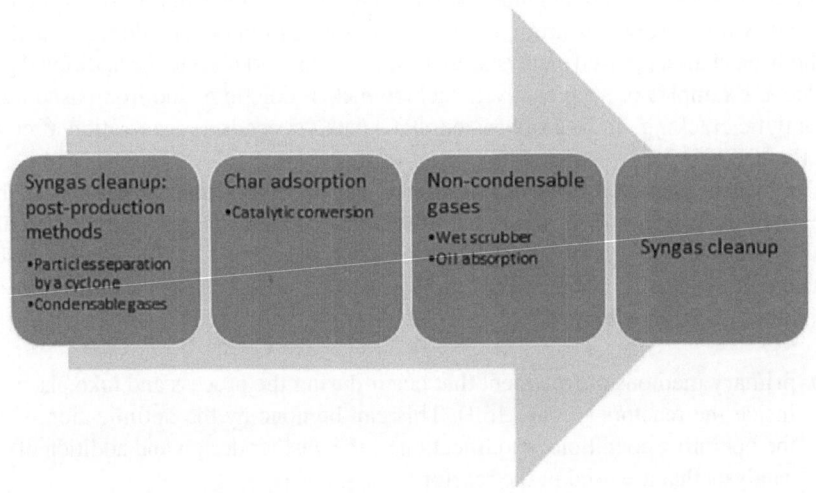

FIGURE 15.2   Post-production methods of syngas cleanup.

Syngas cleaning of contaminates inside of reactors depends on the gasifier type and its efficiency. There are several types of gasifiers that are used for syngas production (Figure 15.2) such as fixed, fluidized bed, entrained flow reactor, rotary kiln reactor, and plasma reactor. The fixed bed gasifiers are classified as updraft gasifier, downdraft gasifier, and crossdraft gasifier while fluidized bed gasifiers are bubbling, circulating, and dual gasifier. Comparison of different gasifier types, operating conditions, products, advantages and disadvantages are presented in Table 15.1. For a good quality syngas to be used as fuels, it needs to be upgraded by methanation, hydrogen separation, Fischer-Tropsch synthesis and hydroprocessing methods; these are known as processes of upgrading syngas to fuels.

TABLE 15.1
Comparison Among Different Types of Gasifiers

Types of gasifier (reactor)	Tar concentration (g/Nm³)	Operating conditions	Products (Wt%)* char, bio-oil, gas	Advantages	Disadvantages	References
Fixed bed						
Updraft	10–150	Gas Flow: Counter current; Feed particle size: 5–100 mm; Gas exit temperature: 200–400 °C	31, 50, 18	Simplest design; Can used varying feed quality	Can be produced low quality syngas	[22, 23, 24]
Downdraft	0.01–6	Gas Flow: Co-current; Feed particle size: 20–100 mm; Gas exit temperature: 700 °C		Can be produced higher quality syngas than updraft reactor	Requires increased biomass combustion to maintain temperature	[22, 23, 24]
Crossdraft	0.01–0.1	Gas Flow: Cross current; Feed particle size: 5–20 mm; Gas exit temperature: 1250 °C				

(Continued)

TABLE 15.1
Continued

Types of gasifier (reactor)	Tar concentration (g/Nm³)	Operating conditions	Products (Wt%)* char, bio-oil, gas	Advantages	Disadvantages	References
Fluidised Bed						
Bubbling	1.5–9	Gas Flow: Counter current; Feed particle size: < 20 mm; Temperature: 750–900 °C	10, 59, 28	Simplest design	Low density syngas produced	[22, 23, 24]
Circulating	9–10			High density syngas: Easy solid particles captured	Difficulty maintaining reactor temperature	[22, 23, 24]
Dual Bed	10 (average)			Can be produced highest quality syngas	Complicated design	[22, 23, 24]

* Note: the total is less than 100%, which is attributed to operation losses.

15.4 EQUIPMENT OF SYNGAS CLEANING

The typical devices used for the collection and removal of particles from waste gases include gravity separators, electrostatic precipitators, fabric filters and wet scrubbers. Table 15.2 provides a guideline for the selection of appropriate air pollution control separation equipment for treatment of gas streams. The efficiency of treatment depends on device type and the goal of treatment. Figure 15.3 provides examples of typical devices used in separation technology and fluid cleaning.

The following is a brief description of the main emission control equipment that is used in the stream of waste gases.

Gravity Separators: Gravity settling chamber (Figure 15.3) is a simple device that is usually used as a preliminary technique in various filter systems to prevent

TABLE 15.2
Selection of the Appropriate Gas Cleaning Equipment (Modified After Cheremisinoff, 2000) [25]

Particle size	Solids particulates	Liquid particulates	Combination
Submicron: <1 micron	Cartridge filter	Scrubber	Scrubber
Invisible: 2–50 microns	Coarse (Grit) filter, Cyclone	Cyclone, baghouse	Mist and fume, filters
Tiny: 51–850 microns	Gravity settler, Cyclone	Gravity settler	Fume filters, Scrubber
Small: 850 microns	Gravity settler	Cyclone	Fume filters, Scrubber

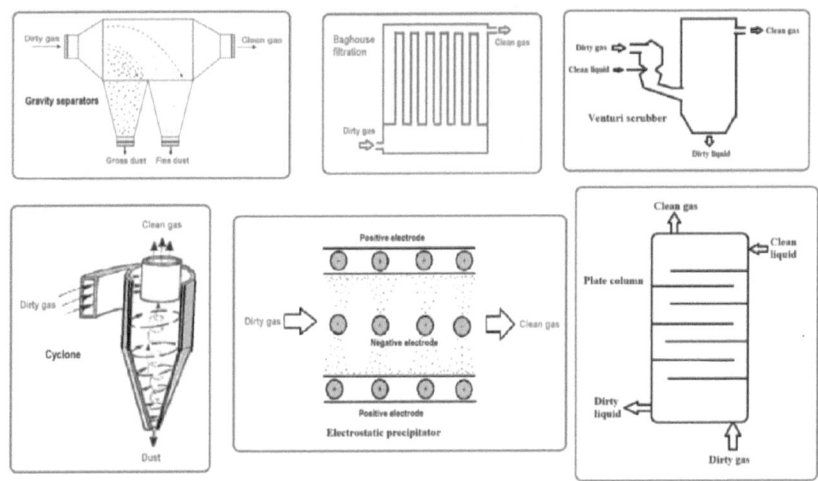

FIGURE 15.3 Examples of typical devices used in waste particulate emission control [9].

entrainment. It is normally integrated with other equipment and used for recovery materials, and it is not restricted for dust content, particulate size more than PM_{50}, but also down to PM_{10}. The efficiency of pollutant removal depends on particle size and feed concentration; it usually has a range between 10–90%. The application limit of flow rate is up to 100,000 Nm^3/h, but it is effective at low gas velocities of less than 200m/min [26] in case of particle size in the range of 50–100 µm for a removal efficiency of 90%.

Cyclone (dry and wet): Cyclone (Figure 15.3) is a gravity separator supported by centrifugal forces that the waste gas stream enters tangentially to make a vortex and rotates in a helical path down. The particles drop down to the bottom and gas flows up through the central inner. The cyclone is usually used after processes of crushing, grinding, spray drying and calcining operations and can be used for recovery. It is often used as pre-cleaners for electrostatic precipitator or fabric filters and suitable for flue gas treatment.

The application limits of temperature can reach more than 1200°C, depending on vessel material type. It has an advantage that it can be operated at high temperatures that exceed 500 °C. It can operate at a flow rate up to 100,000 Nm^3/h for a single unit or up to 180,000 Nm^3/h for multiple units with a gas concentration up to 16000 g/Nm^3. Typically, the exit gas particulate concentrations of cyclones are in the range of 200–300 mg/m^3. The efficiency of pollutant removal depends on the particulate size; it is more effective for removing particles larger than 15µm, but less effective to finer particles, e.g., PM 80–99%; PM_{10} 60–95; PM_5 80–95%; $PM_{2.5}$ 20–70%.

Electrostatic precipitator (ESP) (dry and wet): ESP is similar to gravity or centrifugal separator (Figure 15.3), but in ESP, the separation treatment is done by electric field (electrostatic force) with high voltage. When the waste gas passes through a high voltage electric field, which provides the particles (solid or liquid) with an electrostatic charge, the discharge electrode will have a negative charge while the collecting plate will have a positive charge. The principal cleaning of dust by an electrostatic precipitator has a potential of 50 kV [27].

ESP process consists of three steps: particle charging, particle collection and removal of the collected dust. The ESP can be used to remove particulate matter in industrial applications, chemical manufacturing, refineries, incineration such as utility and industrial power boilers, flue gas streams exiting from waste incineration, cement kilns and glass furnaces, catalytic crackers, paper mills, metal processing, and a wide variety of other processes.

ESP is mainly used to treat small particles in the range from 0.1 to 1 micron [28]. Depending on the type of device, the treatment capacity could reach a very high flow rate of gas volumes between 10^4 and 4×10^6 $m^3 h^{-1}$ [27]. Typically, the exit gas particulate concentrations of ESP are in the range of 5–25 mg/m^3. The device can be operated at temperatures up to 700 °C for dry or less than 90 °C for wet matter. For the control of dioxin emissions under the lowest possible level, the temperature of operation must be below 200 °C [29]. The efficiency of treatment is typically 80 to 90% and, sometimes, it could be more than 90% for particulate size of 0.1 μm, depending on the type of device. The single-field ESP can reduce the particle concentration to below 150 mg/Nm^3, while the two-field ESP is more efficient.

Scrubber: scrubber is a typical device used to remove particles or fly ash and toxic gases from flue gases streams. Depending on the properties of waste gas streams and the process required, there are three types of scrubbers: wet, semi-dry and dry scrubber.

Based on the flow rate and operation parameters, there are also three types of scrubber geometries; these are crossflow, counterflow and co-flow scrubbers [30]. The venturi scrubber is an example of co-flow scrubber. In addition, there are many designs of scrubbers such as a fluidized-bed scrubber, a packed scrubber, a tower scrubber and a spray scrubber. A brief description of the main types of scrubbers is provided below.

1-Wet scrubber: Wet scrubber is a spray scrubber; it is the most typical device used for air pollution control in industry (Figure 15.3). It is used for the removal of toxic gases or particles from waste gases. It can also be used for treatment of a single target pollutant or as a multipurpose removal. The theory of the wet scrubber is based upon the absorption, the mass transfer from gaseous phase into liquid phase (acidic gases in an alkaline liquid phase, as a caustic solution), with or without reaction, which depends on the alkaline solid used such as calcium, magnesium, or sodium gypsum slag.

When using two-stage wet scrubbing, in the first stage, water or acidic solution is used as a scrubber medium to remove HF and HCl. After air injection in the second stage, calcium carbonate suspension is used to remove SO_2 as calcium sulphate. The HCl and calcium sulphate can be recovered. On other hand, two-stage wet scrubbing is used without material recovery to separate HCl and HF ions before desulphurization.

The efficiency of the wet scrubber process for acid gas removal is very high; it exceeds 95% for hydrogen chloride. For abatement of heavy metals as lead, the efficiency is about 99%, and for cadmium it is about 92% [7].

2-Semi-dry scrubber: the semi-dry scrubber uses a spray absorption process where the droplets of calcium hydroxide cool and neutralize the hot flue gas. When the waste gas passes through the scrubber tower, the reaction takes place between calcium hydroxide and acid gases. It is used in conjunction with fabric filters. It also can be used to treat heavy metals as mercury and cadmium and organic micropollutants as dioxins when activated carbon is added to the calcium hydroxide and furans.

3-Dry Scrubbers: the dry scrubber method is an absorption process for the treatment of acid gases that pass through a fine-grained alkaline powder as dry calcium hydroxide ($Ca(OH)_2$). It is sprayed onto the gases. The reaction takes place at about 160 °C in the dry state and the products (e.g., calcium chloride, calcium sulphate with hydrogen chloride and sulphur dioxide,) are removed from the flue gas stream by a filter. Like the semi-dry scrubber, it could be also used to treat heavy metals.

Fabric filters: Fabric filter or baghouse collectors (Figure 15.3) is a device used to separate particulate and acid gases (such as SO_2 and HCl) from gas stream, usually used after the scrubbers or the electrostatic precipitator. It is also used to remove organic micropollutants such as dioxins and furans, and heavy metals such as mercury. The mechanism of operation is as follows: the waste gas is introduced through a porous layer of fabric (medium) or fabric bags and it retains the particulates on the medium surface. After collection, the dust cake is removed from the medium or settled by gravity in the device's housing, and the gas passes from small pores of filter so it becomes clean.

The filtering process and bag-cleaning methods depend on the type of fabric filters, specifically, the composing materials and their resistance to acid or alkali gas attack and the structure geometry. Typically, the exit gas particulate concentrations of fabric filters are less than 5 mg/m³. The efficiency can reach 99% which depends on the type of filter and the condition of its operation. However, the efficiency of acid gases removal is between 85% and 90% [31]. High collection efficiency can be obtained in the case of small particle size in the range of 0.1–0.01μm [31].

The application of the above-mentioned devices is given in Figure 15.4a and b which illustrates the working principles of the systems used for gas treatment produced by the incinerators of MSW. The dry system and simple wet system (Figure 15.4a and b) can remove the particles of HCl, HF and heavy metals with a medium control level of efficiency.

The semidry system with dioxin removal and the selective catalytic reduction (SCR), and the advanced wet system with the selective non catalytic reduction (SNCR), limestone scrubber, and dioxin filter (Figure 15.4c and d) can be used with advanced control level of efficiency to remove HCl, HF, SO_2, NO_x, dioxins, and metals.

The selective non catalytic reduction (SNCR) process at high temperature or a selective catalytic (usually titanium dioxide (TiO_2)) reduction (SCR) at low temperature can be applied to remove NO_x. These are known as De-NOx processes. The De-NOx processes can convert nitrogen oxides to non-polluting nitrogen gas as N_2 and water vapor. The SNCR has better removal efficiency for the incineration plant, technically and economically, when ammonia and steam are injected into the furnace (Figure 15.4d). Furthermore, catalysts can be employed to reduce gaseous pollutants using thermal incinerator systems [27].

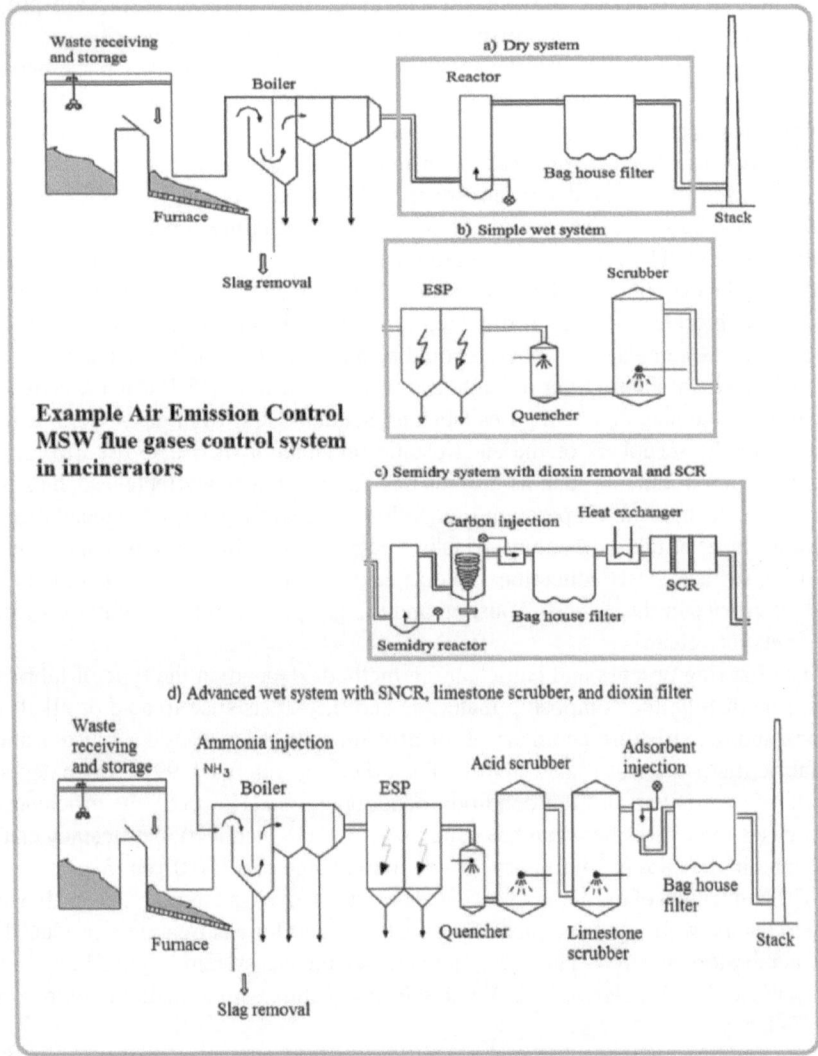

FIGURE 15.4   Illustration of MSW incinerators flue gas treatment systems, (a) dry system, (b) simple wet system, (c) semidry system with dioxin removal and SCR and (d) advanced wet system with SNCR, limestone scrubber and dioxide filter.

15.5   CONCLUSION

In this chapter, several methods and technologies of syngas cleanup have been illustrated and discussed. Syngas produced from pyrolysis/gasification processes can be cleaned up by: (a) primary methods of treatment that occur during the process and takes place inside the reactors, and (b) secondary methods of treatment that occur after the pyrolysis or gasification of process is done and takes place by mechanical or physical processes and reforming processes.

In conclusion, a gas cleanup system depends on: (1) the type and quantity of contaminants in the gaseous effluent; (2) the emission limits to the atmosphere; (3) the required treatment efficiency. The finer fly ash particles can be collected from the flue gases by electrostatic precipitators (ESPs). Scrubber and baghouse, respectively, can remove air pollutants and particulates from gas streams. In addition, electrostatic precipitation, fabric filtration, scrubbers, and mechanical collectors or baghouses can remove most particulate control. In the modern incinerator systems, the efficiency of the emissions control system complies with the environmental standards.

NOTES

1   **Calculation of the flue gas volume**: Rand et al. [8] presented a method for calculating the flue gas volume; it depends on temperature, pressure, and composition of the gas, and based on standard or normal condition of T & P (T = 0 °C, P = atmosphere). For more information about calculations, we recommended the book by C. C. Lee and Shun Dar Lin (2007): "Handbook of Environmental Engineering Calculations", 2nd edition, McGraw-Hill, ISBN: 0071475834.
2   $ngTEQ/Nm^3$ = the concentration in nanogram of toxicity equivalency quantity presented in a cubic meter at a temperature of 273 K and a pressure of 1 atmosphere.

REFERENCES

[1] Al Arni, Saleh, Barbara Bosio, Elisabetta Arato "Syngas from sugarcane pyrolysis: An experimental study for fuel cell applications," Renewable Energy, 2010, 35: 29–3. Available online at: http://dx.doi.org/10.1016/j.renene.2009.07.005
[2] Arni, Saleh "An experimental investigation for gaseous products from sugarcane by fast pyrolysis," Energy Education Science and Technology, 2004, 13(2): 89–96.
[3] Arni, Saleh "Hydrogen-rich gas production from biomass via thermochemical pathways," Energy Education Science and Technology, 2004, 13(1): 47–54.
[4] Rand T., Haukohl J. and Marxen U. "Municipal solid waste incineration: Requirements for a successful project," World Bank technical paper no. 462, June 2000.
[5] Jayarama Reddy P *Energy Recovery from Municipal Solid Waste by Thermal Conversion Technologies*. CRC Press/Balkema, Taylor & Francis Group, London, 2016.
[6] Bilitewski, Bernd, Georg Härdtle, and Klaus Marek. *Waste Management*. Springer Science & Business Media, Heidelberg, 1996.
[7] Williams, P.T., *Waste Treatment and Disposal*, 2nd ed., John Wiley & Sons Ltd, Hoboken, NJ, 2005.
[8] Al Arni, Saleh S and Elwaheidi M., *Concise Handbook of Waste Treatment Technologies*. CRC Press Taylor & Francis Group, London, 2021.
[9] Gaur R.C. *Basic Environmental Engineering*, New Age International (P) Ltd., Publishers, New Delhi, 2021.
[10] Manahan S.E. *Fundamentals of Environmental Chemistry*, 2nd Edition. CRC Press LLC, Boca Raton, 2001.
[11] Lee C.C. and Shun D.L. *Handbook of Environmental Engineering Calculations,* 2nd Edition. McGraw-Hill Companies, London, 2007.
[12] Joseph P.R., John S.J. and Louis T. *Handbook of Chemical and Environmental Engineering Calculations*. John Wiley & Sons, Inc., New York, 2002.
[13] Harrison K.W., Dumas R.D., Nishtala S.R. and Barlaz M.A. "A life cycle inventory model of municipal solid waste combustion," Journal of the Air and Waste Management Association, 2000, 50: 993–1003.

[14] Yaomin J., Ling G., Osvaldo D.F. and María C. Veiga and Christian Kennes, "Bioprocesses for the removal of nitrogen oxides," in *Air Pollution Prevention and Control: Bioreactors and Bioenergy*, edited by Christian Kennes and M. C. Veiga, John Wiley & Sons, Ltd, Hoboken, NJ, 2013.

[15] Williams P.T. "Pyrolysis of waste tyres: A review", Waste Management, 2013, 33: 1714–1728. http://dx.doi.org/10.1016/j.wasman.2013.05.003

[16] Worrell W.A. and Vesilind P.A., *Solid Waste Engineering*. 2nd Edition, Cengage Learning, Stanford, CA.

[17] Wei-Shan L., Guo-Ping C.-C., Lin-Chi W., Wen-Jhy L., Kuen-Yuh W. and Perng-Jy T. Emissions of polychlorinated dibenzo-p-Dioxins and dibenzofurans from stack gases of electric arc furnaces and secondary aluminum smelters," Journal of the Air & Waste Management Association, 2005, 55(2): 219–226. http://dx.doi.org/10.1080/10473289.2 005.10464613

[18] Woodard F., 2001, *Industrial Waste Treatment Handbook*. New York: Butterworth–Heinemann.

[19] Chen D., Yin L., Wang H., He P. "Pyrolysis technologies for municipal solid waste: A review," Waste Management, 2014, 34: 2466–2486.

[20] Shen Y., Wang J., Ge X. and Chen M. "By-products recycling for syngas cleanup in biomass pyrolysis – An overview," Renewable and Sustainable Energy Reviews, 2016, 59: 1246–1268. http://dx.doi.org/10.1016/j.rser.2016.01.077

[21] Basu P. *Biomass Gasification, Pyrolysis and Torrefaction: Practical Design and Theory*. 3rd Edition, Elsevier Inc, London, 2018.

[22] Garcia-Nunez J.A., Pelaez-Samaniego M.R., Garcia-Perez M.E., Fonts I., Abrego J., Westerhof R.J.M. and Garcia-Perez M. Historical developments of pyrolysis reactors: A review. Energy & Fuels, 2017, 31(6).

[23] Witham M.I.G. Biochar based Catalysts for the Cleaning of Syngas Produced from Solid-Fuels Gasification. Thesis of Doctor of Philosophy Degree. Department of Chemical Engineering, Curtin University, 2017.

[24] Cheremisinoff, N.P. *Handbook of Chemical Processing Equipment*. Butterworth-Heinemann, New York, 2000.

[25] Hocking M.B., *Handbook of Chemical Technology and Pollution Control*, 3rd Edition, Butterworth-Heinemann, New York, 2005

[26] Nesaratnam S.T. and Taherzadeh S. (Editors), *Air Quality Management*, John Wiley & Sons Ltd, Hoboken, NJ, 2014.

[27] Schnelle K.B. and Brown C.A. *Air Pollution Control Technology Handbook*, CRC Press LLC, London, 2002.

[28] Georgieva K. and Varma K., World bank technical guidance report, "Municipal Solid Waste Incineration", The International Bank for Reconstruction and Development/the World Bank, Washington, DC, 1999.

[29] De Nevers N. *Air Pollution Control Engineering*. 2nd Edition, McGraw-Hill, London, 2000.

[30] Claffey D., Claffey M. and Childress J., "Fabric filter collectors", in *Air Pollution Control Equipment Selection Guide* editor, Kenneth C. Schifftner, 2nd Edition, CRC Press, Taylor & Francis Group, London, 2014.

[31] Theodore, Louis, *Air Pollution Control Equipment*, John Wiley & Sons, Hoboken, NJ, 2008, Inc.

Index

Note: Page numbers in *italics* refer to figures, those in **bold** refer to tables.

3D printed photocatalytic feed spacer, 106
3Rs (reduce, reuse, and recycle), 58, 62–64
4Rs (reduce, reuse, recycle, and recover), 58, 62–64

A

absorption, treatment for gas effluent, 241–242
acidic pollutants, 290–291, 294
acid rain, 291
acid scrubbers, 241, 254
activated carbon, 245–246
activated persulfate, 205
AdBlue, 279
additives, 60, 197, 246
adsorption, treatment for gas effluent, 245–246
advanced oxidation processes (AOP) as pre-treatment for POME, 168–173
    Fenton, 170–173, *171*, *172*
    ozonation, 168–170, *169*
advanced oxidation processes (AOP) for removal of PFAS, 203–219
    comparisons of processes, 213, **213**
    conventional treatment methods, 204–206
    Fenton-based reaction, *206*, 206–209, *208*
    introduction, 203–206
    ozonation process, 209, *210*, 211
    photocatalytic process, *211*, 211–213
    references, 215–219
    summary/conclusion, 214–215
advanced oxidation process (AOP) for PPCPs, 115–132
    degradation and reaction rate constants of PPCPs, 118, **119**, 120, *120*
    future photocatalytic technologies, 127
    hydroxyl radical production, 115–116, 118
    introduction, 115–116
    photocatalytic membrane reactors, *125*, 125–127, *126*
    references, 128–132
    summary/conclusion, 127–128
    UV irradiation, 116
    UV irradiation for PPCP degradation, 116–118, **117**
    UV/titanium dioxide for PPCP degradation, 116, *121*, 121–125, *122*, **123**

advanced technologies for syngas cleanup, 289–304
    abatement techniques, 292–298
    equipment for syngas cleaning, **298**, 298–301, *299*, *302*
    introduction, 289–290
    methods of gas cleaning, 293–295
    methods of syngas cleaning, 295, *296*, 297, **297–298**
    references, 303–304
    summary/conclusion, 302–303
    treatment technologies for recovery and destruction, 292–293, 294, 299
    types of waste gas, 289, 290–292
Advanced Thermal Processes Methods (ATPM), 44
Advanced Thermal Treatment Technologies (ATTT), 44
aeration
    in food waste composting, 8–9
    in membrane reactors, 134
agriculture waste, 3, 180
AI-based waste management, 33–35, *34*
air-conditioning systems, 256–259, *257*, **258**
air gap membrane distillation (AGMD), 138
air pollution, treatments for, *see* integrated treatments for industrial gas effluent
air-reuse, from gas effluent, 240, 254, 255
alcoholic wastewater, and electrochemical treatment, **80**
algae, microalgae-MBR (MMBR), *142*, 145, **147**
aluminum electrodes
    electrochemical process, *82*, 82–83, **84**
    electrocoagulation process, 183–186, *185*, **187**–**188**, 192
ammonia
    anaerobic ammonium oxidation (ANAMMOX) process, 146–147
    in C/N ratio in food waste composting, 9
    as reducing agent in catalytic processes, 269–270, 278
anaerobic ammonium oxidation (ANAMMOX) process, 146–147
anaerobic baffled reactor (ABR), 165
anaerobic digestion of POME, 162, *165*, 165–168, *166*, *167*, 170
    treatment stages, 165, 167
    types of digesters, 165–166
anaerobic fluidized bed reactor (AFBR), 165
anaerobic MBR (AnMBR), 134, *142*, 146–147, **148**

anaerobic sequencing batch reactor (ASBR), 165, 166
anaerobic treatment, 162, 164, 166, 167
anion exchange resins, 204–205
anodic oxidation (AO), 85, **86**
anoxic-aerobic membrane bioreactor (A/O MBR), 137
antibiotic resistance genes, 7
AOP, *see* advanced oxidation processes (AOP)
aquatic environments, pollution in, 96, 163–164, 179, 226
artificial intelligence in EEW management, 33–35, *34*
ash content of waste tyres, 61, **61**
ash metals particulate, 292
Australia, regulations on waste tyre management, 58
autochthonous microorganisms, 10
automated machines, with AI, 38
automotive industry, *see* catalysts in automotive industry

B

baffles in MBR systems, 142
baghouse collectors, *299*, 301
base scrubbers, 241, 254
batteries, in e-waste stream management, 31–32
biochar, 16, 41
biochemical processes, 41, 222
biodegradable organic waste, 2, 14–15
bio-electrochemical membrane bioreactor (BEC-MBR), *142*, 145–146, **148**
biofilms, 143–144, 146, 150
biofilters
    in food waste composting, 11
    in treatment for gas effluent, *246*, 246–248, 249, 256
biogas production
    by anaerobic digestion treatment of POME, 162, *165*, 165–168, *166*, *167*, 170
    by anaerobic MBR, 146
biological processes, in treatment of gas effluent, *246*, 246–248, 254, 294
biomass
    pyrolysis production of syngas, 295
    as term for food waste, 3
bio-oil, as product of pyrolysis, 49, **50–51**
bioreactors, *see* membrane bioreactors (MBRs) for wastewater treatment
bioscrubbers, *246*, 246–248, 256
biotrickling filters, *246*, 246–248, 255, 256
biowaste, as term for food waste, 3
bismuth-based catalyst, 233
blacklists of electronic parts, 24
BOD
    biochemical oxygen demand, 74, 162

biological oxygen demand, **74**, 134, 180, 222, 226
Brazil, regulation on waste tyre disposal, 69
brewery effluent, and electrocoagulation treatment, **188**, 193, **194**
bulking agents in food waste composting, 9
burning process
    open burning, 294
    of waste tyres, 66
    see also combustion; pyrolysis of solid wastes
butadiene, 59, 64

C

carbon, in C/N ratio in food waste composting, 9
carbon-based catalysts, 234
catalysis, *see* photocatalysis for oily wastewater; photocatalytic techniques
catalysts in automotive industry, 267–288
    De-NOX reactions and kinetics, 268–271, *270*, *272*
    design of automotive SCR systems, 278–281, *279*
    introduction, 267–268
    references, 283–288
    SCR (selective catalytic reduction), 269–271, *272*
    SCR system disadvantages, 281–282
    SCR systems, catalyst types, *see* catalyst types in SCR systems
    summary/conclusion, 283
    system reliability predictions, 282–283
catalyst types in SCR systems, 272–277, *273*, **273–274**
    cerium-added (Ce) catalysts, 277
    copper-based (Cu) catalysts, **274**, 275
    copper catalysts, 271, *272*
    manganese-based (Mn) catalysts, 276–277
    vanadium-based (V) catalysts, 272–273, **273–274**, 275
catalytic converters in cars, 268, 282
catalytic-oxidative conversion of VOCs, 243
catalytic post-combustion, 242
cavitation, 195
central air-conditioning systems, 256–259, *257*, **258**
ceramic membranes, 140, 231
char, product of
    gasification of waste tyres, 68
    pyrolysis of solid wastes, 46, 47, 49–50, **50–51**
chemical oxidative scrubbers, 241
chemical oxygen demand (COD), 74, 141, 162, 222, 226
chemical risk in EEW, regulatory frameworks of, 27, **27–28**
China
    laws on recycling car batteries, 31

production of activated carbon, 245
regulatory frameworks of chemical risk in
EEW, **27**
Circular Economy, 3
C/N ratio in food waste composting, 9
coated membranes, 101–102, *102*, 231
cobalt batteries, in e-waste stream management,
31–32
coffee processing, **181**, **188**, 193
coking wastewater, and electrochemical treatment,
80, **86**
collection, *see* waste collection
combustion
of fuel, in automotive engines, 268
thermal, as treatment for gas effluent, 242
as thermal conversion process, 44, 45–46, *46*, 66
see also burning process
complete autotrophic nitrogen removal over nitrite
(CANON), 146
composting of food wastes, 1–22
characteristics of food waste, 6–8
gaseous emissions, 10–11
introduction, 1–3
microbiology, 9–10
microplastics, 7–8
pollutants, 6–7, 12–13
process conditions (aeration, porosity,
and C/N ratio), 8–9
quality of compost (respiration indices and
maturity), 11–13, **13**
references, 17–22
summary/conclusion, 16–17
suppressor effect on plant pathogens, 16
terminology, 3–6, *4–5*
used as landfill cover, 15–16
used as organic amendment to soil, 3, 13–14
used as soil bioremediation, 14–15
concentration-driven membrane technology, *136*,
137–138
condensation, as treatment for gas effluent, 241
contaminants, PFAS, 203–204, *see also* pollutants
contaminants of emerging concern (CECs), 96, 137
continuous stirred tank reactor (CSTR), 165, 166
copper-based catalysts, **274**, 275, 277, *277*
corrosive pollutants, 290–291
Covid pandemic, 257, 267
cyclones (dry and wet), 299, *299*

D

dairy processing, **181**, **187**, 193
databases
Clarivate Web of Science, 76
Lens.org, 166
Substances of Concern In Products (SCIP),
24, *25*

decentralized wastewater treatment, 148–149
Democratic Republic of Congo (DRC), cobalt
extraction for batteries, 32
De-NOx processes, 268–271, *270*, *272*, 301
design, ecological, of electronics, 31, 32–33
devices, electronics, 31, 33
devulcanization of rubber tyres, 63
dielectric barrier discharge, 243
diesel engines, 268, 281, 283
Diesel Exhaust Fluid (EDF), 279
Diesel Particulate Filter (DPF), 282
dioxins, 291, 294
dip-coating method for membranes, 102–103
direct contact membrane distillation (DCMD), 138
disease suppression by food waste compost, 16
disposal of EEW, 23
dissolved oxygen (DO), 74, 83, 164, 226
dry scrubbers, 294, 301
dry-wet phase inversion method for mixed matrix
membranes, 102
dusts, 290, 294
dyed wastewater
electrochemical treatment, 88
from photocatalysis of oily wastewater, 230–231
photocatalytic membrane treatment, 104–105
dynamic respiration index, in food waste
composting, 11–12, **13**

E

ecological design of electronics, 31, 32–33
ecological sustainability, 35–36, **36–37**
EEW (electronic and electric waste), *see*
integrated management of electronic and
electric waste (EEW)
electrical-driven membrane technology, *136*,
138–139
electricity production via microbial
electrochemical technology, 145
electric vehicles, 31
electric waste, *see* integrated management of
electronic and electric waste (EEW)
electrochemical oxidation of PFAS, 205
electrochemical-peroxidation (ECP) hybrid
process, **194**, 195
electrochemical treatment of municipal
wastewater (MWW), 73–94
anodic oxidation (AO), 85, **86**
combined processes, 86–88, **87**
composition and sources of MWW, 73–74, *74*
electrocoagulation (EC), *82*, 82–85, **84**
electrodes, aluminum, *82*, 82–83, **84**
electrodes, Boron-doped diamond (BDD),
78–79, **80–81**, 85
electrodes, horizontal bipolar electrodes
(BPEs), 83

electrodes, iron, 79, **80–81**, 82–83, **84**
electrodes, platinum, 78, 79
electro-Fenton, 77–79, **80–81**
introduction, 73–77
literature review, *76*, 76–77, *77*, **84**, **86**, **87**
references, 89–94
summary/conclusion, 88–89
technologies, 74–76, **75**
electrocoagulation (EC) in electrochemical
    treatment of municipal wastewater
    (MWW), *82*, 82–85, **84**
electrocoagulation (EC) in food wastewater
    treatment, 179–201
  advantages and disadvantages, 182, **183**
  application, 193–194
  aspects, ecological, 197–198
  aspects, economic, 196–197
  aspects, technical, 198–199
  colloids, 183
  composition of food wastewater, 180, **181**
  EC process, 183–186, *184, 185*
  hybrid technologies, **194**, 194–196
  introduction, 179–183
  operating parameters, *see* electrocoagulation
    operating parameters
  references, 199–201
  summary/conclusion, 199
electrocoagulation operating parameters,
    186–193, **187–188**
  conductivity and supporting electrolyte, 190
  current density, 190–191
  electrode configuration, *191*, 191–192
  electrode material, 192
  electrolysis time, 189
  initial pH, 186, *189*
  inter electrode distance, 192–193
  mixing speed, 190
  power supply, 191
electrode configurations, *191*, 191–192
electro-Fenton, 77–79, **80–81**
electroflotation (EF), 180
electronic and electric waste (EEW), *see*
    integrated management of electronic and
    electric waste (EEW)
electrospinning for membranes, 103
electrostatic precipitator (ESP, dry and wet), 293,
    294, *299*, 299–300
emerging, defined, 96
emerging pollutants (EP) in wastewater, 95–98,
    97, 222, *see also* photocatalytic membrane
    for emerging pollutants (EP) treatment
end-of-life design of EEW, 33
energy, waste conversion into, 41, 42, 44–45, *46*,
    66–68
environmentally eco-friendly
    catalytic technologies for gaseous waste, 267–268

photocatalytic membranes, 108
environmental regulations on pollutants
  on EEW, 27, **27–28**
  on fertilizers and composting, 12, 13
  on POME, 162, 164, **164**
  on waste gas and syngas, 289, 291
  on waste tyre management, 58–59, 68–69
environmental sustainability management of EEW,
    35–36, **36–37**
Equivalent Passenger Units (EPU), 58
European Automobile Manufacturers' Association
    (ACEA), 279
European Chemicals Agency (ECHA), 24
European Commission, Air Quality Directive, 268
European Community, Waste Incineration
    Directive, 291
European Environmental Agency, on emerging
    pollutants, 96, 137
European Union
  Green Deal strategy, 3
  Landfill Directive, 2
  laws on recycling car batteries, 31
  regulation on fertilizers and composting,
    12, 13
  regulations on waste tyre management, 58, 69
  regulatory frameworks of chemical risk in
    EEW, **27–28**
eutrophication, 162, 163, 179, 180, 268
excimer technology, 244
expanded granular sludge bed (EGSB), 165

F

fabric filters, *299*, 301
fast pyrolysis, **43**, 46–47, **48**, **50–51**
Fenton, Henry J., 77
Fenton process
  AOP for removal of PFAS, *206*, 206–209,
    213, **213**
  AOP process as pre-treatment for POME,
    170–173, *171, 172*
  electro-Fenton in treatment of municipal
    wastewater (MWW), 77–79, **80–81**
  historical beginnings, 77–78
  modified Fenton PFOA oxidation process, 207
  modified photo-Fenton process, 207–208, *208*
Fenton's reagent, 209
fermentation, 41, 168
fertilizers, 3, 12, 14, 88, 197
Fick's diffusion law, 242
fishery processing, **181**
flares, 293
flash pyrolysis, 47–48, **48**
flat sheet (FS) MBRs, 134, 140, **147**
flue gas, 268, 289, 290, 291, 292, 300–301,
    302, 303n1

food wastes, composting of, *see* composting of food wastes
food wastewater treatment
    available technologies, 180, **181**
    see also electrocoagulation (EC) in food wastewater treatment
"forever chemicals," 214
forward osmosis (FO) MBR combined process, 137
free air space (FAS) in food waste composting, 9
Fuel Gas Desulphurisation (FGD), 294
furans, 291, 294

G

gallium oxide, 212
garden waste, 6
gaseous emissions
    from automotive engines, *see* catalysts in automotive industry
    from food waste composting, 10–11, 15
    syngas, *see* syngas
    treatments for, *see* integrated treatments for industrial gas effluent
gas hourly space velocities (GHSV), 272
gasification process
    production of syngas, 295, *296*, 297, **297–298**, *see also* advanced technologies for syngas cleanup
    thermal conversion of solid waste via pyrolysis, 44, 46
    tyre recycling, 68
gasifiers, 295, *296*, 297, **297–298**
gas separation process, 137–138
Germany, TA-Luft (Technical Instructions on Air Quality Control), 249
glass hollow fiber membrane, 106
granular activated carbon (GAC), 204–205
graphene catalyst, 234
graphene oxide (GO) membranes, 141
gravity separators, 298–299, *299*
greenhouse gases (GHG)
    emissions numbers, 240
    from food waste composting, 10–11
    see also methane production

H

halogenated organics, 7
hazardous substances and chemicals, also known as emerging pollutants (EP), 96
heat recovery in treating gas effluent, 242, 254
heat treatment processes, 44
heavy metals
    in flue gas from incinerators, 292
    in food waste composting, 12, 14, 15

as origin source for emerging pollutants (EP), 96
    photocatalysis restriction on, 234
    photocatalytic membrane treatment, 104
Henry partition coefficients, 241
HEPA filters, 258
hollow fiber (HF) MBRs, 134, 140, **147–148**
Hong Kong, regulations on waste tyre management, 58
household waste, as term for food waste, 4
human health, 12, 23–24, 96, 108, 204, 222, 226, 227, 239, 294
hydraulic retention time (HRT), 145, 166
hydro-pyrolysis, 47, **48**
hydrothermal carbonization (HTC), 45
hydrothermal gasification (HTG), 45
hydrothermal liquefaction (HTL), 45
hydrothermal processes, 45
hydrous pyrolysis, 47
hydroxyl radical production
    in AOP for PFAS, 205, 206, 212
    in AOP for POME, 168–169, 170
    in AOP for PPCPs, 115–116, 118
    in photocatalysis of oily wastewater, 223–224, 230

I

image recognition, *34*
immersed membrane bioreactors (iMBR), 134, 148
immobilization techniques, 231–232
incineration, thermal conversion of solid waste via pyrolysis, 44, 45–46, *46*
incinerators, industrial, 289, 295, 301, *302*
indium oxide, 212
industrial waste air, *see* integrated treatments for industrial gas effluent
industrial wastewater
    electrochemical treatment, 87–88
    vs. domestic wastewater, 141–142
    see also wastewater
inner plasma catalyst (IPC), 243
Institute for Health Metrics and Evaluation, on air pollution, 239
integrated/hybrid membrane technology, 139–140
integrated management of electronic and electric waste (EEW), 23–39
    artificial intelligence in EEW management, 33–35, *34*
    classification: UNU-Keys, 28–30, **29–30**
    ecological design, 31, 32–33
    environmental sustainable management, 35–36, **36–37**
    global production, 25–27, *26*
    introduction, 23–24, *24*, *25*
    management of e-waste streams, batteries example, **31**, 31–32

references, 38–39
regulatory frameworks of risk, 27, **27–28**
sorting of, 34–35
summary/conclusion, 38
integrated treatments for industrial gas effluent,
    239–265
    absorption, 241–242
    adsorption, 245–246
    air conditioning systems, 256–259, *257*
    biological processes, *246*, 246–248
    combined treatment processes, 253–259
    condensation, 241
    conventional techniques, 240–248
    introduction, 239–240
    membrane processes, 242
    non-thermal plasma (NTP), 243–244
    odor emissions, 248–249, **251–253**
    oxidative catalysis, 243
    parameters for suitability of processes,
        248–249, *249*, **250**
    references, 259–265
    summary/conclusion, 259
    thermal combustion, 242
    thermal plasma, 244–245
    types of compounds, 255
    UV oxidation, 244
    waste air conditions, high contaminated,
        254–256
    waste air conditions, low contaminated,
        256–259, **258**
integration of AOPs for palm oil mill effluent
    (POME), 161–177
    anaerobic digestion systems and biogas
        production, 162, *165*, 165–168, *166*, *167*
    AOP, 168
    AOP, Fenton, 170–173, *171*, *172*
    AOP, ozonation, 168–170, *169*
    characteristics of POME and effects on
        environment, 162–164, **163**, **164**, **181**
    electrocoagulation process, 180, **187**, **194**
    introduction, 161–162
    open ponding treatment, 164
    published research, *166*, 166–167, *167*
    references, 173–177
    summary/conclusion, 173
International Energy Agency (IEA), on air
    pollution, 240
International Union of Pure and Applied Chemistry
    (IUPAC), on photocatalysis, 98
ion exchange membrane bioreactor (IEMBR),
    138–139
iron electrodes
    electrochemical process, 79, **80–81**,
        82–83, **84**
    electrocoagulation process, 183–186,
        187–188, 192

J

Japan, regulations on waste tyre management, 59

K

Keggin cation, 186
Korea, regulations on waste tyre management,
    58–59

L

landfills
    anodic oxidation, 85
    disposal of waste tyres, 68–69
    food waste compost as cover, 15–16
    life cycle of EEW, **37**
latex, 59
laws, *see* environmental regulations on pollutants
life cycle of electronics, 35–36, **36–37**
lignocellulosic compounds, 168, 169
lithium batteries, in e-waste stream management,
    31–32

M

machine learning, 35, 283
magnetic fields, 255
Malaysia, and POME, 161, 180
    regulations, 162, 164, **164**
manganese-based catalysts, 276–277, 282
MBRs, *see* membrane bioreactors (MBRs) for
    wastewater treatment
meat processing, **181**
membrane aerated biofilm reactor (MABR), 138,
    142, 144–145, **147**
membrane-biofilm reactor (MBB-MBR), 138,
    143–144, **147**
membrane bioreactors (MBRs) for wastewater
    treatment, 133–159
    anaerobic MBR (AnMBR), 134, *142*,
        146–147, **148**
    bio-electrochemical membrane bioreactor
        (BEC-MBR), *142*, 145–146, **148**
    challenges and future directions, 149–150
    concentration-driven membrane technology,
        136, 137–138
    conventional MBR configurations, 134, *142*,
        142–145
    electrical-driven membrane technology, *136*,
        138–139
    integrated/hybrid membrane technology,
        139–140
    introduction, 133–134, *135*
    membrane aerated biofilm reactor (MABR),
        138, *142*, 144–145, **147**

membrane-biofilm reactor (MBB-MBR), 138,
  143–144, **147**
membrane materials and modules, *140*,
  140–141, **147–148**
membrane types, 134, 136–137, 231
microalgae-MBR (MMBR), *142*, 145, **147**
moving bed biofilm reactor (MBBR), *142*,
  143–144
novel MBRs, *142*, 145–148
pressure-driven membrane technology,
  135–137, *136*
references, 150–159
removal mechanisms, 153
single-chamber MBR: simultaneous biological
  nitrogen removal MBR (SBNR-MBR),
  142, 143
small-scale MBR applications, 148–149
thermal-driven membrane technology,
  136, 138
two-chamber MBR: sequencing anoxic/oxic
  MBR system, *142*, 142–143, **147**
wastewater types: industrial *vs.* domestic,
  141–142
membrane distillation, 106, 138
membranes
  fouling in MBRs, 134, 139, 141, 144
  materials and modules, *140*, 140–141,
    147–148
  for photocatalysis, *see* photocatalytic
    membrane for emerging pollutants (EP)
    treatment
  photocatalytic membrane reactors (PMRs),
    101, 116, *125*, 125–127, *126*, 222
  in treatments for gas effluent, 242
  types of, 134, 136–137, 231
membrane separation technology, 100
metallic membranes, 140
metal oxide catalysts, 243
  rare-earth, 282
methane production
  from anaerobic digestion of POME, 162–164,
    165–167, 170
  emissions from food waste composting,
    10–11, 15–16
  see also biogas production
methanotrophs, 15
microalgae-MBR (MMBR), *142*, 145, **147**
microbial electrochemical technology, 145
microbiology, in food waste composting, 9–10
microbioplastics (MBPs), 7–8
microfiltration (MF) membranes, 136–137
microplastics, 7–8
micropollutants in the environment, sources and
  routes of, 97, *97*
mixed matrix membranes, 101–102, *102*
models in AI technology, 35

monolithic designs for catalysts, 272, *273*
moving bed biofilm reactor (MBBR), *142*,
  143–144
multi-tubular (MT) MBRs, 134
municipal solid waste (MSW)
  composition and classifications of, 42, **43**,
    43, 44
  flue gas produced by incinerators, 289,
    290–291
  generation rates and definitions, 1–3
  thermal conversion via pyrolysis, *see* pyrolysis
    of solid wastes
municipal waste
  composting of, *see* composting of food wastes
  as term for food waste, 4
municipal wastewater (MWW)
  composition and sources of, 73–74, *74*
  industrial *vs.* domestic, 141–142
  treatment of organic compounds, *see*
    electrochemical treatment of municipal
    wastewater (MWW)
  see also wastewater

N

nanofiltration (NF) membranes, 136–137, 139
nanoparticles, 196
natural rubber, *59*, 59–60, **60**
nitrogen
  C/N ratio in food waste composting, 9
  in POME, 163, **163**
  removal from domestic wastewater, 142–143,
    144, 146
nitrogen oxides
  reactions in treating flue gases, 268–271,
    270, *272*
  types of, as emissions in flue gases, 291
nitrogen starvation response (NSR), 269
nitrous oxide emissions, from food waste
  composting, 10–11
non-selective catalytic reduction (NSCR), 269
non-thermal plasma (NTP), 243–244, 259
Norway, regulatory frameworks of chemical risk
  in EEW, **28**
nut processing, **188**, 193–194, **194**

O

odor emissions, 248–249, **251–253**, 295
off-the-road tyres, 57
oil, *see* bio-oil; palm oil mill effluent (POME)
oilfield wastewater, **86**
oil refineries, 222
oily wastewaters, 222, 225–227, *228*, *see also*
  photocatalysis for oily wastewater
olfactometer, 249

olive oil mill wastewater, **84**, **181**, 193
open burning, 294
open ponding treatment for POME, 164
organic compounds in MWW, *see* electrochemical treatment of municipal wastewater (MWW)
organic fraction of MSW (OFMSW), 2, 3–6, *4–5*, *see also* composting of food wastes
organic materials, 11, 42, *43*, 44
organic waste, as term for food waste, 3
Organization for Economic Co-operation and Development (OECD), municipal waste defined, 2
oxidative catalysis, 243, 254
oxidative scrubbers, 241
ozonation process
    AOP as pre-treatment for POME, 168–170, *169*
    AOP for removal of PFAS, 209, *210*, 211, 213, **213**

P

palm oil mill effluent (POME)
    characteristics and effects on environment, 161–164, **163**, **164**, **181**
    in oily wastewater, 227
    see also integration of AOPs for palm oil mill effluent
PANI membranes, 106–107
paper mill wastewater, **80**, **84**, 85
particulate materials (dusts), 290, 294
PCDD (polychlorinated dibenzodioxins), 291–292
PCDF (polychlorinated dibenzofurans), 291
per- and polyfluoroalkyl substances (PFAS), *see* advanced oxidation processes (AOP) for removal of PFAS
peroxi-electrocoagulation hybrid process, 195
personal care products, as origin source for emerging pollutants (EP), 96, *see also* pharmaceuticals and personal care products (PPCPs)
pesticides
    electrochemical treatment, **81**, **84**, **86**
    as origin source for emerging pollutants (EP), 96
    photocatalytic membrane treatment, 104
    as pollutants in food wastes, 7, 12, 15
petroleum refinery wastewater
    electrochemical treatment, **86**
    photocatalysis of oily waste, 227, 231
PFAS (per- and polyfluoroalkyl substances), *see* advanced oxidation processes (AOP) for removal of PFAS
PFOA (perfluorooctanoic acid), 204, 207–208, 209, 211, 212
PFOS (perfluoro-octane sulfonate), 204, 209, 211
pharmaceutical products

electrochemical treatment in MWW, 79, **80**, **86**
    as origin source for emerging pollutants (EP), 96
    photocatalytic membrane treatment, 104, 105
pharmaceuticals and personal care products (PPCPs), *see* advanced oxidation process (AOP) for PPCPs
photocatalysis, history of term, 98
photocatalysis for oily wastewater, 221–237
    application of, 225–229, *228*
    heterogeneous photocatalysis, 221, 222, 224, 234
    introduction, 221–223
    miscellaneous catalysts, 224, 227, *228*, 233–234
    pinciple of photocatalysis, 223–225, *224*
    references, 235–237
    restrictions, 234
    summary/conclusion, 234–235
    titanium oxide catalyst, 222–223, 224–225, **225**, **229**, 229–232
    zinc oxide catalyst, 222–223, 224–225, **225**, **229**, 232–233
photocatalytic membrane for emerging pollutants (EP) treatment, 95–113
    challenges, 107–108
    emerging pollutants in wastewater, 95–98, *97*, 222
    fabrication of photocatalytic membranes, 101–103, *102*
    membrane technology, 100–101
    performance of photocatalytic membrane on EPs, 103–107, *105*
    photocatalysis mechanism, 98–100, *99*
    references, 109–113
    summary/conclusion, 108–109
photocatalytic membrane reactors (PMRs), 101, 116, *125*, 125–127, *126*, 222
photocatalytic techniques
    for PPCPs, *see* advanced oxidation process (AOP) for PPCPs
    for removal of PFAS, *211*, 211–213, **213**
photocatalytic water treatment, published research, 122, *122*
photo-electrocoagulation hybrid process, **194**, 195
photo-Fenton process for removal of PFAS, 207–208, *208*
physical scrubbers, 241
plasma arc technology, 44–45, 68
plasma gases for degradation of air pollutants, 244
plastics as pollutants in food wastes, 7–8, 12–13
pollutants
    acidic and corrosive, 290–291, 294
    in air, *see* gaseous emissions
    in food wastes, 6–8, 12–13
    from incinerators, 289
    in water, *see* wastewater
    see also contaminants

polychlorinated biphenyls (PCBs), 12
polycyclic aromatic hydrocarbons (PAHs), 12
polyfluoroalkyl substances (PFAS)
    as pollutants in food wastes, 7, 12
    see also advanced oxidation processes (AOP)
        for removal of PFAS
polymer-based membranes, 140–141, 231
POME, see palm oil mill effluent (POME)
porosity, in food waste composting, 8–9
post plasma catalyst, 243
precipitators, electrostatic, 293, 294, *299*,
    299–300
prefiltration-membrane process, 139
pre-oxidation catalyst, 278, *279*
pressure-driven membrane technology,
    135–137, *136*
pre-treatment
    AOP for POME, see integration of AOPs for
        palm oil mill effluent (POME)
    in membrane bioreactor, 136, 139, 141
    of waste tyres, 67, *67*
process designs in automotive industry, 267
PTFE (polytetrafluoroethylene), 125–126, 138,
    141, 291
PVDF (polyvinylidene difluoride), 104, 125–126,
    137, 138, 140, 149
pyrolysis of solid wastes, 41–55
    composition and classifications of solid waste,
        42, **43**, *43*, 44
    gasification, 44, 46
    incineration, 44, 45–46, *46*
    introduction, 41–42
    process types and mechanisms, 47–49, **48**
    products of pyrolysis processes, 49–50,
        50–51
    reactor types, 49
    references, 52–55
    summary/conclusion, 51
    thermal conversion processes, *44*, 44–47
pyrolysis of waste tyres, 66–67, *67*, *68*
pyrolysis production of syngas, 289, 295, see also
    advanced technologies for syngas cleanup

R

rare-earth metal oxides, 282
raw materials, **36**, 59, 62
reactor types
    for electrochemical treatment in MWW, 83, 87
    fixed bed photoreactors, 126–127
    photocatalytic membrane reactors (PMRs),
        101, 116, *125*, 125–127, *126*
    for pyrolysis of solid wastes, **43**, 49
    slurry batch photoreactor, 126–127
    see also membrane bioreactors (MBRs) for
        wastewater treatment

reclamation of crumb rubber, 63
recuperators, 242
recycling EEW
    batteries, 31–32
    ecological design of electronics, 33
recycling waste tyres, 57–71
    burning process, 66
    components of tyres, 59–61, **60**, **61**
    estimates of waste tyres, 58
    gasification process, 68
    introduction, 57–58
    landfill disposal, 68–69
    pyrolysis process, 66–67, *67*, *68*
    recycling processes, 62–64, *65*
    references, 69–71
    regulations, 58–59, 68–69
    reusing tyres, 62–63, **64**
    summary/conclusion, 69
    thermochemical conversion into useful energy,
        66–68
regenerators, 242
regulations, see environmental regulations on
    pollutants
respiration indices, in food waste composting,
    11–12, **13**
reverse osmosis (RO) membranes, 105, 136–137,
    139–140
rubber, in tyres, *59*, 59–60, **60**, 63, **64**
rubber crumb, reclamation of, 63
rubber tree, 59

S

scrubbers, treatment for gas effluent, 241,
    293, 294
    geometries of, 300–301
selective catalytic reduction (SCR), 269–271,
    272, 301
    system reliability predictions, 282–283
    systems disadvantages, 281–282
    typical catalysts used in SCR systems,
        272–277
selective non-catalytic reduction (SNCR),
    269, 301
semiconductor photocatalysis, 98, 100
semi-dry scrubbers, 294, 301
separate collection, *4*, 6, 12, **31**, 34–35
sequencing anoxic/oxic MBR system, *142*,
    142–143, **147**
sequencing batch reactor (SBR), 147
side-stream membrane bioreactors (sMBR), 134
silicone membranes, 242
simultaneous biological nitrogen removal MBR
    (SBNR-MBR), *142*, 143
simultaneous nitrification and denitrification
    (SND), 143, 149

single-chamber MBR, *142*, 143
  simultaneous biological nitrogen removal
    MBR (SBNR-MBR), 143
slaughterhouse wastewater, **86**, **187**, 193, **194**
slow pyrolysis, **43**, 47–48, **50–51**
small-scale MBR applications, 148–149
smuts, 290
soil amendment, 3, 13–15
soil bioremediation, 14–15
solids retention time (SRT), 143
solid waste materials (SWM), 41, *see also*
    pyrolysis of solid wastes
sono-electrocoagulation hybrid process, **194**,
    195–196
soot, 290
spin coating method for membranes, 103
stacks (industrial incinerators), 289, 295, *302*
static respiration index, in food waste composting, 11
Stockholm Convention on Persistent Organic
    Pollutants, 204
Strategic Approach International Chemical
    Management (SAICM), 27
Submerged Membrane Electro-Bioreactor, 83, 87
Substances of Concern In Products (SCIP), 24, *25*
substances of very high concern (SVHC), 24
sugar mills, 45, **181**, **187**
sulfate radicals, 205
supercritical water gasification (SCWG), 45
suppressor effect, of food waste compost, 16
sustainable management, environmental, of
        electronic and electric waste (EEW),
        35–36, **36–37**
sweeping gas membrane distillation (SGMD), 138
syngas
    as fuel, 291, 295, 297
    as product of gasification of waste tyres, 68
    as product of pyrolysis of solid waste, 46, 47,
        49–50, **50–51**
    see also advanced technologies for syngas
        cleanup
synthesis gas, 42, 47
synthetic rubber, 59–60, **60**

T

tannery wastewater, **81**, **84**
textile wastewater, **80**, **84**
thermal combustion
    open burning, 294
    treatment for gas effluent, 242
    see also burning process; combustion
thermal conversion processes
    combustion/incineration, 44, 45–46, *46*, 66
    gasification, 44, 46, 68
    hydrothermal processes, 45
    plasma arc technology, 44–45

pyrolysis, *see* pyrolysis of solid wastes
    torrefaction process, 45
thermal-driven membrane technology, *136*, 138
thermal oxidation, 293
thermal plasma, 244–245
thermal treatment, 42, 45, 294
thermochemical conversion, 41–42, 66–68, 289
thermochemical processes, 45–47, 67
three-way catalysts (TWCs), 281
tires, *see* tyres
titanium dioxide
    in AOP for PPCPs, 116, *121*, 121–125,
        122, **123**
    in AOPs for PFAS, 212
    in photocatalysis for oily wastewater,
        222–223, 224–225, **225**, **229**, 229–232
torrefaction process, 45
Total Diet Study (TDS), 204
total dissolved solids (TDS), 74, 163
total nitrogen (TN), 74
total phosphate (TP), 74
total suspended solids (TSS), 74, 137, 163, 164
transmembrane pressure (TMP), 134
tungsten trioxide catalyst, 233–234
two-chamber MBRs, *142*, 142–143, **147**
    iMBR for small-scale applications, 148–149
    MBB-MBR, 144
    sequencing anoxic/oxic MBR system, *142*,
        142–143, **147**
tyre derived aggretate, 63
tyre pyrolysis oil (TPO), 67
tyres
    retreaded, 63
    rubber aggregate, crumbs, 63, **64**
    rubber components, 59–61, **60**, **61**
    see also recycling waste tyres

U

ultrafiltration (UF) membranes, 136–137,
        139–140
ultrarapid pyrolysis, 47, **48**
ultrasonication oxidation methods, 205
ultrasound irradiation, 195
United Nations International Children's
        Emergency Fund (UNICEF), on clean
        water access, 128
United Nations University (UNU), 29–30
United Nations Water (UN-Water), 95
United States, regulatory frameworks of chemical
        risk in EEW, **28**
United States Environmental Protection Agency
        (EPA), on water contamination, 96
UNU keys, 29–30, **29–30**
up-flow anaerobic sludge blanket (UASB),
        165, 166

up-flow anaerobic sludge fixed-film (UASFF), 165
urea, as reducing agent in catalytic processes,
        269–270, 278, 281
UV (ultraviolet) light
    in modified photo-Fenton process, 207–208, *208*
    oxidation process as treatment for gas effluent,
        244, 255
    for PPCPs treatment, *see* advanced oxidation
        process (AOP) for PPCPs

V

vacuum membrane distillation (VMD), 138
vacuum pyrolysis
    of solid wastes, **43**, 47–48, **48**, **50–51**
    of waste tyres, 67
vanadium-based catalysts, 272–273, **273–274**, 275
ventilation systems, 257
vermicomposting, 11
volatile organic compounds (VOCs)
    from food waste composting, 10–11
    in industrial gas effluent, *see* integrated
        treatments for industrial gas effluent
    oxidative catalysis, 243
    in waste gas, 291, 294, 295

W

waste air, *see* gaseous emissions; integrated
        treatments for industrial gas effluent;
        syngas
waste collection
    of EEW, 26, 31, 34–35, **37**

of food wastes, 2, 7, 12, **13**
    from syngas, 298, 300, 301
waste conversion into energy, 41, 42, 44–45, *46*,
        66–68
waste electrical and electronic equipment
        (WEEE), 28, *see also* integrated
        management of electronic and electric
        waste (EEW)
Waste Framework Directive (WFD), 24
waste management, 34–35, 58–59
waste tyres, *see* recycling waste tyres
wastewater
    emerging pollutants (EP), 95–98, *97*, 222
    from food processing, *see* electrocoagulation
        (EC) in food wastewater treatment
    removal of PFAS by AOPs, *see* advanced
        oxidation processes (AOP) for removal
        of PFAS
    types of: industrial *vs.* domestic, 141–142
    see also municipal wastewater (MWW)
water-silicone oil emulsions, 241
wet scrubbers, 293, 294, 300
winery wastewater, **86**
World Health Organization (WHO)
    on air pollution, 239, 268
    on clean water access, 95, 128

Z

zeolites, 125, 245, 275, 278, 294
zinc oxide catalyst, in photocatalysis for oily
        wastewater, 222–223, 224–225, **225**, **229**,
        232–233